Trigonometric Functions

Sines and tangents, read down.
Cosines and cotangents, read up.

∠	sin		tan		∠	∠	sin		tan		∠
0°	.0000	.0175	.0000	.0175	89°	45°	.7071	.7193	1.000	1.036	44°
1°	.0175	.0349	.0175	.0349	88°	46°	.7193	.7314	1.036	1.072	43°
2°	.0349	.0523	.0349	.0524	87°	47°	.7314	.7431	1.072	1.111	42°
3°	.0523	.0698	.0524	.0699	86°	48°	.7431	.7547	1.111	1.150	41°
4°	.0698	.0872	.0699	.0875	85°	49°	.7547	.7660	1.150	1.192	40°
5°	.0872	.1045	.0875	.1051	84°	50°	.7660	.7771	1.192	1.235	39°
6°	.1045	.1219	.1051	.1228	83°	51°	.7771	.7880	1.235	1.280	38°
7°	.1219	.1392	.1228	.1405	82°	52°	.7880	.7986	1.280	1.327	37°
8°	.1392	.1564	.1405	.1584	81°	53°	.7986	.8090	1.327	1.376	36°
9°	.1564	.1736	.1584	.1763	80°	54°	.8090	.8192	1.376	1.428	35°
10°	.1736	.1908	.1763	.1944	79°	55°	.8192	.8290	1.428	1.483	34°
11°	.1908	.2079	.1944	.2126	78°	56°	.8290	.8387	1.483	1.540	33°
12°	.2079	.2250	.2126	.2309	77°	57°	.8387	.8480	1.540	1.600	32°
13°	.2250	.2419	.2309	.2493	76°	58°	.8480	.8572	1.600	1.664	31°
14°	.2419	.2588	.2493	.2679	75°	59°	.8572	.8660	1.664	1.732	30°
15°	.2588	.2756	.2679	.2867	74°	60°	.8660	.8746	1.732	1.804	29°
16°	.2756	.2924	.2867	.3057	73°	61°	.8746	.8829	1.804	1.881	28°
17°	.2924	.3090	.3057	.3249	72°	62°	.8829	.8910	1.881	1.963	27°
18°	.3090	.3256	.3249	.3443	71°	63°	.8910	.8988	1.963	2.050	26°
19°	.3256	.3420	.3443	.3640	70°	64°	.8988	.9063	2.050	2.145	25°
20°	.3420	.3584	.3640	.3839	69°	65°	.9063	.9135	2.145	2.246	24°
21°	.3584	.3746	.3839	.4040	68°	66°	.9135	.9205	2.246	2.356	23°
22°	.3746	.3907	.4040	.4245	67°	67°	.9205	.9272	2.356	2.475	22°
23°	.3907	.4067	.4245	.4452	66°	68°	.9272	.9336	2.475	2.605	21°
24°	.4067	.4226	.4452	.4663	65°	69°	.9336	.9397	2.605	2.747	20°
25°	.4226	.4384	.4663	.4877	64°	70°	.9397	.9455	2.747	2.904	19°
26°	.4384	.4540	.4877	.5095	63°	71°	.9455	.9511	2.904	3.078	18°
27°	.4540	.4695	.5095	.5317	62°	72°	.9511	.9563	3.078	3.271	17°
28°	.4695	.4848	.5317	.5543	61°	73°	.9563	.9613	3.271	3.487	16°
29°	.4848	.5000	.5543	.5774	60°	74°	.9613	.9659	3.487	3.732	15°
30°	.5000	.5150	.5774	.6009	59°	75°	.9659	.9703	3.732	4.011	14°
31°	.5150	.5299	.6009	.6249	58°	76°	.9703	.9744	4.011	4.331	13°
32°	.5299	.5446	.6249	.6494	57°	77°	.9744	.9781	4.331	4.705	12°
33°	.5446	.5592	.6494	.6745	56°	78°	.9781	.9816	4.705	5.145	11°
34°	.5592	.5736	.6745	.7002	55°	79°	.9816	.9848	5.145	5.671	10°
35°	.5736	.5878	.7002	.7265	54°	80°	.9848	.9877	5.671	6.314	9°
36°	.5878	.6018	.7265	.7536	53°	81°	.9877	.9903	6.314	7.115	8°
37°	.6018	.6157	.7536	.7813	52°	82°	.9903	.9925	7.115	8.144	7°
38°	.6157	.6293	.7813	.8098	51°	83°	.9925	.9945	8.144	9.514	6°
39°	.6293	.6428	.8098	.8391	50°	84°	.9945	.9962	9.514	11.43	5°
40°	.6428	.6561	.8391	.8693	49°	85°	.9962	.9976	11.43	14.30	4°
41°	.6561	.6691	.8693	.9004	48°	86°	.9976	.9986	14.30	19.08	3°
42°	.6691	.6820	.9004	.9325	47°	87°	.9986	.9994	19.08	28.64	2°
43°	.6820	.6947	.9325	.9657	46°	88°	.9994	.9998	28.64	57.29	1°
44°	.6947	.7071	.9657	1.000	45°	89°	.9998	1.000	57.29	∞	0°
	cos		cot		∠		cos		cot		∠

COLLEGE ALGEBRA

THIRD EDITION
REVISED

COLLEGE ALGEBRA

GORDON FULLER

Texas Technological University

D. VAN NOSTRAND COMPANY

New York Cincinnati Toronto London Melbourne

D. VAN NOSTRAND COMPANY Regional Offices:
New York Cincinnati Millbrae

D. VAN NOSTRAND COMPANY International Offices:
London Toronto Melbourne

Copyright © 1974 by Litton Educational Publishing, Inc.

Library of Congress Catalog Card Number 69—15871

ISBN: 0-442-22509-1

Published by D. VAN NOSTRAND COMPANY
450 West 33rd Street, New York, N.Y. 10001

Published simultaneously in Canada by
VAN NOSTRAND REINHOLD LTD.

10 9 8 7 6 5 4

PREFACE TO THE THIRD EDITION

This edition of *College Algebra*, like the previous ones, has been planned for college freshmen who are not sufficiently undergirded in algebra to begin the study of analytic geometry and calculus. The new edition, however, has been "modernized" through emphasis on the basic concepts, procedures, and terminology. This approach, together with precise definitions and language, makes possible a judicious amount of rigor which in turn facilitates the mastery of the essential techniques employed in the conventional subject matter of elementary algebra.

The concept of a set is introduced in the first chapter. The treatment, though brief, suffices for the use of the concept in the chapters which follow. Set notation and set operations appear at various places where they tend to promote understanding and help with simplifications.

The field axioms of real numbers and certain theorems based on the axioms are treated in Chapter 2. This discussion makes way for justifying the fundamental operations on algebraic expressions. Although students at this stage have been using the axioms and theorems, many do not appreciate the basic nature of the axioms since they do not realize that the theorems stem from the axioms. We have emphasized the dependence of the theorems on the axioms by stating formally all the necessary theorems and referring freely to the axioms in making the proofs. The coverage of the material in the chapter will, of course, depend on the class at hand. As a minimum accomplishment, however, the students, by knowing the axioms and understanding the statements contained in the theorems, should begin to envision elementary algebra as a logical structure, thus gaining some insight into the nature of mathematics.

Several chapters treated at the level of intermediate algebra follow Chapter 2. Then, in Chapter 10, we introduce the order axioms of real numbers and derive certain theorems based on the axioms. This paves the way for solving inequalities. Included in this discussion are inequalities of one and two variables and also included are systems of inequalities.

All the topics of the previous editions, with considerable upgrading in treatment, are taken up in this edition. The new material provides a basis for the improved treatment of the conventional material of the earlier editions.

We list here certain other notable features of the book:

1. Functions and relations are defined as sets of ordered pairs of numbers with the distinction between the two concepts clearly pointed out.

2. A careful discussion is given of operations on equations which lead to equivalent equations, and also on systems of equations which yield equivalent systems.

3. We have achieved a more satisfactory presentation of probability by proceeding from certain axioms and using the idea of sets and subsets.

4. Complex numbers are treated quite adequately. Although the form (a, b) is introduced briefly, the main discussion involves the rectangular form $a + bi$ and the polar form $r(\cos \theta + i \sin \theta)$. Attention is given to the fact that complex numbers obey the field axioms of Chapter 2.

5. An introduction to the algebra of matrices and proofs of the most important properties of determinants of matrices appear in Chapter 16.

6. Logarithms are treated rather fully with emphasis on the theoretical aspects.

7. The exposition throughout is in clear and simple language. Each new topic is amply illustrated with solved problems, these examples having been planned carefully to make evident the principles involved and to obviate any troublesome points. In fact, the organization and exposition are such that the material is largely self-teaching, thus affording more time for the teacher to help with difficult steps and to deepen the student's insight by emphasizing the underlying reasons.

8. As with the text proper, meticulous care has been used in planning the exercises, each of which has an abundance of problems. Answers to the odd-numbered problems are included in the text, and answers to the even-numbered ones are given in the instructor's manual.

9. All the necessary numerical tables—powers and roots of numbers, common logarithms of numbers, and values of trigonometric functions—are bound with the book.

10. The book is flexible enough for the selection of material to accommodate classes of varying degrees of preparation. Some students may need to spend considerable time on Chapters 3 through 8, while others, in a short time, can successfully review these chapters (or even omit them) and have ample time to cover most of the remaining chapters, thus securing a promising foundation for the study of calculus and other areas of mathematics. Although three class meetings a week for a semester is the most common practice, ample material has been provided for students who would be better served by four or five class meetings a week.

CONTENTS

COLLEGE ALGEBRA

Chapter 1

SETS AND OPERATIONS

I-I SETS

The idea of a set is a fundamental concept in mathematics. Although a set is classed as an undefined term, we gain an intuitive understanding by describing a set as a specified collection (or aggregate) of objects. Thus the totality of trees in a park, the collection of chairs in a room, and the list of families in a town are examples of sets. Each object of a set is called an *element* or *member* of the set. A set may be specified by listing its elements or by describing the set so that it can be determined if any given object is, or is not, a member of the set.

Many kinds of sets arise in mathematical investigations. We mention a few examples of sets, using a capital letter to stand for each set.

A = the set of all angles having measures between 0° and 90°.
B = the set of vowels in the English alphabet.
C = the set of positive integers less than 7.*
D = the set of all numbers between 3 and 5.
E = the set of counties in Texas.
F = the positive odd integers less than 100.

We note that the elements of the sets B, C, E, and F may be listed. But we cannot list the elements of sets A and D. If the elements of a set can be listed from the first through the last, we say the set is *finite*; otherwise the set is *infinite*. It is customary to specify an infinite set by enclosing a description of the set within braces. Although a finite set can be indicated in this way, it is sometimes more convenient simply to list the elements. We may specify the sets above with the following notation.

$A = \{\theta | 0° < \theta < 90°\}$
$B = \{a, e, i, o, u\}$
$C = \{1, 2, 3, 4, 5, 6\}$
$D = \{x | 3 < x < 5\}$

---

* We assume the existence of the numbers 1, 2, 3, 4, and so on. This set of numbers is called the *counting numbers*, the *natural numbers*, or the *positive integers*.

$E = \{x | x \text{ is a county in Texas}\}$
$F = \{1, 3, 5, 7, \ldots, 99\}$

The letter θ in set A stands for any unspecified member of the set and the symbols $<$ and $>$ mean respectively "is less than" and "is greater than." Hence the set consists of all angles whose measures are greater than $0°$ and less than $90°$. Similarly, x in set D stands for any unspecified member of the set which consists of all numbers greater than 3 and less than 5. The vertical bar in sets A, D, and E may be read "such that," and the three dots in set F indicate that the missing odd integers are members of the set. With these explanations, we see that the members of each set A to F are definitely spelled out.

We express the fact that θ is a member of set A, and x is a member of set D, by writing

$$\theta \in A \quad \text{and} \quad x \in D$$

The symbol $\in$ may be read "is a member of, is an element of," or "belongs to"; and the symbol $\notin$ is read "is not a member of." The letters θ and x, as used here, are variables, as is borne out by the following definition.

Definition 1-1 A symbol, usually a letter, which may stand for any member of a specified set of objects is called a *variable*. If the set has only one member the symbol is called a *constant*.

1-2 RELATED SETS

In this section we shall point out certain ways in which two sets may be related.

Definition 1-2 Two sets A and B are said to be *equal* ($A = B$) if each element of set A is an element of set B and each element of set B is an element of set A.

The sets $A = \{x, y, z\}$ and $B = \{y, z, x\}$, for example, are equal.

Definition 1-3 If it is possible to pair each element of a set A with exactly one element of a set B and each element of B with exactly one element of A, then we say the elements of the sets can be arranged in a *one-to-one correspondence*.

Definition 1-4 Two sets A and B are said to be *equivalent* ($A \leftrightarrow B$) if their elements can be put into a one-to-one correspondence.

According to this definition the infinite set of positive integers and the infinite set of positive even integers are equivalent. The equivalence of the sets $A = \{1, 2, 3, \ldots\}$ and $B = \{2, 4, 6, \ldots\}$ becomes evident from the following method of pairing the elements:

$$(1, 2), (2, 4), (3, 6), (4, 8), (5, 10), \ldots$$

Definition 1-5 If each element of a set A is an element of set B, then A is called a *subset* of B. If A is a subset of B and if B has one or more elements not belonging to A, then A is a *proper* subset of B.

To indicate that A is a subset of B, we write $A \subseteq B$, which is read "A is contained in B." This notation does not tell us if A is or is not a proper subset. If, however, A is a proper subset, we may emphasize that relationship by writing $A \subset B$.

We write here some of the subsets of $\{a, b, c, d\}$.

$$\{a\} \subset \{a, b, c, d\}$$
$$\{b, c\} \subset \{a, b, c, d\}$$
$$\{a, c, d\} \subset \{a, b, c, d\}$$
$$\{a, b, c, d\} \subseteq \{a, b, c, d\}$$

As illustrated by the last subset appearing here, a set is a subset of itself; this is a consequence of the very definition of a subset.

Thus far we have mentioned sets which contain one or more elements. Conditions may be specified, however, such that a set has no element. For example, the set of blondes in a group of five brunettes has no element. A set which contains no element is called the *null*, or *empty*, set. The symbol $\emptyset$ is used to stand for the empty set. Thus,

$$\{x | x = 1 \text{ and } x = 2\} = \emptyset$$
$$\{x | x \text{ is an integer between 5 and 6}\} = \emptyset$$

Since $\emptyset$ has no element, we can conclude that each element of $\emptyset$ belongs to any set A, and therefore $\emptyset \subseteq A$. In other words, the empty set is a subset of every set.

EXERCISE 1-1

Enclose within braces (a) a list of the elements of each set and (b) a description of the set.

1. The positive integers less than 7.
2. The positive even integers between 1 and 13.
3. The days of the week.
4. The months of the year having 30 days.
5. The first three presidents of the United States.
6. The thirteen original states of the United States.
7. The books of the Pentateuch.

Tell if the two sets are equal, equivalent, or if one is a subset of the other.

8. $A = \{4, 7, 11\}$, $B = \{7, 4, 11\}$.
9. $C = \{a, e, i, o, u\}$, $D = \{u, v, x, y\}$.

10. $C = \{x|x$ is a number between 1 and 20 and divisible by 4$\}$
      $D = \{t|t$ is a positive integer less than 20$\}$.
11. $S = \{x|x$ is a lady president of the United States$\}$
      $T = \{x|x$ is a positive integer less than 1$\}$.

12. Given $\{5, 7, 8\} = \{x, 8, 7\}$, find x.
13. Find a set of values for a, b, c such that $\{a, b, c\} = \{0, 1, 4\}$.
14. Write all the subsets of (a) $\{1\}$, (b) $\{u, v\}$.
15. Write the eight subsets of $\{1, 2, 3\}$.
16. If $E = \{2, 5, 8, 11\}$ and $F = \{8, 11, 14, 2\}$, tell which of the following is true:
   (a) $8 \in E$, (b) $8 \subset F$, (c) $8 \subset E$, (d) $\varnothing \in F$, (e) $\varnothing \subset F$, (f) $\{8\} \in E$, (g) $\{8\} \subset E$, (h) $E \leftrightarrow F$,
   (i) $\{2, 5, 8\}$ is a proper subset of E.
17. If $A \subset B$ and $B \subset A$, show that A is (or is not) equal to B.
18. By using definitions concerning sets in Sec. 1-2, show that the following are true for sets A, B, and C:
   (a) $A = A$ (equality is reflexive).
   (b) If $A = B$, then $B = A$ (equality is symmetric).
   (c) If $A = B$ and $B = C$, then $A = C$ (equality is transitive).

1-3   OPERATIONS ON SETS

We have seen that new sets, called subsets, can be obtained from a given set. In this section we shall consider operations by which new sets can be obtained from two given sets.

Definition 1-6   The set consisting of the totality of elements under consideration in a particular discussion is called the *universal set*. The universal set is commonly denoted by U.

Definition 1-7   The set of all elements which belong to a given universal set U and do not belong to a given subset A of U is called the *complement* of A.

Denoting the complement of a set A by A', we may write

$$A' = \{x|x \in U \text{ and } x \notin A\}$$

Suppose, for example, that we have under consideration the sets $A = \{a, b, c\}$ and $B = \{c, d, f, g\}$. Then we could choose $U = \{a, b, c, d, f, g\}$ and have

$$A' = \{d, f, g\} \quad \text{and} \quad B' = \{a, b\}$$

Definition 1-8   The *union* of two sets A and B, denoted by $A \cup B$, is defined to be the set composed of all elements which belong to A or to B or to both A and B.

It follows from this definition that if $x \in A$ then $x \in A \cup B$ and if $x \in B$, then $x \in A \cup B$. Hence we have

$$A \cup B = \{x|x \in A \text{ or } x \in B\}$$

Definition 1-9 The *intersection* of two sets A and B, denoted by $A \cap B$, is defined to be the set consisting of all elements which belong to A and also belong to B.

We express this definition symbolically by writing

$$A \cap B = \{x \mid x \in A \text{ and } x \in B\}$$

EXAMPLE 1. If $A = \{u, v, w, x, y\}$ and $B = \{a, v, w, y\}$, then

$$A \cup B = \{u, v, w, x, y, a\}$$

and

$$A \cap B = \{v, w, y\}$$

EXAMPLE 2. If $C = \{5, 7, 11\}$ and $D = \{\text{sun, moon}\}$, then

$$C \cup D = \{5, 7, 11, \text{sun, moon}\}$$

and

$$C \cap D = \varnothing$$

Since C and D have no common element, their intersection is the empty set $\varnothing$.

Definition 1-10 Two sets which have no common element are called *disjoint sets*.

Figures 1-1 and 1-2 furnish pictorial illustrations of the union and intersection of two sets. We let A stand for the set of all points inside the larger circle and B the set of all points inside the smaller circle. Pictures of this kind are called Venn diagrams in honor of the English logician John Venn (1834–1883).

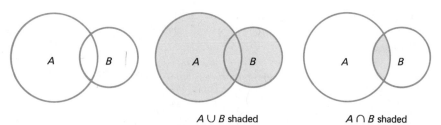

$A \cup B$ shaded $A \cap B$ shaded

Fig. 1-1

Definition 1-11 If x and y denote two objects (alike or different) with x specified as the first object and y the second object, then the symbol (x, y) is called an *ordered pair*.

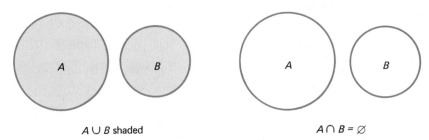

$A \cup B$ shaded $A \cap B = \emptyset$

Fig. 1-2

Definition 1-12 If X and Y are sets, the set of all ordered pairs (x, y) such that $x \in X$ and $y \in Y$ is called the *product set*, or Cartesian product, of X and Y; or, in symbols,

$$X \times Y = \{(x, y) | x \in X \text{ and } y \in Y\}$$

EXAMPLE 3. If $X = \{a, b\}$ and $Y = \{1, 2, 3\}$, the product set is

$$X \times Y = \{(a, 1), (a, 2), (a, 3), (b, 1), (b, 2), (b, 3)\}$$

EXAMPLE 4. If $S = \{(a, b)\}$, then

$$S \times S = \{(a, a), (a, b), (b, a), (b, b)\}$$

EXERCISE 1-2

If $A = \{0, 1, 2, 3, 4, 5\}$, $B = \{0, 1, 3, 4\}$, and $C = \{4, 5, 6, 7\}$, enclose within braces the elements of the following sets.

1. $A \cup B$	**2.** $A \cap B$	**3.** $A \cup C$	**4.** $B \cap B$
5. $C \cup C$	**6.** $A \cup \emptyset$	**7.** $B \cap \emptyset$	**8.** $A \cup (B \cup C)$
9. $A \cup (B \cap C)$	**10.** $A \cap (B \cap C)$	**11.** $A \cap (B \cup C)$	**12.** $A \cup (B \cap \emptyset)$

If $A = \{p, q, r, s\}$, $B = \{r, s, t, u\}$, and $C = \{t, u, v, w\}$, determine if the following equations are true by examining each member of the equations.

13. $A \cap (B \cup C) = (A \cap B) \cup (A \cap C)$
14. $A \cup (B \cap C) = (A \cup B) \cap (A \cup C)$
15. $A \cap (\emptyset \cup C) = (A \cup B) \cap (A \cup C)$

Prove that if A and B are any sets, then each of the following equations is true.

16. $A \cup A = A$ **17.** $A \cap A = A$
18. $A \cap B = B \cap A$ **19.** $A \cup B = B \cup A$

Draw Venn diagrams to illustrate the following theorems.

20. *Theorem.* If $A \subset B$, then $B \cup A = B$.
21. *Theorem.* If $B \cup A = B$, then $A \subset B$.
22. *Theorem.* If $A \cap B = A$, then $A \subset B$.
23. *Theorem.* If $A \subset B$, then $A \cap B = A$.

24. If $X = \{1\}$ and $Y = \{2, 3\}$, find $X \times Y$ and $Y \times X$.
25. If $P = \{1, 2\}$ and $Q = \{2, 3\}$, find $P \times Q$ and $Q \times P$.
26. If $S = \{a, b, c\}$ and $T = \{1, 2\}$, find $S \times T$.
27. If $U = \{J, F, M, A, S, O, N\}$, $A = \{J, M, S\}$ and $B = \{O, N\}$, find A' and B'. Then verify that $(A \cup B)' = A' \cap B'$.
28. If $U = \{J, F, M, A, S, O, N\}$, $A = \{J, F\}$, and $B = \emptyset$, find A', and $\emptyset'$. Then verify that $(A \cup \emptyset)' = A' \cap \emptyset'$.

Show by means of Venn diagrams that the following equations are true if A, B, and C are subsets of a universal set U.

29. $A \cup B = B \cup A$
30. $A \cup (B \cup C) = (A \cup B) \cup C$
31. $A \cap (B \cup C) = (A \cap B) \cup (A \cap C)$
32. $(A \cap B) \cup (A' \cup B') = U$
33. $(A \cup B)' = A' \cap B'$
34. $(A \cap B) \cup (A \cap B') = A$
35. $(A')' = A$

Chapter 2

THE REAL NUMBERS

2-I A DEDUCTIVE SYSTEM

We recall that a theorem of high school plane geometry consists of a hypothesis and conclusion, and that the proof of the theorem is accomplished by a process of logical reasoning which shows that the conclusion is a consequence of the hypothesis. The hypothesis of the theorem rests on earlier theorems, and, in turn, each of the earlier theorems depends on preceding theorems. In this manner, then, the theorems of plane geometry extend back to some first theorem which, clearly, is not a consequence of other established theorems. Accordingly, there must be at least one theorem whose proof rests on an unproved statement or statements. The unproved statements are called *axioms* or *postulates* and are assumed to be true.

There are also numerous definitions of terms in plane geometry. The first of the succession of definitions must describe a term with words whose meanings are assumed to be known. Such words are called *undefined terms*. The meaning of the words "point" and "line," for example, are assumed to be known and are classed as undefined terms.

The undefined terms and axioms constitute the foundation of plane geometry. Inasmuch as the theorems depend on this foundation, plane geometry is called a *deductive*, or *axiomatic, system.*

Algebra, like geometry and many other areas of mathematics, is a deductive system. This system is based on a set of axioms and undefined terms which we shall introduce and study in this chapter. In the treatment we shall consider a set of undefined elements called *real numbers*, and indicate the set by $R = \{a, b, c, \ldots\}$. The set shall have two basic, undefined operations called *addition* and *multiplication*. The symbol $+$ denotes the operation of addition and the symbols $\cdot$ or $\times$ denote multiplication. Thus, if a and b stand for real numbers, we write $a + b$ for addition and $a \cdot b$, $a \times b$, or just ab to indicate multiplication. The result of addition is called the *sum* and of multiplication, the *product.*

Later we shall introduce numbers different from the set of real numbers. In the meantime, for brevity, we shall use the word *number* for real number.

The equals sign in the number system denotes the relation of identity. Thus $a = b$ means that a and b are symbols for the same number. A statement of equality, such as $a = b$, is called an *equation*.

For the equality relation, we assume the following properties for all elements $a, b, c, \ldots$ of R.

1. $a = a$ (*reflexive* property)
2. If $a = b$, then $b = a$ (*symmetric* property)
3. If $a = b$ and $b = c$, then $a = c$ (*transitive* property)
4. If $a = b$ and $c = d$, then $a + c = b + d$ (*additive* property)
5. If $a = b$ and $c = d$, then $ac = bd$ (*multiplicative* property)

The last three properties are sometimes described, respectively, by saying: *Things equal to the same thing are equal to each other; if equals are added to equals, the results are equal; if equals are multiplied by equals, the results are equal.*

The five assumptions seem quite reasonable. The first assumption, stating that a number is equal to itself, appears acceptable. If in the equation $a = b$, we replace a by its equal b and b by its equal a, we obtain $b = a$. Hence Assumption 2 is really an agreement that any quantity may be replaced by an equal quantity; this is called the *principle of substitution*. Assumptions 3, 4, and 5 likewise follow from this principle. To check Assumption 4, we start with the sum $a + c$. Then replacing a by its equal b and c by its equal d, we have $a + c = b + d$.

If in Assumptions 4 and 5, we use $c = c$ instead of $c = d$, we see that $a + c = b + c$ and $ac = bc$. These results give rise to the familiar rules:

1. The same number may be added to both members of an equation.
2. Both members of an equation may be multiplied by the same number.

2-2   FIELD AXIOMS OF REAL NUMBERS

We shall state here six axioms, called *axioms of a field*, for the set of real numbers R. Later (Sec. 10-1) additional axioms will be introduced to complete our study of real numbers.

Axiom 1   *The closure laws.* For any $a, b \in R$,

$$a + b \in R \quad \text{and} \quad a \cdot b \in R$$

This axiom simply states that the operations of addition and multiplication on two numbers which belong to R yield numbers which also belong to R. Hence the set R is said to be *closed* under these operations. The numbers $a + b$ and $a \cdot b$ are called respectively the *sum* and *product* of a and b.

Axiom 2   *The commutative laws.* For any $a, b \in R$,

$$a + b = b + a \quad \text{and} \quad a \cdot b = b \cdot a$$

The commutative axiom tells us that the sum or product of two numbers does not depend on the order in which the addition or multiplication is performed. Thus, $3 + 7 = 7 + 3$ and $3 \cdot 7 = 7 \cdot 3$.

Axiom 3   *The associative laws.* For any $a, b, c \in R$,

$$(a + b) + c = a + (b + c) \quad \text{and} \quad (ab)c = a(bc)$$

The associative axiom tells us that the sum or product of any three numbers is independent of the way in which the numbers are grouped for these operations. In fact, Axioms 1 to 3 permit us to add or multiply any number of real numbers in any chosen order. For two of the several possible ways of finding the product of four numbers, we have

$$2 \cdot 3 \cdot 4 \cdot 5 = (2 \cdot 3) \cdot 4 \cdot 5 = (6 \cdot 4) \cdot 5 = 24 \cdot 5 = 120$$
$$= 2 \cdot (3 \cdot 4) \cdot 5 = 2 \cdot (12 \cdot 5) = 2 \cdot 60 = 120$$

Axiom 4   *The distributive law.* For any $a, b, c \in R$,

$$a(b + c) = ab + ac$$

The distributive axiom states that a particular product is equal to a sum, and, conversely, the sum is equal to a product since equality is symmetric. The name of this axiom seems appropriate because the multiplier a *is distributed to each element of the sum* $(b + c)$. The distributive law, as we shall see, has important consequences; it is basic to the structure of elementary algebra.

A consequence of Axiom 4 is the fact that the product of a number and the sum of three or more numbers is the same as the sum of the products of the first number by each of the numbers forming the sum. Thus, for the sum of three numbers, we have

$$a[b + c + d] = a[b + (c + d)] \qquad \text{Axiom 3}$$
$$= ab + a(c + d) \qquad \text{Axiom 4}$$
$$= ab + (ac + ad) \qquad \text{Axiom 4}$$
$$= ab + ac + ad \qquad \text{Axiom 3}$$

Axiom 5   *The identity elements.* For any $a \in R$, there exists a real number called "zero," denoted by 0, such that

$$a + 0 = a$$

For any $a \in R$, there exists a real nonzero number called "one," and denoted by 1, such that

$$a \cdot 1 = a$$

The number 0 is called the *identity element for addition*, and the number 1 is called the *identity element for multiplication*. Since 0 and 1 belong to R, we have, by Axiom 2,

$$a + 0 = 0 + a = a$$

and

$$a \cdot 1 = 1 \cdot a = a$$

Axiom 6 *The inverse elements.* For any $a \in R$, there exists a real number, denoted by $-a$, such that

$$a + (-a) = 0$$

For any nonzero number $a \in R$, there exists a real number, denoted by $\frac{1}{a}$, or $1/a$, such that

$$a \cdot \frac{1}{a} = 1$$

The number $-a$ is called the *additive inverse of a*, the *negative of a*, or *minus a*. The number $1/a$ is called the *multiplicative inverse of a* or the *reciprocal of a*. By the commutative axiom, we have

$$a + (-a) = (-a) + a = 0$$

$$a \cdot \frac{1}{a} = \frac{1}{a} \cdot a = 1$$

Hence, we see that a is the additive inverse of $-a$, and $a \neq 0$ is the multiplicative inverse of $1/a$. (The symbol $\neq$ is read "not equal to.")

EXERCISE 2-1

Tell which axiom or axioms are used in each of the statements 1 through 13.

1. $4 + 5 = 5 + 4$
2. $6 + 0 = 6$
3. $0 + 5 = 5$
4. $(-2) \cdot 1 = -2$
5. $3 + (-3) = 0$
6. $-3 + [-(-3)] = 0$
7. $(-xy) + xy = 0$
8. $1 \cdot 7 = 7$

9. $-3\left(\dfrac{1}{-3}\right) = 1$

10. $4\left(\dfrac{1}{4}\right) = 1$

11. $3(5 + 4) = 3 \cdot 5 + 3 \cdot 4$

12. $x + y = 1 \cdot x + 1 \cdot y$

13. $\dfrac{1}{x + 2y}(x + 2y) = 1$

Find the additive inverse and also the multiplicative inverse of each of the following numbers.

14. 5   **15.** 1   **16.** -2

17. $\frac{2}{3}$ **18.** $-a$ **19.** $(2x + y)$

20. $\dfrac{2x}{5y}$ **21.** $-\dfrac{3a}{2b}$ **22.** $-\dfrac{5}{c}$

23. Assuming that $d \neq 0$, give the value of x if (a) $dx = d$, (b) $d + x = d$, (c) $(1/d) \cdot x = 1$.

Show that the following statements are true.

24. $(b + c)a = ab + ac$ **25.** $(ab)c = (ca)b$

26. $(b + c + d)a = ab + ac + ad$ **27.** $(a + 3) + b = 3 + (a + b)$

28. $-a + (a + b) = b$ **29.** $\dfrac{1}{a}(ab) = b$

2-3 IMMEDIATE CONSEQUENCES OF THE FIELD AXIOMS*

We shall next establish certain theorems which follow directly from the field axioms. The theorems represent basic properties of the set of real numbers and will be applied in later work. Most of the steps in the proofs of the theorems will be justified by referring to the appropriate axioms or previously established theorems. For brevity, the axioms will be referred to by the numbers assigned them in the preceding section. In cases where references are missing, the student should supply the reasons. The student may also establish the theorems which are stated without proofs.

Theorem 2-1 The identity element for multiplication is unique.

Proof. We wish to show that 1 is the only identity element for multiplication. Suppose we assume that some number b is an identity element for multiplication. Then we have

$$1 = 1 \cdot b$$
$$= b \cdot 1 \qquad\qquad \text{Axiom 2}$$
$$= b \qquad\qquad \text{Axiom 5}$$

Hence we see that any identity element for multiplication must be equal to 1, and the theorem is proved. ∎

Theorem 2-2 The identity element for addition, 0, is unique.

* The coverage of the material in this section will vary from class to class. And in occasional instances it may be preferable for the students to continue assuming that the theorems are true.

Theorem 2-3 The multiplicative inverse of any nonzero number $a \in R$ is unique.

Proof. Let us assume that some number b is a multiplicative inverse of a real number a different from zero. Then we have

$$a \cdot b = 1 \qquad\qquad \text{hypothesis}$$

$$(a \cdot b)\frac{1}{a} = 1 \cdot \frac{1}{a} \qquad\qquad \text{multiplying by } \frac{1}{a}$$

$$\frac{1}{a}(a \cdot b) = \frac{1}{a} \cdot 1 \qquad\qquad \text{Axiom 2}$$

$$\left(\frac{1}{a} \cdot a\right)b = \frac{1}{a} \qquad\qquad \text{Axioms 3 and 5}$$

$$1 \cdot b = \frac{1}{a} \qquad\qquad \text{Axioms 2 and 6}$$

$$b \cdot 1 = \frac{1}{a} \qquad\qquad \text{Axiom 2}$$

$$b = \frac{1}{a} \qquad\qquad \text{Axiom 5}$$

Hence any multiplicative inverse of a is equal to $1/a$, and the theorem is proved. ∎

Theorem 2-4 The additive inverse of any $a \in R$ is unique.

Theorem 2-5 For any $a \in R$,

$$-(-a) = a$$

Proof.

$$(-a) + [-(-a)] = 0 \qquad\qquad \text{Axiom 6}$$

and also

$$a + (-a) = 0 \qquad\qquad \text{Axiom 6}$$

$$(-a) + a = 0 \qquad\qquad \text{Axiom 2}$$

These equations reveal that $-(-a)$ and a are both additive inverses $-a$. We conclude, then, by Theorem 2-4, that $-(-a) = a$. ∎

Theorem 2-6 For any $a \in R$, $a \cdot 0 = 0$.

Proof.

$$a \cdot 0 = a \cdot 0 + 0 \qquad \text{Axiom 5}$$
$$= a \cdot 0 + \{a \cdot 0 + [-(a \cdot 0)]\} \qquad \text{Axiom 6}$$
$$= (a \cdot 0 + a \cdot 0) + [-(a \cdot 0)] \qquad \text{Axiom 3}$$
$$= a(0 + 0) + [-(a \cdot 0)] \qquad \text{Axiom 4}$$
$$= a \cdot 0 + [-(a \cdot 0)] \qquad \text{Axiom 5}$$
$$= 0 \quad \blacksquare \qquad \text{Axiom 6}$$

Theorem 2-7   If $a, b \in R$ and $ab = 0$, then either $a = 0$ or $b = 0$.

Proof. If $a = 0$, the theorem is certainly verified. Hence we assume that $a \neq 0$. Then $1/a \in R$ and

$$\frac{1}{a}(ab) = \frac{1}{a}(0) = 0 \qquad \text{Theorem 2-6}$$

But

$$\frac{1}{a}(ab) = \left(\frac{1}{a} \cdot a\right)b \qquad \text{Axiom 3}$$
$$= b\left(a \cdot \frac{1}{a}\right) \qquad \text{Axiom 2}$$
$$= b \qquad \text{Axiom 6}$$

Since $b = (1/a)(ab)$ and $(1/a)(ab) = 0$, we have, by the transitive property, $b = 0$. $\blacksquare$

Theorem 2-8   If $a, b \in R$, then

$$(-a)b = -(ab)$$

Proof.

$$ab + (-a)b = ba + b(-a) \qquad \text{Axiom 2}$$
$$= b[a + (-a)] \qquad \text{Axiom 4}$$
$$= b \cdot 0 \qquad \text{Axiom 6}$$
$$= 0 \qquad \text{Theorem 2-6}$$

Hence $(-a)b$ is the additive inverse of ab. Since $-(ab)$ is also the additive inverse of ab, we have, by Theorem 2-4,

$$(-a)b = -(ab) \quad \blacksquare$$

Corollaries $(-a)b = a(-b)$ and $(-1)b = -b$.

Theorem 2-9 For any $a, b \in R$,

$$(-a)(-b) = ab$$

This theorem follows quite readily from Theorems 2-8 and 2-5, and we leave the proof to the student.

Definition 2-1 The operation of subtraction is defined in terms of addition by the equation

$$a - b = a + (-b)$$

We call $a - b$ the "difference of a and b" or "b subtracted from a" or "a minus b." We remark that in this definition the symbol $-$, called the *minus sign*, is used in two different ways. In the difference $a - b$, the minus sign denotes the operation of subtraction, and $-b$ in $a + (-b)$ is the negative of b.

Theorem 2-10 If $a, b, c \in R$, then

$$a(b - c) = ab - ac$$

Proof.

$$
\begin{array}{ll}
a(b - c) = a[b + (-c)] & \text{Definition 2-1} \\
\quad = ab + a(-c) & \text{Axiom 4} \\
\quad = ab + (-c)a & \text{Axiom 2} \\
\quad = ab + [-(ca)] & \text{Theorem 2-8} \\
\quad = ab - ca & \text{Definition 2-1} \\
\quad = ab - ac \quad \blacksquare & \text{Axiom 2}
\end{array}
$$

Definition 2-2 For any $a, b \in R$, $b \neq 0$, the fraction a/b is defined by the equation

$$\frac{a}{b} = a\left(\frac{1}{b}\right)$$

We call a/b "the quotient of a divided by b"; a is the numerator and b is the denominator of the fraction. We have defined subtraction in terms of addition and here we define division in terms of multiplication. That is, the quotient of a divided by b is equal to a multiplied by the reciprocal of b.

Theorem 2-11 For any $b \in R$, $b \neq 0$,

$$\frac{1}{-b} = -\frac{1}{b}$$

Proof.

$$-b\left(\frac{1}{-b}\right) = 1 \qquad\qquad \text{Axiom 6}$$

and

$$-b\left(-\frac{1}{b}\right) = b\left(\frac{1}{b}\right) = 1 \qquad\qquad \text{Theorem 2-9}$$

The theorem then follows from the fact that $-b$ has only one multiplicative inverse. ∎

Theorem 2-12   For any $a, b \in R,\ b \neq 0$,

$$\frac{a}{-b} = -\frac{a}{b} \quad \text{and} \quad \frac{-a}{b} = -\frac{a}{b}$$

Proof.

$$\frac{a}{-b} = a\left(\frac{1}{-b}\right) \qquad\qquad \text{Definition 2-2}$$

$$= a\left(-\frac{1}{b}\right) \qquad\qquad \text{Theorem 2-11}$$

$$= -\frac{1}{b}(a)$$

$$= -\left(\frac{1}{b} \cdot a\right) \qquad\qquad \text{Theorem 2-8}$$

$$= -\frac{a}{b} \qquad\qquad \text{Definition 2-2}$$

Similarly,

$$\frac{-a}{b} = -\frac{a}{b} \qquad ∎$$

Theorem 2-13   For any $a, b \in R,\ b \neq 0$,

$$\frac{-a}{-b} = \frac{a}{b}$$

Proof.

$$\frac{-a}{-b} = -a\left(\frac{1}{-b}\right) \qquad\qquad \text{Definition 2-2}$$

$$= -a\left(-\frac{1}{b}\right) \qquad \text{Theorem 2-11}$$

$$= a \cdot \frac{1}{b} \qquad \text{Theorem 2-9}$$

$$= \frac{a}{b} \quad \blacksquare \qquad \text{Definition 2-2}$$

Theorem 2-14   For any nonzero numbers $a, b \in R$,

$$\frac{1}{a} \cdot \frac{1}{b} = \frac{1}{ab}$$

Proof.

$$ab\left(\frac{1}{a} \cdot \frac{1}{b}\right) = \left(a \cdot \frac{1}{a}\right)\left(b \cdot \frac{1}{b}\right)$$

$$= 1 \cdot 1 = 1$$

But

$$ab\left(\frac{1}{ab}\right) = 1$$

Since ab has only one multiplicative inverse, we conclude that

$$\frac{1}{a} \cdot \frac{1}{b} = \frac{1}{ab} \quad \blacksquare$$

Theorem 2-15   For any $a, b, c, d \in R$ except $b = d = 0$,

$$\frac{a}{b} \cdot \frac{c}{d} = \frac{ac}{bd}$$

This theorem may be applied repeatedly to show that the product of three or more fractions is equal to the product of the numerators divided by the product of the denominators.

Theorem 2-16   For any $a, b, c \in R$ except $b = c = 0$,

$$\frac{ac}{bc} = \frac{a}{b}$$

Proof.

$$\frac{ac}{bc} = \frac{a}{b} \cdot \frac{c}{c} \qquad \text{Theorem 2-15}$$

$$= \frac{a}{b} \quad \blacksquare$$

This theorem establishes the fact that the value of a fraction is not changed when the numerator and denominator are divided by, or multiplied by, the same nonzero number.

Theorem 2-17   For any $a, b, c \in R$, $c \neq 0$,

$$\frac{a}{c} + \frac{b}{c} = \frac{a + b}{c}$$

This theorem is an immediate consequence of the definition of division and the distributive axiom. The theorem may be extended to show that the sum of three or more fractions with the same denominator is equal to the sum of the numerators divided by the common denominator.

Theorem 2-18   For any $a, b, c, d \in R$ with $b \neq 0$ and $d \neq 0$,

$$\frac{a}{b} + \frac{c}{d} = \frac{ad + bc}{bd}$$

Proof.

$$\frac{a}{b} + \frac{c}{d} = \frac{ad}{bd} + \frac{cb}{db} \qquad\qquad \text{Theorem 2-16}$$

$$= \frac{ad + bc}{bd} \quad \blacksquare \qquad\qquad \text{Theorem 2-17}$$

For convenience we sometimes express the quotient of two fractions, like a/b divided by c/d, by the notation

$$\frac{a}{b} \div \frac{c}{d} \quad \text{or} \quad \frac{a/b}{c/d}$$

Theorem 2-19   For any $a, b, c, d \in R$ with b, c, and d nonzero numbers

$$\frac{a}{b} \div \frac{c}{d} = \frac{a}{b} \cdot \frac{d}{c}$$

Proof.

$$\frac{a}{b} \div \frac{c}{d} = \frac{a/b}{c/d}$$

$$= \frac{(a/b)(d/c)}{(c/d)(d/c)} \qquad\qquad \text{Theorem 2-16}$$

$$= \frac{(a/b)(d/c)}{1} \qquad\qquad \text{Theorem 2-15}$$

$$= \frac{a}{b} \cdot \frac{d}{c} \quad \blacksquare$$

From this theorem, we state the following rule.

To divide one fraction by another fraction, multiply the dividend by the inverted divisor.

Theorem 2-20   If any $a, b, c \in R$ and if $a + c = b + c$, then $a = b$.

To prove the theorem, add $-c$ to both sides of the first equation and apply the necessary axioms.

Theorem 2-21   If $a, b, c \in R$, $c \neq 0$, and if $ac = bc$, then $a = b$.

To prove the theorem, start by multiplying both sides of the first equation by $1/c$.

These theorems are sometimes called, respectively, the *cancellation law of addition* and the *cancellation law of multiplication*.

EXERCISE 2-2

1. Let a be a nonzero number and use the equations $a \cdot 0 = 0$ and $a \cdot 1 = a$ to point out why $1 \neq 0$.
2. Let $a = -1$ in the distributive axiom and show that $-(b + c) = -b - c$.
3. Show that $-(a - b) = -a + b$ and that $-(a + b - c - d) = -a - b + c + d$.

Show that the following statements are correct. The letters stand for any real numbers except that no denominator shall be equal to 0.

4. $-0 \cdot 0 = 0$.

5. If $a + b = 0$, then $b = -a$.

6. If $x + c = a$, then $x = a - c$.

7. If $a = b$, then $-a = -b$.

8. If $\dfrac{a}{b} = x$, then $a = bx$.

9. If $x = y$, then $x/y = 1$.

10. If $\dfrac{a}{b} = \dfrac{c}{d}$, then $ad = bc$.

11. If $ad = bc$, then $a/b = c/d$.

12. Prove that the identity element for addition is unique.
13. Prove Theorems 2-20 and 2-21.

2-4   RATIONAL AND IRRATIONAL NUMBERS

Thus far in our study of real numbers, we have not given explicit attention to any particular subsets of the set R. We shall now discuss two special subsets of the real number system. We continue to assume the existence of the natural numbers $1, 2, 3, \ldots$, which are also called the counting numbers or the positive integers. Then, by the additive inverse axiom, we have the negatives of the natural numbers $-1, -2, -3, \ldots$. The natural numbers, their negatives, and zero constitute the integers of the real number system. The set of integers give rise to a new set of numbers as here defined.

Definition 2-3 Any number which can be expressed as the quotient of two integers (division by zero excluded) is called a *rational number.*

According to this definition numbers like $\frac{17}{3}$, $-\frac{5}{8}$, $\frac{3}{4}$ are rational numbers. Likewise any integer, being equal to itself divided by 1, is a rational number. There are real numbers, however, which are not equal to the quotient of two integers. These numbers constitute an important subset of the real number system and are called *irrational numbers.* It can be proved, for example, that the numbers $\sqrt{2}$, $\sqrt{6}$, and π are not expressible as the quotient of two integers, and are therefore irrational. We recall that the symbols $\sqrt{2}$ and $\sqrt{6}$ stand for the numbers whose squares are 2 and 6. We shall prove that $\sqrt{2}$ is irrational. First, however, we need to discuss a few preliminary ideas.

A real number a is said to be an *even integer* if it can be expressed as $a = 2n$, with n an integer. Thus, 8, -6, 0 are even integers since $8 = 2(4)$, $-6 = 2(-3)$, $0 = 2(0)$. A number b which is equal to $2n + 1$, n an integer, is called an *odd integer.* As examples, 1, -7, 13 are odd integers because $1 = 2(0) + 1$, $-7 = 2(-4) + 1$, $13 = 2(6) + 1$.

Since any integer p is either even or odd, there is some integer n such that

$$p = 2n$$

or

$$p = 2n + 1$$

Hence

$$p^2 = 4n^2 = 2(2n^2)$$

or

$$p^2 = 4n^2 + 4n + 1 = 2(2n^2 + 2n) + 1$$

From the last results, we discover that the square of an even integer is an even integer, and the square of an odd integer is an odd integer. Accordingly, an integer p is an even integer if p^2 is an even integer, and is an odd integer if p^2 is an odd integer. We are now ready to prove the following theorem.

Theorem 2-22 The number $\sqrt{2}$ is an irrational number.

Proof. To establish the theorem we need to show that the number $\sqrt{2}$ is not equal to the quotient of two integers. We proceed by supposing that there is some rational number $p/q = \sqrt{2}$. Further, we assume that the fraction p/q is in lowest terms; that is, p and q have no common factor other than 1 and -1. Then, squaring, we get

$$\frac{p^2}{q^2} = 2 \quad \text{and} \quad p^2 = 2q^2$$

The last equation reveals that p^2 is divisible by 2, and consequently $p = 2n$ for some integer n. Then, substituting, we obtain

$$4n^2 = 2q^2 \quad \text{and} \quad 2n^2 = q^2$$

The second equation here shows that q^2 is divisible by 2, which makes q an even integer. Thus on the basis of our assumption that the $\sqrt{2}$ is rational, we have shown that p and q are both even integers. This is contrary to our agreement that the fraction p/q is in lowest terms. Hence we conclude that the number $\sqrt{2}$ is not a rational number. ∎

In an analogous fashion we could prove that numbers like $\sqrt{3}$, $\sqrt{6}$, and $\sqrt{10}$, for example, are irrational numbers.

We note that the set of rational numbers and the set of irrational numbers are proper subsets of the system of real numbers. The rational numbers contain the set of integers as a subset and the noninteger rational numbers as a subset.

We state the following properties of the system of real numbers.

Add

Property 1   The sum, difference, and product of any two integers is an integer.

Property 2   The sum, difference, and product of any two rational numbers is a rational number. The quotient of any rational number divided by a nonzero rational number is a rational number.

We accept Property 1 as a postulate; we do not prove the property. Property 2 follows at once from Theorems 2-15, 2-18, and 2-19 if we let the letter symbols stand for integers.

2-5   THE REAL NUMBER AXIS

The representation of all real numbers by points on a line is a basic concept of mathematics. To establish this representation, we first choose a direction on a line as positive (to the right in Fig. 2-1) and select a point of the line, which

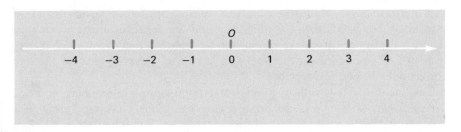

Fig. 2-1

we call the *origin*, to represent the number zero. Next we mark points at distances 1, 2, 3, and so on, units to the right of the origin to represent the natural numbers. In the same way we locate points to the left of the origin to represent the negatives of the natural numbers. We now have a one-to-one correspondence between the set of integers and the points of the line thus far located.

Numbers whose values are between two consecutive integers have their points between the points associated with those integers. Thus the number $\frac{1}{4}$ has its point one-fourth of a unit to the right of the origin. A line segment of length $\frac{1}{4}$ unit can be constructed by using the geometric method of dividing a line segment into four equal parts. Then any rational number of the form $p/4$ has its point at a distance p times one of these smaller units from the origin— to the right if p is a natural number and to the left if p is the negative of a natural number. The points corresponding to certain irrational numbers can be located geometrically. Figures 2-2 and 2-3 exhibit line segments of lengths $\sqrt{2}$ and $\sqrt{3}$, and open the way for locating points corresponding to numbers of the form $(p/q)\sqrt{2}$ and $(p/q)\sqrt{3}$, where p/q is a rational number.

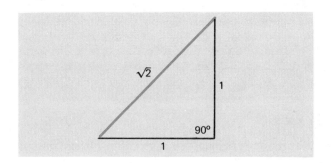

Fig. 2-2

It is impossible to construct geometrically a line segment whose length is equal to an arbitrary irrational number of units. For example, we cannot locate geometrically a point π units from the origin. We assume, however, that there are points of the line π units to the right and left of the origin and we associate these points with the number π and the negative of π. Drawing on the idea of distance from the origin, we assume that every real number corresponds to one point on the line, and, conversely, every point on the line corresponds to one real number.

The line of Fig. 2-1, with its points corresponding to real numbers, is called a *real number axis* or a *real number scale*. The number corresponding to a point on the line is called the *coordinate of the point*. For convenience, we shall sometimes speak of a point as being a number, and vice versa. For example,

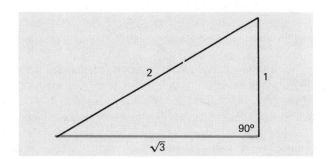

Fig. 2-3

we may say "the point 5" when we mean "the number 5," and "the number 5" when we mean "the point 5."

Thus far in this chapter we have spoken of real numbers and their negatives, but have not mentioned positive numbers and negative numbers. Now, however, we arbitrarily define the numbers corresponding to points on the number axis to the right of the origin as *positive numbers* and those to the left of the origin as *negative numbers*. We note that the positive numbers correspond to points in the chosen positive direction from the origin, and the negative numbers correspond to points in the opposite or negative direction from the origin. In view of this circumstance, we shall consider the coordinate of a point on a number axis to be a *directed distance* from the origin—positive if in the positive direction from the origin, and negative if in the negative direction.

The one-to-one correspondence between the set of real numbers and the points of a line furnish a geometrical representation of the *order* property of real numbers, a topic which we shall study in Chapter 10. Consider a pair of numbers a and b and their corresponding points on a number axis. Either the two points coincide or do not coincide. If the points coincide, we say the numbers a and b are equal. If the point a is in the negative direction from b (to the left in Fig. 2-4), we say a is less than b. If point a is in the positive

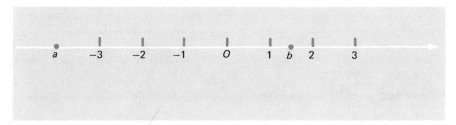

Fig. 2-4

direction from point b, we say that a is greater than b. We express these three possibilities by writing

$$a < b, \quad a = b, \quad a > b$$

where the symbols $<$ and $>$ mean respectively "is less than" and "is greater than." We assume, then, that for any two real numbers one and only one of these relations is true. We note that when $b = 0$, these relations become

$$a < 0, \quad a = 0, \quad a > 0$$

Hence any real number is negative, equal to zero, or positive—one of the three.

Our discussion of real numbers in this section and the previous one enable us to make the following observations: The set of rational numbers and the set of irrational numbers are proper subsets of the system of real numbers. The rational numbers contain the set of integers as a subset and the nonintegral rational numbers as a subset. And, finally, the integers consist of positive integers, zero, and negative integers. This makeup of the system of real numbers is pictured in the accompanying diagram.

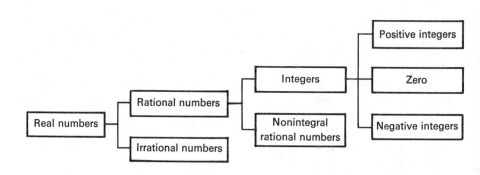

EXERCISE 2-3

Show that each of the following numbers can be expressed as the quotient of two integers and therefore is rational.

1. 0.38 **2.** $5\frac{7}{8}$ **3.** -11.62 **4.** 0

By long division, express each of the following numbers decimally. Note that from some stage on the decimal consists of an endless repetition of a group of one or more digits. Hence we call the decimal a *repeating decimal*. A repeating decimal, as we shall learn in Chap. 11, can be expressed as the quotient of two integers. We conclude then that an irrational number cannot be represented by a repeating decimal.

5. $\frac{5}{6}$ **6.** $\frac{14}{9}$ **7.** $\frac{34}{11}$ **8.** $\frac{7}{4}$

9. $\frac{21}{37}$ **10.** $\frac{8}{9}$ **11.** $\frac{103}{33}$ **12.** $\frac{329}{333}$

13. Paralleling the proof that $\sqrt{2}$ is irrational, prove that $\sqrt{3}$ is irrational.

Place the appropriate symbol "$<$" or "$>$" between each pair of numbers. Also locate the points on a number axis corresponding to the pair of numbers.

14. 4 6 **15.** 3 8 **16.** 7 1

17. 5 3 **18.** 6 -3 **19.** 3 -4

20. -2 0 **21.** -2 -4 **22.** -5 -2

Give the name chosen from the diagram at the end of Sec. 2-5 which best describes each of the following sets of numbers.

23. 1, 4, 8, 9 **24.** 2, 5, 7, 0

25. $-1, -2, -3, -4$ **26.** $-4, -2, -3, -1$

27. $-\frac{2}{3}, 3, \frac{4}{5}, 0$ **28.** 4, $\sqrt{2}, -3, 1$

29. $\sqrt{2}, \sqrt{5}, \sqrt{7}, \sqrt{11}$ **30.** 6, $-4, 2, \sqrt{7}$

Chapter 3

OPERATIONS ON ALGEBRAIC EXPRESSIONS

.

3-1 SOME DEFINITIONS

The field axioms and theorems of the previous chapter furnish the basis for the operations which we shall perform in this chapter. As a preliminary to this study, we introduce certain ideas and definitions.

Addition, subtraction, multiplication, and division are called the *fundamental operations of algebra*. Each of two or more numbers which are multiplied together to form a product is called a *factor* of the product. In the product $7xy$, for example, 7, x, and y are the factors, with x and y the *literal factors*. Usually a numerical factor, such as the 7, is called the *coefficient* of the literal factors. But, more generally, any factor or factors may be regarded as the coefficient of the other factors. Thus in $7xy$, $7x$ is the coefficient of y, and x is the coefficient of $7y$.

A single number or two or more numbers combined with some or all of the fundamental operations of algebra indicated by appropriate signs is called an *algebraic expression*. An algebraic expression may be separated into parts by plus and minus signs. Each of these parts with the sign preceding it is called a *term*. Frequently the sign of the first term of an algebraic expression is omitted: in all such cases the plus sign is understood.

EXAMPLES.

$3a$ is an expression of one term.

$mn - pq + 7$ is an expression with the terms mn, $-pq$, and $+7$.

An expression of one term is called a *monomial*; an expression of two terms is called a *binomial*; and an expression of three terms is called a *trinomial*. The word *multinomial* is used to indicate an expression of two or more terms.

The terms in an algebraic expression with the same literal factor or factors are called *like terms*. The expression $3xy - 3y - 2xy + 4x$, for example, has $3xy$ and $-2xy$ as like terms.

Frequently the concept of the absolute value of a real number is of particular significance. Relative to this concept, we have the following definition.

Definition 3-1 The absolute value of a real number a, denoted by $|a|$, is the real number such that

$|a| = a$ when a is positive or zero.

$|a| = -a$ when a is negative.

According to the definition, the absolute value of every nonzero number is positive and the absolute value of zero is zero. Thus

$$|4| = 4, \qquad |-4| = -(-4) = 4, \qquad |0| = 0$$

3-2 ADDITION AND SUBTRACTION

In arithmetic, positive numbers are added, but in algebra addition is performed on both positive and negative numbers. Addition in this wider system of numbers is sometimes called *algebraic addition*.

EXAMPLE 1.

$$11 + 4 = 15$$

$$-11 - 4 = -(11 + 4) = -15$$

$$11 - 4 = 7$$

$$-11 + 4 = -(11 - 4) = -7$$

These results are obtained graphically in Fig. 3-1. We note, for example, that the point corresponding to 15 is 4 units to the right of the point corresponding to 11. All points located in this manner by two positive numbers are to the right of the origin, and this motivates us to *assume* that the sum of two positive numbers is positive. The results of the four additions are illustrations of the following statements.

1. The sum of two numbers of the same sign (both positive or both negative) is equal to the sum of their absolute values preceded by the common sign.
2. The sum of two numbers of different signs is obtained by subtracting the smaller absolute value from the larger absolute value and taking the sign of the number of larger absolute value.

EXAMPLE 2. Find the sum of $3ab$ and $-7ab$.
Solution. To find the sum of these like terms, we write

$$3ab - 7ab = (3 - 7)ab \qquad \text{distributive law}$$

$$= -4ab$$

We note that the coefficient of the sum is the algebraic sum of the coefficients of the separate terms. This method of adding can be extended to three or more terms.

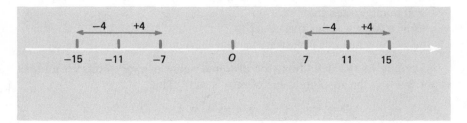

Fig. 3-1

EXAMPLE 3. Combine like terms of the expression

$$11a + 6ac - 4 - 7a + 5ac + 6 - a$$

Solution. Grouping like terms of the expression, we have

$(11a - 7a - a) + (6ac + 5ac) + (-4 + 6)$ associative axiom

$= (11 - 7 - 1)a + (6 + 5)ac + 2$ distributive axiom

$= 3a + 11ac + 2$

EXAMPLE 4. Find the difference if $-5x + 4y + 5z$ is subtracted from $2x + 6y - 3z$.

Solution. We may obtain the difference by finding the sum of the second expression above (the minuend) and the additive inverse of the first expression (the subtrahend). We get the additive inverse of the subtrahend by changing the sign of each of its terms. Thus we have

$$
\begin{array}{r}
2x + 6y - 3z \\
5x - 4y - 5z \\
\hline
7x + 2y - 8z
\end{array}
$$

Alternatively, we may change the signs of the terms of the subtrahend mentally and write

$$
\begin{array}{rl}
2x + 6y - 3z & \text{MINUEND} \\
-5x + 4y + 5z & \text{SUBTRAHEND} \\
\hline
7x + 2y - 8z & \text{DIFFERENCE}
\end{array}
$$

In performing algebraic operations, it is sometimes necessary to place parentheses (or other grouping symbols) about an expression or to remove parentheses from an expression. We illustrate the procedure for introducing and for removing parentheses when a plus sign precedes the parentheses and also when a minus sign precedes the parentheses.

EXAMPLE 5.

$$+(3x - 2y + 4) = 1 \cdot (3x - 2y + 4) \qquad \text{identity element}$$
$$= 1 \cdot 3x + 1(-2y) + 1 \cdot 4 \qquad \text{distributive law}$$
$$= 3x - 2y + 4$$

EXAMPLE 6.

$$-(3x - 2y + 4) = -(3x - 2y + 4) \cdot 1 \qquad \text{identity element}$$
$$= -1 \cdot (3x - 2y + 4) \qquad (-a)b = (-b)a$$
$$= -3x + 2y - 4 \qquad \text{distributive law}$$

As these illustrations show, parentheses preceded by a plus sign may be placed about, or removed from, an expression if no change is made in the signs of the terms of the expression. If a minus sign precedes the parentheses, the signs of all terms must be changed.

Frequently one or more sets of grouping symbols are contained within another. To remove such multiple sets, it is advisable to begin with the innermost symbols.

EXAMPLE 7.

$$\{6x - [3x - (x - y) - 4y] + y\}$$
$$= \{6x - [3x - x + y - 4y] + y\}$$
$$= \{6x - [2x - 3y] + y\}$$
$$= \{6x - 2x + 3y + y\}$$
$$= 4x + 4y$$

EXERCISE 3-1

Perform the indicated operations.

1. $(-14) + 6$ 2. $9 + (-3)$ 3. $(-17) + (-5)$
4. $|-7| + |3| + |-5|$ 5. $|6 - 9| + |10 - 4|$ 6. $|5| + |-5|$
7. $18 + (-17) + (10)$ 8. $(-7) + (-9) + (-13)$ 9. $mn + (-mn) + 6mn$
10. $(-3y) + (-7y + (-6y)$ 11. $6 - (-6)$ 12. $(-20) - 13$
13. $(-x) - (-7x)$ 14. $13 - (-17) - 10$ 15. $(-7) - 1 - (-6)$
16. $(-8a) - (-3a) - (-2a)$ 17. $5a - 6a - 7a$
18. $|4 - 8| - |-6|$ 19. $|14 - 11| - |-8|$

Find the sum of the two expressions in each problem 20 through 24. Next subtract the second expression from the first.

20. $2x + 10y - 16$
$7x - 8y + 10$

21. $5a + 9b - 10c$
$-6a + 7c$

22. $7x + 7y$
$3x - 9y + 6z$

23. $3x + 6xy + 3yz + 4$
$3x - 7xy - 6yz - 13$

24. $4mn - 3m - 5n + 6$
$8mn + 7m - 2n - 8$

Add the three expressions in each problem 25 through 27. Next subtract the third expression from the sum of the first and second.

25. $7a - 3b + 11c; -14a + 10b + 10c; 8a + 8b + 13c$
26. $3xy + 4yz - x; 2x - 4xy + 7yz; 3yz - x + 5xy$
27. $2r - 3rs + 7s; -4s - 3r + 5rs; 2rs + 3s - 8r$

Remove the grouping symbols and simplify by combining like terms.

28. $2a - (b + c) + (a - b - c)$
29. $4x + (y - 3) - (3x + 1)$
30. $(3x - 2y) - (x + 4y - z)$
31. $(x - y) - (2x - 3y) - (-x + y)$
32. $a - (b - c + 4) + (a + c - 2)$
33. $1 - [a - 2b - (3 - a) + 3]$
34. $8a - [(a + 3b) - (3a - 3b)]$
35. $-[x + (3 - x) - (4 + 3x)]$
36. $-\{5a - b - [3b - [c \not= b \mp 2a) \not= 4a] + c\}$
37. $3xy - \{-(2xy + 4x) + [3y - (-xy + x + 2xy)]\}$
38. $-\{a - 2ab + b - [3a + 5ab + 6b - (a - b) + 5]\}$

Place parentheses preceded by a minus sign about the last three terms of each expression. Next place parentheses preceded by a plus sign about the last three terms of each expression.

39. $2 - 5x - 6y + 7z$
40. $a + 2b - 3c - 5$
41. $c - xy + 3 - yz$
42. $1 + 3 - p + q$
43. $-x - 4y - 3z$
44. $r - s + 2t - st$

3-3 MULTIPLICATION

In multiplying two signed numbers, we need to know the sign of the product. To answer this question, we first *assume* that the product of two positive numbers is positive. Then according to Theorems 2-8 and 2-9, we have

$$-a(b) = -(ab)$$

$$(-a)(-b) = ab$$

These equations, being true for any real numbers a and b, are certainly true when a and b are positive numbers. If a and b are positive, $-a$ and $-b$ are negative. Then the first equation reveals that the product of a negative number and a positive number is negative, and the second equation reveals that the product of two negative numbers is positive. So we have a justification for the following rules.

The product of two numbers of like signs is positive. The product of two numbers of unlike signs is negative.

EXAMPLE 1.

$$(-2)(-3)(4) = (6)(4) = 24$$

$$(-2)(3)(4) = (-6)(4) = -24$$

$$(-2)(-3)(-4) = (6)(-4) = -24$$

Products often have the same factor occurring two, three, or more times. A product of this kind may be expressed by writing the factor the proper number of times, but it is more convenient to use a shorthand notation. Thus

$$a \cdot a = a^2 \qquad \text{(read "}a\text{ square")}$$

$$a \cdot a \cdot a = a^3 \qquad \text{(read "}a\text{ cube")}$$

$$a \cdot a \cdot a \cdot a = a^4 \qquad \text{(read "}a\text{ fourth")}$$

These are illustrations of the following definition.

Definition 3-2   If a is any number and n is a positive integer, then a^n denotes the product of n factors each of which is a. That is,

$$a^n = a \cdot a \cdot a \cdot a \cdots a \qquad (n \text{ factors})$$

The quantity a^n is called the *nth power of a*, and is sometimes read "*a* to the *n*th." The number a is the *base* and n is the *exponent*. The first power of a number is usually expressed without an exponent. Thus, by definition, $a^1 = a$.

Theorem 3-1   If a is a real number and m and n are positive integers, then

$$a^m a^n = a^{m+n}$$

Proof. The quantity a^m means the product of m factors each of which is a, and a^n means a is a factor n times. In all, a occurs as a factor $m + n$ times, which we exhibit by writing

$$a^m a^n = \overbrace{(a \cdot a \cdot a \cdot a \cdots a)}^{m \text{ factors}}\overbrace{(a \cdot a \cdot a \cdot a \cdots a)}^{n \text{ factors}} = a^{m+n} \quad \blacksquare$$

As illustrations of this law of exponents, we have

$$2^4 \cdot 2^3 = 2^{4+3} = 2^7$$

$$a^2 a^5 a^7 = a^{2+5+7} = a^{14}$$

$$(2a^3 b^2)(3ab^4) = 2 \cdot 3 \cdot a^3 \cdot a^1 \cdot b^2 \cdot b^4 = 6a^4 b^6$$

EXAMPLE 2. Multiply $x^2 - 3xy - 2y^2$ by $2y^3$.

Solution. Using the distributive law for multiplication and Theorem 3-1, we get

$$2y^3(x^2 - 3xy - 2y^2) = 2y^3(x^2) + 2y^3(-3xy) + 2y^3(-2y^2)$$
$$= 2x^2y^3 - 6xy^4 - 4y^5$$

The intermediate step is written to emphasize the process. The usual procedure is to write only the final result.

The following theorem may be established by repeated applications of the distributive axiom.

Theorem 3-2 The product of two multinomials is the sum of all results obtained by multiplying all the terms of one multinomial by each term of the other.

EXAMPLE 3. Multiply $2x^2 + xy - 3y^2$ by $x^2 - 3xy + y^2$.

Solution. We obtain the product by multiplying the first expression, in turn, by x^2, $-3xy$, and y^2 and then adding the separate results. For convenience in adding the separate products, we arrange the work so that like terms are in columns.

$$
\begin{array}{ll}
2x^2 + \ xy - 3y^2 & \\
\underline{\ x^2 - 3xy + \ y^2} & \\
2x^4 + \ x^3y - 3x^2y^2 & \text{multiply by } x^2 \\
\quad\ - 6x^3y - 3x^2y^2 + \ 9xy^3 & \text{multiply by } -3xy \\
\underline{\qquad\qquad + 2x^2y^2 + \ xy^3 - 3y^4} & \text{multiply by } y^2 \\
2x^4 - 5x^3y - 4x^2y^2 + 10xy^3 - 3y^4 & \text{add}
\end{array}
$$

EXERCISE 3-2

Multiply as indicated and collect like terms.

1. $(4)(-5)(-2)$
2. $(3)(-1)(6)$
3. $(-4)(-5)(-3)$
4. $(5xy)(2xy)$
5. $(-6a^2b)(3ab^2)$
6. $(-2a^2)(-3a)$
7. $(xy)(-2y)(3x)$
8. $(-xy)(2y)(-5x)$
9. $(3x^2y^2)(2xy)(x^2y)$
10. $2x^2y(3xy^2 + x)$
11. $xy(x^3 - y^3)$
12. $-7a^2(4a - 3)$
13. $(x + 2y)(x - 2y)$
14. $(2m - 3n)(m + n)$
15. $(1 - 2x)(4 + 3x)$
16. $(4m - 5)(3m + 6)$
17. $(3x + 5y)(3x - 5y)$
18. $(s^2 - t^2)(s - t)$
19. $(x + 1)(x^2 - x + 1)$
20. $(a - b)(a^2 + ab + b^2)$
21. $(c + 4a)(2c^2 + 5ac - a^2)$
22. $(3a - 2c)(a^2 - 7ac + 8c^2)$
23. $(x^2 - 7)(3x^2 - 5x - 6)$
24. $(b^2 + 3)(2b^2 + 3b - 4)$
25. $(a + b - 1)(a + b + 1)$
26. $(2x - y + 4)(2x + y + 4)$
27. $(a^2 - 3a + 4)(a^2 + 2a - 3)$
28. $(c^2 - 3c + 2)(c^2 - c + 4)$

29. $(3x^2 - x - 6)(2x^2 + x + 4)$ **30.** $(m^2 - 4m + 3)(2m^2 + m - 4)$
31. $(x - 1)(x - 2)(x - 3)$ **32.** $(a + 2)(a - 1)(a + 3)$
33. $(2 - m)(3 - 2m)(1 + m)$ **34.** $(4 - x)(4 + 2x)(4 - x)$

3-4 DIVISION

In this section we shall consider the operation of dividing one algebraic expression by another. First, however, let us discover a rule for determining the sign of the quotient of two signed numbers. Suppose c is the quotient of a number a divided by a nonzero number b. Then

$$\frac{a}{b} = c \quad \text{and} \quad a = bc$$

The last equation tells us that the quotient c is positive if a and b are both positive or both negative, and that c is negative if a and b have unlike signs. Accordingly, we have the following rule.

Rule *The quotient of two numbers of like signs is positive. The quotient of two numbers of unlike signs is negative.*

In defining division (Definition 2-2), we rejected zero as a divisor. The reason for rejecting zero becomes evident when we try to use zero as a divisor. To divide a number a (different from zero) by zero requires that we find a number c such that $a = 0 \cdot c$. This is impossible, of course, because the product of zero and any number is equal to zero. The equation would be satisfied, however, if the dividend a is also zero. But in this case the quotient c could be any number whatever since $0 = 0 \cdot c$. Hence division by zero is never allowed.

To prepare the way for dividing a multinomial by a monomial, we need to establish another law of exponents.

Theorem 3-3 If m and n are positive integers and $a \neq 0 \in R$, then

$$\frac{a^m}{a^n} = a^{m-n} \quad \text{if} \quad m > n$$

$$= 1 \quad \text{if} \quad m = n$$

$$= \frac{1}{a^{n-m}} \quad \text{if} \quad m < n$$

Proof. If $m > n$, we have by Theorem 3-1

$$\frac{a^m}{a^n} = \frac{a^n \cdot a^{m-n}}{a^n}$$

$$= \frac{a^n}{a^n}(a^{m-n})$$

$$= a^{m-n}$$

If $m = n$, we have from the definition of division (Definition 2-2)

$$\frac{a^m}{a^n} = \frac{a^m}{a^m} = a^m\left(\frac{1}{a^m}\right) = 1 \quad \blacksquare$$

We leave the proof of the last part of the theorem to the student.
As illustrations of the theorem, we write

$$\frac{y^5}{y^2} = y^{5-2} = y^3 \quad \text{and} \quad \frac{15x^2}{-3x^4} = -\frac{5}{x^2}$$

EXAMPLE 1. Divide $3x^2y - 6xy^2 + 12x$ by $3x$.

Solution. Theorem 2-17 permits us to divide each term of the multinomial by the monomial. Hence

$$\frac{3x^2y - 6xy^2 + 12x}{3x} = \frac{3x^2y}{3x} - \frac{6xy^2}{3x} + \frac{12x}{3x}$$

$$= xy - 2y^2 + 4$$

We next consider the problem of division when the dividend and divisor are multinomials. We shall, however, restrict the multinomials to a special kind called polynomials.

An algebraic expression in which each term involving x is of the form kx^n, with n a positive integer and k free of x, is called a *polynomial in x*. The polynomial may have one or more terms not involving x. According to this definition

$$3x + 4, \quad \tfrac{5}{4}x^3 - \tfrac{1}{3}x^2 - x - 1, \quad 3x^4 - 5x^2 + 7x$$

are polynomials in x.

The *degree* of each term of a polynomial in x is the exponent of x in that term, and the degree of the polynomial is defined as the greatest exponent of x in the polynomial. For example, the terms of

$$7x^4 + 6x^3 - x^2 - 5x + 3$$

are respectively of degrees four, three, two, one, and zero. Note that we regard the term 3 to be of degree zero.

We may have polynomials in one, two, three, or more letters. The degree of a term with respect to a set of letters is the sum of the exponents of those letters in the term. And the greatest sum thus obtained among the terms of the polynomial determines the degree of the polynomial with respect to the set of letters. The expression

$$2x^3y^2z - 3x^2y + xy - z + 5$$

is a polynomial of degree three in x, of degree two in y, of degree one in z, of degree five in x and y, of degree four in x and z, of degree three in y and z, and of degree six in x, y, and z.

We now describe a method of dividing one polynomial by another. The procedure is quite similar to that of long division in arithmetic.

1. Arrange the dividend and divisor in order of descending powers of a common letter, leaving a gap for any missing power of the letter in the dividend.
2. Divide the first term of the dividend by the first term of the divisor. This gives the first term of the quotient.
3. Multiply the divisor by the first term of the quotient and subtract the result from the dividend.
4. Consider the remainder thus obtained as a new dividend and repeat steps 2 and 3 to find the second term of the quotient and the next remainder.
5. Continue this process until a remainder is obtained which is zero or is of lower degree in the common letter than the degree of the divisor.

If the remainder is zero, the division is exact and the result may be expressed as

$$\frac{\text{dividend}}{\text{divisor}} = \text{quotient}$$

If the remainder is not zero, we express the result as

$$\frac{\text{dividend}}{\text{divisor}} = \text{quotient} + \frac{\text{remainder}}{\text{divisor}}$$

EXAMPLE 2. Divide $3x - 6x^2 + 18$ by $2x + 3$.

Solution. We write the dividend and divisor in order of descending powers of x and carry out the steps listed above.

$$
\begin{array}{r}
-3x + 6 \qquad \text{QUOTIENT} \\
2x + 3 \overline{)\,-6x^2 + 3x + 18} \quad \text{DIVIDEND} \\
-6x^2 - 9x \\
\hline
12x + 18 \\
12x + 18 \\
\hline
0 \quad \text{REMAINDER}
\end{array}
$$

DIVISOR

Hence we have

$$\frac{-6x^2 + 3x + 18}{2x + 3} = -3x + 6$$

EXAMPLE 3. Divide $6a^4 - 41a^2 + 3a + 6$ by $2a^2 - 4a - 3$.

Solution. We carry out the steps listed above using an arrangement which facilitates the operations.

$$
\begin{array}{l}
\;3a^2 +\;\;6a\;-\;4 \qquad\qquad \text{QUOTIENT}\\
\text{DIVISOR}\quad 2a^2 - 4a - 3\overline{)6a^4 - 41a^2 +\;\;3a\;+\;\;6}\quad \text{DIVIDEND}\\
\;6a^4 - 12a^3 -\;\;9a^2\\
\overline{\;12a^3 - 32a^2 +\;\;3a}\\
\;12a^3 - 24a^2 - 18a\\
\overline{\;-\;\;8a^2 + 21a\;+\;\;6}\\
\;-\;\;8a^2 + 16a + 12\\
\overline{\;5a\;-\;\;6}\quad \text{REMAINDER}
\end{array}
$$

Hence

$$
\frac{6a^4 - 41a^2 + 3a + 6}{2a^2 - 4a - 3} = 3a^2 + 6a - 4 + \frac{5a - 6}{2a^2 - 4a - 3}
$$

A problem in division can be checked by using the relation

$$
\text{dividend} = (\text{quotient})(\text{divisor}) + \text{remainder}
$$

due mult of 3

EXERCISE 3-3

Perform the indicated divisions.

1. $\dfrac{4x^4}{2x^2}$

2. $\dfrac{-36x^2y^3}{-6x^2y}$

3. $\dfrac{8ab}{-4ab}$

4. $\dfrac{14x^2 - 21x^3}{7x^2}$

5. $\dfrac{15a^2b^3 - 9ab}{3ab}$

6. $\dfrac{9x^2 - 6x + 3}{-3}$

7. $\dfrac{m^2n - mn^2 - mn}{mn}$

8. $\dfrac{m^4n^3 + 2m^2n^4 - m^2n^2}{m^2n^2}$

9. $\dfrac{5x^2y^3 - 10x^4y^4}{5x^2y^3}$

10. $\dfrac{x^2 - 7x + 6}{x - 1}$

11. $\dfrac{3x^2 + x - 24}{x + 3}$

12. $\dfrac{x^3 - 2x^2 - 32}{x - 4}$

13. $\dfrac{8 - 6x + x^3}{x + 2}$

14. $\dfrac{12x^2 - 7x - 10}{3x - 4}$

15. $\dfrac{2x^3 - 20x - 5}{2x + 6}$

16. $\dfrac{a^3 - b^3}{a - b}$

17. $\dfrac{a^3 + b^3}{a + b}$

18. $\dfrac{a^4 - b^4}{a^2 - b^2}$

19. $\dfrac{x^3 + 10x^2 + 12x + 27}{x^2 + x + 3}$

20. $\dfrac{4 - 7x + 10x^2 + 12x^3}{2x^2 + x + 1}$

21. $\dfrac{2x^4 - 5x^3 + 11x - 2x^2 - 6}{x^2 - 3x + 2}$

22. $\dfrac{4x^4 - 2x^3 + 6x^2 + x + 3}{1 + x + 2x^2}$

23. $(x^4 + x^2y^2 + y^4) \div (x^2 - xy + y^2)$

24. $(2m^4 - 3m^3n + 3m^2n^2 + mn^3 - n^4) \div (2m - n)$

25. $(2m^4 + 5m^3n - 2m^2n^2 + 10mn^3 - 3n^4) \div (m^2 + 3mn - n^2)$

26. $(x^5 - 5x^4 + 7x^3 + x^2 - 8x + 7) \div (x - 2)$

27. $(6 + 10x + 3x^2 + x^3 + x^4 + x^5) \div (x^2 + x + 2)$

28. $(2x^5 + 5x^2 + 1 - 5x^4 - 4x^3 - 5x) \div (2x - 1)$

29. $(2x^5 - 5x^4 + 6x^3 + 4x^2 - 11x + 4) \div (2x^2 - 3x + 1)$

30. $(6x^5 + x^4 + 10x^3 + 4x + 3) \div (2x^2 + x + 3)$

3-5 SPECIAL PRODUCTS

In many problems certain kinds of algebraic expressions occur over and over as factors to be multiplied. Consequently it is worthwhile to learn to write the products speedily. We shall list some formulas for multiplying which will greatly condense the procedure used in Sec. 3-3. Each formula may be verified by carrying out the multiplication indicated on the left side. The student should memorize the formulas and apply them until proficiency is acquired in obtaining products of factors of these types, and, conversely, in supplying the factors from the products. It is important to understand that the letters in the formulas may stand for any quantity.

Product of the sum and difference of two numbers

$$(x + y)(x - y) = x^2 - y^2 \tag{1}$$

Square of a binomial

$$(x + y)^2 = x^2 + 2xy + y^2 \tag{2}$$

$$(x - y)^2 = x^2 - 2xy + y^2 \tag{3}$$

Product of two binomials with like terms

$$(ax + by)(cx + dy) = acx^2 + (ad + bc)xy + bdy^2 \tag{4}$$

The cube of a binomial

$$(x + y)^3 = x^3 + 3x^2y + 3xy^2 + y^3 \tag{5}$$

$$(x - y)^3 = x^3 - 3x^2y + 3xy^2 - y^3 \tag{6}$$

With the exception of formula (4), the student might find it helpful to express the formulas verbally. We suggest the following for formulas (1) and (3).

The product of the sum and difference of two numbers is equal to the square of the first minus the square of the second.

The square of the difference of two numbers is equal to the square of the first minus twice the product of the numbers plus the square of the second.

EXAMPLE 1. Perform the multiplication $(3a + 4b)(3a - 4b)$.

Solution. We apply formula (1) with $x = 3a$ and $y = 4b$. Thus

$$(3a + 4b)(3a - 4b) = (3a)^2 - (4b)^2 = 9a^2 - 16b^2$$

EXAMPLE 2. Find the product $(x - 2y + z)(x - 2y - z)$.

Solution. Grouping terms and applying formulas (1) and (3), we have

$$[(x - 2y) + z][(x - 2y) - z] = (x - 2y)^2 - z^2$$
$$= x^2 - 4xy + 4y^2 - z^2$$

EXAMPLE 3. Write by inspection the product of $3x - 4y$ and $2x + 5y$.

Solution. We write the binomials and describe a procedure. The product of the first terms of the binomials is $(3x)(2x) = 6x^2$, and of the second terms $(-4y)(5y) = -20y^2$.

$$(3x - 4y)(2x + 5y) \qquad -8xy + 15xy = 7xy$$

As indicated in the diagram, the product of the two inner terms is $(-4y)(2x) = -8xy$ and the product of the two outer terms is $(3x)(5y) = 15xy$, which gives $7xy$ as the sum of these like terms. Hence we have

$$(3x - 4y)(2x + 5y) = 6x^2 + 7xy - 20y^2$$

The middle term of this product is called the *cross product*. Being the sum of two terms, it is the least simple to find. However, the idea is to determine mentally each term of a product of this kind and to write the result at once.

EXAMPLE 4. Find the cube of $2m - 5n$.

Solution. We may use formula (6) with $x = 2m$ and $y = 5n$ or, alternatively, formula (5) with $x = 2m$ and $y = -5n$. Choosing the latter, we have

$$(2m - 5n)^3 = (2m)^3 + 3(2m)^2(-5n) + 3(2m)(-5n)^2 + (-5n)^3$$
$$= 8m^3 - 60m^2n + 150mn^2 - 125n^3$$

EXERCISE 3-4

Write the products by inspection.

1. $(m + 4)(m - 4)$
2. $(3 + 4x)(3 - 4x)$
3. $(x^2 + y^2)(x^2 - y^2)$
4. $(ab + cd)(ab - cd)$
5. $(3a^2 - 11b^2)(3a^2 + 11b^2)$
6. $(6 - 3b^2)(6 + 3b^2)$
7. $(r + 4s)^2$
8. $(x - 3)^2$
9. $(2x - 3y)^2$
10. $(mn^2 + 4)^2$
11. $(-4x^2 - 3y^2)^2$
12. $(a^3 - b^3)^2$
13. $[(x - 11) + y]^2$
14. $(2 - m - 2n)^2$
15. $(x + 5)(x + 3)$
16. $(2x - 1)(x - 1)$
17. $(3m + 2)(2m + 1)$
18. $(2a + 3b)(5a + b)$
19. $(7a + b)(a - 6b)$
20. $(3x^2 - 2)(4x^2 + 2)$
21. $(2r^2 - 3s^2)(3r^2 + 2s^2)$
22. $(2r^2 - 3s^2)(3r^2 + s^2)$

23. $(mn - 2y)(mn + y)$ **24.** $(2xy + z^2)(3xy - 2z^2)$
25. $(x - 1)^3$ **26.** $(x + 3)^3$
27. $(2x + 3y)^3$ **28.** $(5 - x)^3$
29. $(-a - b)^3$ **30.** $(x^3 + y^3)^3$

Treat the factors in each problem as the sum and difference of two quantities and then find the product.

31. $(a + b + 1)(a + b - 1)$ **32.** $(x - y + 3)(x - y - 3)$
33. $(1 - 3m + n)(1 - 3m - n)$ **34.** $(a + b + 1)(a - b - 1)$
35. $(x + 5y + 1)(x - 5y + 1)$ **36.** $(x^2 + 1 - x)(x^2 + 1 + x)$

REVIEW EXERCISE

In each of the following problems substitute the given values for the variables. Then simplify the result by performing the indicated operations (Secs. 3-1 and 3-2).

1. $|x| - y + |z - 2|; x = -2, y = 3, z = 1$
2. $x - |y| - |z - 1|; x = -2, y = -3, z = 4$
3. $|x - y| + |2y| - |3z|; x = 3, y = -4, z = 3$
4. $|-5x| + |y - z| + |3z|; x = 4, y = 0, z = 2$
5. $|3x - 1| - (3y + 6) - |x - 2y|; x = 0, y = -4, z = 6$
6. $|3 - 2x| + |2 - 3y| - 2(y + z); x = -3, y = -2, z = 4$
7. $2x - 3 - |3x - 2y| + |4 - 5z|; x = 1, y = 2, z = -2$
8. $|3x - z| - |2x - z| - |y - z|; x = -1, y = 3, z = -1$

Perform the indicated multiplications (Sec. 3-3).

9. $(3a^2b^3)(2ab)$ **10.** $(-3a^2b)(4ab^3)$
11. $(7ab^4)(-2a^2b^2)$ **12.** $(2ab^2)(-5a^3b^3)$
13. $(6a^2b)(ab^2)(-2ab^3)$ **14.** $(4a^3b)(-3ab^3)(-5a^2b^2)$
15. $(3a^3b)(2a^2b^2)(-ab)$ **16.** $(ab^4)(a^4b)(4a^3b^3)$
17. $2b^2(a^2 + 3ab - 2a^2b^3)$ **18.** $-3a^2b(3a + 4ab^2 - b^2)$
19. $-5a^2(a^3 - 3ab^2 - 3a^2b^2)$ **20.** $-ab^4(2a^2 - 4ab - b^3)$
21. $(a + b)(2a - b)$ **22.** $(3a + b)(2a - 3b)$
23. $(3a - 2b)(4a + 3b)$ **24.** $(a^2 - b^2)(a^2 - 3b^2)$
25. $(b + 2a)(b^2 - ab + a^2)$ **26.** $(a^2 - 7)(3a^2 - 5a - 6)$
27. $(a^2 + 3a + 2)(a^2 - a - 4)$ **28.** $(2a^2 - a + 7)(3a^2 + a - 4)$

Perform the indicated divisions (Sec. 3-4).

29. $\dfrac{6x^4}{2x^2}$ **30.** $\dfrac{-25x^3y^2}{-5xy}$ **31.** $\dfrac{14x^4y^3}{-7x^3y}$

32. $\dfrac{21x^2 - 14x^3}{7x^2}$ **33.** $\dfrac{9x^2y^3 - 6xy}{3xy}$ **34.** $\dfrac{4xy - x^2y^2}{xy}$

35. $\dfrac{x^2 - 8x + 7}{x - 1}$ **36.** $\dfrac{x^2 + 3x + 2}{x + 1}$ **37.** $\dfrac{2x^2 - x - 6}{x - 2}$

38. $\dfrac{12x^2 - x - 19}{3x - 4}$ **39.** $\dfrac{8x^2 + 10x - 5}{4x - 1}$ **40.** $\dfrac{x^3 - 2x^2 + 96}{x + 4}$

41. $\dfrac{2x^3 - 9x - 30}{x - 3}$ **42.** $\dfrac{2x^3 - 6x^2 + 7}{2x - 4}$ **43.** $\dfrac{x^3 - 10x^2 + 12x + 27}{x^2 - x + 3}$

44. $\dfrac{x^4 + x^2y^2 + y^4}{x^2 + xy + y^2}$ **45.** $\dfrac{3x^4 - 5x^3 + 12x^2 - 8x - 3}{x^2 - 2x + 4}$

46. $\dfrac{4x^4 + x^3 - 4x^2 + 6x - 3}{x^2 + x - 1}$ **47.** $\dfrac{4x^4 - 2x^3 - 8x^2 + 3x + 1}{2x - 3}$

Write the products by inspection (Sec. 3-5).

48. $(5 + x)(5 - x)$ **49.** $(3x^2 + 4)(3x^2 - 4)$
50. $(x + yz)(x - yz)$ **51.** $(4 - x)^2$
52. $(2x - 3)^2$ **53.** $(x^2 + 2y^2)^2$
54. $(xy^2 - 3)^2$ **55.** $(2x - 3y + 2)^2$
56. $(x^2 + x - 1)^2$ **57.** $(x - 6)(x + 5)$
58. $(2x + 1)(x - 5)$ **59.** $(2x + 7y)(3x - 4y)$
60. $(3x^2 + 2)(2x^2 - 1)$ **61.** $(2x + 7y)(7x - 4y)$
62. $(x + 2)^3$ **63.** $(x - 3)^3$
64. $(x^3 - y^3)^3$ **65.** $(xy - 3z)^3$

Treat the factors in each problem as the sum and difference of two quantities and then find the product (Sec. 3-5).

66. $(x - y + 1)(x - y - 1)$ **67.** $(x + y - 3)(x + y + 3)$
68. $(1 - 2x + y)(1 - 2x - y)$ **69.** $(x^2 + 1 - x)(x^2 + 1 + x)$
70. $(x - y + 1)(x + y - 1)$ **71.** $(x + 5y - 1)(x - 5y + 1)$

Chapter 4

FACTORING AND OPERATIONS ON FRACTIONS

4-1 SIMPLE TYPES OF FACTORING

A considerable part of Chapter 3 dealt with finding the products of factors. In this chapter we shall consider the problem of finding the factors when the product is given. This process, called *factoring*, is used extensively in performing algebraic operations. Factoring at this stage is usually restricted to polynomials and factors whose terms have rational coefficients. We say that a polynomial is completely factored when none of its factors can be factored. Each factor, then, being expressible only as 1 times itself or minus 1 times its negative, is said to be *prime*.

In this section we take up polynomials whose factors are quite easily determined.

Type 1. *Common factors.* If each term of a polynomial has a common factor, the distributive law enables us to express the polynomial as the product of two factors, one of which is the common factor.

EXAMPLE 1. $4x^3y^2 + 2x^2y^3 - 6x^2y^2 = 2x^2y^2(2x + y - 3)$

EXAMPLE 2. $x(a + b) + y(a + b) = (a + b)(x + y)$

Type 2. *The difference of two squares.* By reading the special product formula $(x + y)(x - y) = x^2 - y^2$ from right to left, we obtain the factoring formula

$$x^2 - y^2 = (x + y)(x - y)$$

EXAMPLE 1. $4a^2 - 9b^2 = (2a)^2 - (3b)^2 = (2a + 3b)(2a - 3b)$

EXAMPLE 2. $4ax^2 - 16ay^2 = 4a(x^2 - 4y^2) = 4a(x + 2y)(x - 2y)$

Note: As here illustrated, the removal of a common factor, if any, should be the first step in factoring.

EXAMPLE 3. $(a + b)^2 - (c - 2d)^2 = (a + b + c - 2d)(a + b - c + 2d)$

Type 3. *Perfect square trinomial.* From the formulas for the square of a binomial (Sec. 3-5), we have the factoring formulas

$$x^2 + 2xy + y^2 = (x + y)^2$$
$$x^2 - 2xy + y^2 = (x - y)^2$$

EXAMPLE 1.

$$25x^2 + 30x + 9 = (5x)^2 + 2(5x)(3) + (3)^2$$
$$= (5x + 3)^2$$

EXAMPLE 2.

$$9x^2 - 12xy + 4y^2 = (3x)^2 - 2(3x)(2y) + (2y)^2$$
$$= (3x - 2y)^2$$

Type 4. *The sum and difference of two cubes.* By multiplying we may verify the formulas

$$x^3 + y^3 = (x + y)(x^2 - xy + y^2)$$
$$x^3 - y^3 = (x - y)(x^2 + xy + y^2)$$

which enable us to factor the sum and difference of two-cubes.

EXAMPLE 1.

$$27x^3 + 64y^3 = (3x)^3 + (4y)^3$$
$$= (3x + 4y)(9x^2 - 12xy + 16y^2)$$

EXAMPLE 2.

$$2x^4y - 54xy^4 = 2xy(x^3 - 27y^3)$$
$$= 2xy(x - 3y)(x^2 + 3xy + 9y^2)$$

EXERCISE 4-1

Factor each expression completely.

1. $x^3 - x^2y + 2xy^3$
2. $10x^2y + 6xy^3 - 2x^3y$
3. $3x^2 - 3xy + 6xy^2$
4. $3(a + 3) + x(a + 3)$
5. $m(a + 2b) - n(a + 2b)$
6. $2x(1 - y) + 3a(1 - y)$
7. $9x^2 - y^2$
8. $x^2 - 25$
9. $1 - 49y^2$
10. $4a^2 - 9b^2c^2$
11. $2ax^2 - 50ay^2$
12. $m^3 - 16m$
13. $(a - 1)^2 - b^2$
14. $(a - b)^2 - 9c^2$
15. $4x^2y^2 - (z - 1)^2$
16. $x^2 + 4x + 4$
17. $x^2 - 6x + 9$
18. $4x^2 + 12x + 9$
19. $16a^2 + 8ab + b^2$
20. $4a^2x^2 - 4ax + 1$

21. $3ax^2 + 6ax + 3a$

22. $8x^3 + y^3$

23. $a^3 - 64b^3$

24. $1 - 8b^3$

25. $27p^3q^3 + 8$

26. $a^4 + 8ab^3$

27. $2xy^4 - 2x^4y$

28. $8(x + 1)^3 + y^3$

29. $(x + 2y)^3 - 8y^3$

30. $54x^5y - 2x^2y^4$

31. $x^4 - 16$

32. $81 - y^4$

33. $a^4 - 625b^4$

34. $a^6 - b^6$

35. $64x^6 - y^6$

36. $x^6 + 64y^6$

37. $3a^7b^2 - 3a^3b^2$

38. $2a^5b^2 - 162ab^2$

39. $(2m - 1)^2 - (n - 1)^2$

40. $2a^3b^3 + 12a^3b^2 + 18a^3b$

41. $4a^4b^2 - 12a^3b^3 + 9a^2b^4$

42. $(2a - b)^2 - (a + 3b)^2$

43. $\dfrac{a^2}{9} - \dfrac{ab}{3} + \dfrac{b^2}{4}$

44. $x^2 + \dfrac{b}{a}x + \dfrac{b^2}{4a^2}$

4-2   GENERAL SECOND-DEGREE TRINOMIAL

We now consider the case of factorable trinomials which are not perfect squares. By reversing the order of writing the members of the formula for the product of two binomials with like terms (Sec. 3-5), we have

$$acx^2 + (ad + bc)xy + bdy^2 = (ax + by)(cx + dy)$$

To factor a trinomial of this kind, we need to find rational values for a, b, c, and d from the known values ac, $ad + bc$, and bd. If such rational values exist, we say the trinomial is factorable. Factoring for this case is less simple than factoring perfect-square trinomials. The following examples, however, show a procedure.

EXAMPLE 1. Factor the polynomial $x^2 - 5x + 6$.

Solution. If x is used as the first term of each factor, the product will contain x^2. The product of the other terms must be 6. This value is the product of 1 and 6, of -1 and -6, of 2 and 3, and of -2 and -3. Observing the middle term of the given trinomial, we see that the numbers whose product is 6 must have the sum -5. Only the pair -2 and -3 have the required sum. Hence

$$x^2 - 5x + 6 = (x - 2)(x - 3)$$

EXAMPLE 2. Factor the expression $8x^2 - 2xy - 15y^2$.

Solution. For the first terms of the factors, the pair x and $8x$ and the pair $2x$ and $4x$ are the only possibilities with positive coefficients. Then the second terms must have $-15y^2$ as their product. The choices for this product are y and $-15y$, $-y$ and $15y$, $3y$ and $-5y$, and $-3y$ and $5y$. But the sum of the cross products (the first term of each factor times the second term of the other) must yield $-2xy$. By trying the various combinations it may be determined that $2x - 3y$ and $4x + 5y$ are the factors. Hence

$$8x^2 - 2xy - 15y^2 = (2x - 3y)(4x + 5y)$$

EXERCISE 4-2

Factor each expression completely.

1. $m^2 + 6m + 8$	**2.** $m^2 - 9m + 18$	**3.** $m^2 - 9m + 20$
4. $m^2 + 5m - 14$	**5.** $m^2 - 4mn - 21n^2$	**6.** $m^2 - 5mn - 36n^2$
7. $m^2 + 11mn + 18n^2$	**8.** $2x^2 + x - 3$	**9.** $3x^2 + 4x - 4$
10. $4x^2 + 23x - 6$	**11.** $10x^2 - 13x - 3$	**12.** $4x^2 - 4x - 15$
13. $21 - 25x - 4x^2$	**14.** $3x^2y^2 - 11xy + 8$	**15.** $6x^2 - 11xy - 7y^2$
16. $4x^2y^2 - 5xy^3 - 6y^4$	**17.** $a^4 - 7a^2 + 12$	**18.** $6a^4 - 5a^2 - 6$
19. $2a^4 - a^2 - 1$	**20.** $4a^4 - 17a^2 + 4$	**21.** $3a^4 - 10a^2 - 8$
22. $4a^4 - 37a^2 + 9$	**23.** $9a^4 - 13a^2 + 4$	
24. $(m + n)^2 - (m + n) - 2$	**25.** $3(x - y)^2 - 7(x - y) + 2$	
26. $4x^4y - 2x^3y^2 - 20x^2y^3$	**27.** $12x^3y - 27x^2y^2 + 6xy^3$	
28. $x^6 - 7x^3 - 8$	**29.** $8x^6 - 7x^3 - 1$	

4-3 FACTORING BY GROUPING

Sometimes a polynomial whose terms contain no common factor can be separated into groups of terms which have a common factor. And some polynomials not in the form of the difference of two squares may be so expressed by a proper grouping of terms.

EXAMPLE 1.

$$3x + 3y + ax + ay = (3x + 3y) + (ax + ay)$$
$$= 3(x + y) + a(x + y)$$
$$= (x + y)(3 + a)$$

EXAMPLE 2.

$$x^2 - y^2 - 2x + 2y = (x^2 - y^2) + (-2x + 2y)$$
$$= (x + y)(x - v) - 2(x - y)$$
$$= (x - y)(x + y - 2)$$

EXAMPLE 3.

$$4 - x^2 + 2xy - y^2 = 4 - (x^2 - 2xy + y^2)$$
$$= 4 - (x - y)^2$$
$$= (2 + x - y)(2 - x + y)$$

EXAMPLE 4. Factor $x^4 - 13x^2 + 4$.

Solution. The given trinomial would be a perfect square if the middle term were $-4x^2$. We obtain the perfect square by adding $9x^2$. If we add $9x^2$, we must subtract $9x^2$. Thus we have

$$x^4 - 13x^2 + 4 = (x^4 - 4x^2 + 4) - 9x^2$$
$$= (x^2 - 2)^2 - (3x)^2$$
$$= (x^2 - 2 + 3x)(x^2 - 2 - 3x)$$

The device of adding and subtracting a term, in a problem of this kind, works only if (a) the expression to be factored becomes a perfect square by the addition, not the subtraction, of a term, (b) the term added, and also subtracted, is a perfect square.

$\vee$ *EXERCISE 4-3*

Factor each expression completely.

1. $ax + ay + 4x + 4y$
2. $4x + 12 + xy + 3y$
3. $ax - ay + 2cx - 2cy$
4. $12xy + 9cy + 3cx + 4x^2$
5. $3xy + 9ax - ay - 3a^2$
6. $m^3 - 2m^2 - 3m + 6$
7. $4 - 2m - 4m^2 + 2m^3$
8. $am^3 - 3m^2 - am + 3$
9. $x^2 + 2x + 1 + ax + a$
10. $x^2 - 4xy + 4y^2 - 3x + 6y$
11. $x^2 - y^2 + 5x + 5y$
12. $x^2 - 4y^2 - 3x + 6y$
13. $x^2 - 2xy + y^2 - 1$
14. $x^2 - y^2 - 8x + 16$
15. $m^2 - 9 + 12n - 4n^2$
16. $25 - m^2 - 2mn - n^2$
17. $x^2 - 2x + 1 - y^2 - 2yz - z^2$
18. $x^2 + 4x + 4 - y^2 + 2yz - z^2$
19. $x^2 - 8x - y^2 + 4y + 12$
20. $x^3 + y^3 + x + y$
21. $a^3 + b^3 - a^2 - 2ab - b^2$
22. $x^4 + 4$
23. $64a^4 + 1$
24. $x^4 - 3x^2 + 1$
25. $x^4 + x^2y^2 + y^4$
26. $36x^4 + 11x^2 + 1$
27. $4a^4 + 8a^2b^2 + 9b^4$

4-4 ALGEBRAIC FRACTIONS

In Chapter 2 we defined a fraction and proved some theorems which provide the basis for certain operations on fractions. The fraction, as there defined, involved literal numbers. Fractions of this kind are sometimes called *algebraic fractions* to distinguish them from the fractions of arithmetic.

We may think of a fraction as having three signs associated with it—the sign preceding the fraction, the sign of the numerator, and the sign of the denominator. With the agreement that when one of these three signs is missing the plus sign is understood, we have from Theorems 2-12 and 2-13

$$\frac{a}{b} = \frac{-a}{-b} = -\frac{-a}{b} = -\frac{a}{-b}, \qquad b \neq 0$$

These equations tell us that the value of a fraction is not altered (a) when the sign of the numerator and the sign of the denominator are changed, (b) when the sign of the fraction and the sign of either the numerator or denominator are changed. In other words, *any two of the three signs associated with a fraction may be changed without altering the value of the fraction.*

We call attention also to Theorem 2-16 which is expressed symbolically by

$$\frac{a}{b} = \frac{ac}{bc}, \qquad b \cdot c \neq 0$$

Reading from left to right and from right to left, we see that the numerator and denominator of a fraction may be multiplied or divided by any nonzero number without altering the value of the fraction.

If the numerator and denominator of a fraction are in factored form, the signs of the terms of any two of the factors, or the signs of the terms of one of the factors and the sign of the fraction, may be changed without altering the value of the fraction. We leave the proof of this statement to the student.

In dealing with algebraic fractions, we shall agree that the literal number or numbers in a denominator may not have values which make the denominator equal to zero.

4-5 REDUCTION TO LOWEST TERMS

A fraction is said to be in *lowest terms* or in *simplest form* if the numerator and denominator have no common factor except ± 1. Hence we may determine if a fraction is in lowest terms by expressing the numerator and denominator as products of their prime factors. Any factor appearing in both the numerator and denominator can then be readily removed by division.

EXAMPLE 1. Reduce the fraction $\dfrac{x^2 - 5x + 6}{x^2 + x - 6}$ to lowest terms.

Solution.

$$\frac{x^2 - 5x + 6}{x^2 + x - 6} = \frac{(x - 2)(x - 3)}{(x - 2)(x + 3)} = \frac{x - 3}{x + 3}$$

The common factor is removed by dividing both the numerator and denominator by $x - 2$.

EXAMPLE 2. Reduce the fraction $\dfrac{x^2 - 9x + 20}{25 - x^2}$ to lowest terms.

Solution.

$$\frac{x^2 - 9x + 20}{25 - x^2} = \frac{(x - 5)(x - 4)}{(5 - x)(5 + x)}$$

$$= \frac{(5 - x)(4 - x)}{(5 - x)(5 + x)}$$

$$= \frac{4 - x}{5 + x}$$

Observe that we obtained a common factor in the numerator and denominator by reversing the signs of the terms in the two factors of the numerator.

The process of eliminating a common factor from the numerator and denominator of a fraction is called *multiplicative cancelling*. This operation is sometimes used erroneously by striking out equal quantities which are not factors. The fraction

$$\frac{(2x + y) + 5}{(2x + y)(2x - y)}$$

for example, has $2x + y$ as a *factor* of the denominator but $2x + y$ is a *term* of the numerator, not a factor. Striking out the equal quantities here would mean that the denominator is divided by $2x + y$ and that $2x + y$ is subtracted from the numerator. This is a serious error and should be avoided. The fraction as given is in lowest terms.

EXERCISE 4-4

Reduce each fraction to lowest terms.

1. $\dfrac{25xy^2}{75x^2y}$

2. $\dfrac{24ab^2c}{18a^2bc^2}$

3. $\dfrac{3^2a^6b}{18a^4b^6}$

4. $\dfrac{2x}{x^2 + 2x}$

5. $\dfrac{ax + 3x}{a^2 + 3a}$

6. $\dfrac{3x + 2}{9x^2 - 4}$

7. $\dfrac{(x - 2)^2}{x^2 - 4}$

8. $\dfrac{a^2 - ab}{ab - b^2}$

9. $\dfrac{x - 1}{x - x^2}$

10. $\dfrac{6x - 6y}{3x - 3y}$

11. $\dfrac{b^2 - a^2}{(a - b)^2}$

12. $\dfrac{3x^2 + 15xy}{2x^2 + 10xy}$

13. $\dfrac{6abc - 18ab}{3a^2bc - 9a^2b}$

14. $\dfrac{5x^3y - 5x^2y^2}{x^2y^2 - xy^3}$

15. $\dfrac{a^2 - b^2}{2a^2 + ab - b^2}$

16. $\dfrac{3x^2 - 11x + 6}{3x^2 + x - 2}$

17. $\dfrac{4x^2 + 7x - 2}{1 - 16x^2}$

18. $\dfrac{3 + 2x - x^2}{3 + 5x + 2x^2}$

19. $\dfrac{9 - 6x + x^2}{9 - 9x + 2x^2}$

20. $\dfrac{a^2 - ab + b^2}{a^3 + b^3}$

21. $\dfrac{a^2 - 2ab + b^2}{a^3 - b^3}$

22. $\dfrac{x^3 - y^3}{x^3 + x^2y + xy^2}$

23. $\dfrac{x^2 - y^2 + x + y}{x^2 - y^2}$

24. $\dfrac{4a^2 - b^2}{4a^2 - b^2 + 2a - b}$

25. $\dfrac{y^3 - y^2 + y - 1}{y^3 + y^2 + y + 1}$

26. $\dfrac{a^3 - 2a^2 - a + 2}{a^3 + 2a^2 - a - 2}$

27. $\dfrac{c^3 + 3c^2 + 2c + 6}{c^3 - c^2 + 2c - 2}$

4-6   MULTIPLICATION AND DIVISION OF FRACTIONS

Theorems 2-15 and 2-19 furnish procedures for multiplying and dividing fractions. These theorems are expressed by the equations

$$\frac{a}{b} \cdot \frac{c}{d} = \frac{ac}{bd} \qquad b, d \neq 0$$

$$\frac{a}{b} \div \frac{c}{d} = \frac{a}{b} \cdot \frac{d}{c} \qquad b, c, d \neq 0$$

These theorems permit us to make the following statements.

1. The product of two or more fractions is equal to the product of the numerators divided by the product of the denominators.
2. The quotient of two fractions is equal to the dividend multiplied by the inverted divisor.

It is usually desirable to reduce the product, or the quotient, of fractions to lowest terms. That being the case, the best procedure is to write each fraction in factored form. Common factors of the numerator and the denominator can then be easily removed by division.

EXAMPLE 1. Multiply $\dfrac{2x^2 - x - 3}{x^2 - 1}$ by $\dfrac{x^2 - 2x + 1}{3x^2 - x - 2}$.

Solution. Factoring the numerators and denominators of the fractions, we have

$$\frac{(2x - 3)(x + 1)}{(x + 1)(x - 1)} \cdot \frac{(x - 1)(x - 1)}{(x - 1)(3x + 2)} = \frac{2x - 3}{3x + 2}$$

EXAMPLE 2. Find the product of

$$(x^2 - y^2) \cdot \frac{x^2 + 3xy + 2y^2}{2x^2 - 3xy + y^2} \cdot \frac{2x - y}{x^2 + 2xy + y^2}$$

Solution. We write $x^2 - y^2$ in fractional form and factor all numerators and denominators. Thus, we obtain

$$\frac{(x + y)(x - y)}{1} \cdot \frac{(x + 2y)(x + y)}{(2x - y)(x - y)} \cdot \frac{2x - y}{(x + y)^2} = x + 2y$$

EXAMPLE 3. Divide $\dfrac{a^3 - 1}{a^2 - 9}$ by $\dfrac{a^2 + a + 1}{a^2 - 2a - 3}$.

Solution. To find the quotient we invert the divisor and then multiply. Thus we get

$$\frac{a^3 - 1}{a^2 - 9} \cdot \frac{a^2 - 2a - 3}{a^2 + a + 1} = \frac{(a - 1)(a^2 + a + 1)}{(a + 3)(a - 3)} \cdot \frac{(a + 1)(a - 3)}{a^2 + a + 1}$$

$$= \frac{(a - 1)(a + 1)}{a + 3}$$

EXAMPLE 4. Perform the indicated operations:

$$\frac{x^2 - 7x + 12}{x^2 - 6x + 8} \cdot \frac{x^2 + 3x - 10}{x^2 - 10x + 21} \div \frac{x + 5}{x - 7}$$

Solution. Inverting the divisor and factoring, we have

$$\frac{(x-3)(x-4)}{(x-2)(x-4)} \cdot \frac{(x+5)(x-2)}{(x-7)(x-3)} \cdot \frac{x-7}{x+5} = 1$$

EXERCISE 4-5

Perform the indicated operations and express each result in lowest terms.

1. $\dfrac{28y}{15x} \cdot \dfrac{5x}{7y}$

2. $\dfrac{4x^2}{3a^2b} \cdot \dfrac{9a^3b}{2x}$

3. $8ab^2 \cdot \dfrac{3}{4ab}$

4. $\dfrac{64x^2}{39y} \cdot \dfrac{13y^2}{48x^2}$

5. $\dfrac{17x^2}{27ab} \cdot \dfrac{9ab^2}{34x}$

6. $\dfrac{5a^4x^3}{8b^2y} \cdot \dfrac{4by}{a^3x^4}$

7. $\dfrac{7bc}{9a} \div \dfrac{14c}{a^2}$

8. $\dfrac{10xy^3}{3z^3} \div \dfrac{5x^2y}{6z^3}$

9. $64x^3 \div \dfrac{8x^2y^3}{3}$

10. $\dfrac{x^2+x}{y^2-1} \cdot \dfrac{y+1}{x+1}$

11. $\dfrac{a+b}{x-2y} \cdot \dfrac{2y-x}{2b+2a}$

12. $\dfrac{1-a}{1-x^3} \cdot \dfrac{1-x^2}{a^2-1}$

13. $\dfrac{x-y}{a-b} \div \dfrac{y-x}{2a-2b}$

14. $\dfrac{2x-12}{3x} \div (x-6)$

15. $\dfrac{x^2-y^2}{x+2y} \div \dfrac{x+y}{2x+4y}$

16. $\dfrac{a^3+b^3}{x^3+y^3} \cdot \dfrac{x+y}{a+b}$

17. $\dfrac{4-x^2}{8-x^3} \cdot \dfrac{2-x}{2+x}$

18. $\dfrac{a^3-b^3}{x^2-y^2} \cdot \dfrac{x-y}{a-b}$

19. $\dfrac{30-7x-x^2}{24-2x-x^2} \cdot \dfrac{x^2-x-12}{x^2+7x-30}$

20. $\dfrac{7x+21y}{x^2-9y^2} \cdot \dfrac{x^2-6xy+9y^2}{x^2-2xy-3y^2}$

21. $\dfrac{2bx-b}{2ax-14a} \cdot \dfrac{x^2-49}{6x^2-7x+2}$

22. $\dfrac{a^2-4}{2a^2-5a+2} \cdot \dfrac{2a-1}{a^3+8}$

23. $\dfrac{x^2-6x+9}{x^2+8x+7} \div \dfrac{x^2-x-6}{x^2-x-2}$

24. $\dfrac{2xy+2y^2}{3y^2-xy-4x^2} \div \dfrac{3x^2+3xy}{9y^2-16x^2}$

25. $\dfrac{a^3+2a^2-a-2}{a^2+3a+9} \div \dfrac{a^2+3a+2}{a^3-27}$

26. $\dfrac{x^4+x^2+1}{4x^2-1} \div \dfrac{x^2+x+1}{2x-1}$

27. $\dfrac{4x^2-1}{2x^2-5x-3} \cdot \dfrac{x^2+x-12}{4x^2-8x+3} \cdot \dfrac{2x^2+3x-9}{x^2+7x+12}$

28. $\dfrac{(x-3)^3}{6x^2-19x+3} \cdot \dfrac{6x^2+35x+36}{x^2-6x+9} \cdot \dfrac{x+1}{x^2+37x+36}$

29. $\left(\dfrac{x^2}{1-x^2} \div \dfrac{2x}{1-x}\right) \div \dfrac{1-x}{1+x}$

30. $\dfrac{x^2}{1-x^2} \div \left(\dfrac{2x}{1-x} \div \dfrac{1-x}{1+x}\right)$

31. $\dfrac{3a^2-a-10}{8a^2-2a-3} \cdot \dfrac{10a^2+a-2}{3a^2+20a+25} \div \dfrac{5a^2+8a-4}{12a^2+11a-15}$

32. $\dfrac{x^2+4xy-12y^2}{x^2+7xy+6y^2} \cdot \dfrac{x^2-6xy-7y^2}{x^2-xy-12y^2} \div \dfrac{x^2-9xy+14y^2}{x^2-xy-12y^2}$

33. $\dfrac{3x^2+4xy+y^2}{4x^2+2xy-2y^2} \cdot \dfrac{3x^2+13xy-10y^2}{x^2-4y^2} \div \dfrac{9x^2-12xy+4y^2}{2x^2-5xy+2y^2}$

4-7 ADDITION OF FRACTIONS

The sum of two or more fractions with the same denominator is equal to the fraction whose numerator is the algebraic sum of the numerators and whose denominator is the common denominator (Theorem 2-17). The result obtained in this manner is usually called addition even though a minus sign precedes some of the fractions. To add fractions which have different denominators, it is necessary first to express the fractions as equivalent fractions with a common denominator (Theorem 2-18). In view of this circumstance, we shall introduce the idea of a common multiple of polynomials.

A polynomial which is exactly divisible by two or more polynomials is called a *common multiple* of the latter polynomials. The common multiple determined in the following way is called the *lowest common multiple* (*L.C.M.*).

1. Factor each of the given polynomials.
2. Form a product by using all the different prime factors of the polynomials, giving each prime factor the largest exponent which that factor has in any of the polynomials.

EXAMPLE 1. Find the L.C.M. of the polynomials

$$6x^2 + 3x - 3, \qquad 7x^2 - 14x + 7, \qquad 12x^2 - 12$$

Solution. Factoring each polynomial, we have

$$6x^2 + 3x - 3 = 3(2x^2 + x - 1) = 3(2x - 1)(x + 1)$$

$$7x^2 - 14x + 7 = 7(x^2 - 2x + 1) = 7(x - 1)^2$$

$$12x^2 - 12 = 12(x^2 - 1) = 2^2 \cdot 3(x + 1)(x - 1)$$

Selecting the proper power of each prime factor, we get

$$2^2 \cdot 3 \cdot 7(2x - 1)(x + 1)(x - 1)^2 = 84(2x - 1)(x + 1)(x - 1)^2$$

which is the lowest common multiple of the given polynomials.

The L.C.M. may be expressed in factored form or the actual multiplication may be carried out. The factored form, however, is usually more convenient.

In adding fractions, we shall use the L.C.M. of the various denominators for the common denominator. It is then called the *lowest common denominator* (L.C.D.). Having found the L.C.D., we obtain the sum of the fractions by the following operations.

1. Multiply the numerator and denominator of each fraction by the quotient obtained by dividing the L.C.D. by the denominator of that fraction.
2. Add the resulting numerators, being careful to reverse the signs of all terms of the numerator of any fraction preceded by a minus sign. This gives the numerator of the sum of the fractions; the denominator is the common denominator.

EXAMPLE 1. Add the fractions $\dfrac{2}{(x-3)(x+4)} + \dfrac{3}{(3-x)(2x+1)}$.

Solution. The factor $3-x$ in the second denominator is the negative of the factor $x-3$ in the first denominator. To handle this situation, we reverse the signs of the terms of $3-x$ and change the sign before the fraction from plus to minus. Hence we obtain

$$\frac{2}{(x-3)(x+4)} - \frac{3}{(x-3)(2x+1)}$$

$$= \frac{2(2x+1)}{(x-3)(x+4)(2x+1)} - \frac{3(x+4)}{(x-3)(x+4)(2x+1)}$$

$$= \frac{4x+2-3x-12}{(x-3)(x+4)(2x+1)}$$

$$= \frac{x-10}{(x-3)(x+4)(2x+1)}$$

EXAMPLE 2. Combine $\dfrac{12x^2}{9x^2-1} - \dfrac{2x}{3x-1} - \dfrac{3x}{3x+1}$ into a single fraction.

Solution. Factoring the denominator of the first fraction, we get

$$\frac{12x^2}{(3x+1)(3x-1)} - \frac{2x}{3x-1} - \frac{3x}{3x+1}$$

$$= \frac{12x^2 - 2x(3x+1) - 3x(3x-1)}{(3x+1)(3x-1)}$$

$$= \frac{12x^2 - 6x^2 - 2x - 9x^2 + 3x}{(3x+1)(3x-1)}$$

$$= \frac{-x(3x-1)}{(3x+1)(3x-1)} = \frac{-x}{3x+1}$$

EXERCISE 4-6

Mult of 3

Find the L.C.M. and leave each result in factored form except for the numerical coefficient.

1. $18x^2y, 20xy^2, 30x^3y$ 2. $60xy^2, 36x^3, 84y^3z^2$
3. $a^2-4, 4a^2-8a, 3a^2+6a$ 4. $9a^2-1, 6a^3-2a^2, 9a+3$
5. $3x^2+5xy-2y^2, 2x^2+5xy+2y^2, 6x^2+xy-y^2$
6. $2x^2-xy-3y^2, x^2-xy-2y^2, 2x^2-7xy+6y^2$

Combine into a single fraction in lowest terms.

7. $\dfrac{2x-3y}{3z} + \dfrac{3x+4y}{3z}$ 8. $\dfrac{x+y}{x-6} - \dfrac{5x-2y}{x-6}$ 9. $\dfrac{x+2}{x-3} + \dfrac{x-2}{3-x}$

10. $\dfrac{a}{5} - \dfrac{a}{10} + \dfrac{7a}{25}$ 11. $\dfrac{1}{x} + \dfrac{1}{y} + \dfrac{1}{z}$ 12. $\dfrac{1}{x} + \dfrac{1}{2x} + \dfrac{1}{3}$

13. $\dfrac{a}{a+5} - \dfrac{2}{a}$

14. $\dfrac{2x}{x-5} + \dfrac{5}{x-2}$

15. $\dfrac{4}{a-9} - \dfrac{3}{a+3}$

16. $\dfrac{3}{a-b} + \dfrac{4}{a+2b}$

17. $\dfrac{2}{y-2} + \dfrac{3}{4-y^2}$

18. $\dfrac{2a-1}{4-a} - \dfrac{a+2}{3a-12}$

19. $\dfrac{2y-b}{1} + \dfrac{1}{2y+b}$

20. $a + 2b - \dfrac{a^2}{a-3b}$

21. $\dfrac{a+3}{a^2+1} + 2a - 4$

22. $\dfrac{2m}{m-3} - \dfrac{4m}{m+6} - 2$

23. $3 - \dfrac{4m}{m+3} + \dfrac{m+2}{m-5}$

24. $\dfrac{2}{x-3} + \dfrac{3}{x-2} + \dfrac{1}{x+1}$

25. $\dfrac{2}{3-x} + \dfrac{3}{1-x} - \dfrac{2}{2-x}$

26. $\dfrac{x+3}{1+x} - \dfrac{x+4}{1-x} + \dfrac{x-5}{2-x}$

27. $\dfrac{x-4}{x-1} - \dfrac{x+5}{x-2} - \dfrac{x+3}{x+2}$

28. $\dfrac{2x}{x-1} + \dfrac{3x}{x+1} - \dfrac{4x^2}{x^2-1}$

29. $\dfrac{16x^2}{4x^2-1} - \dfrac{4x}{2x-1} - \dfrac{6x}{2x+1}$

30. $\dfrac{2x}{3x-1} + \dfrac{3x}{3x+1} - \dfrac{12x^2}{9x^2-1}$

31. $\dfrac{x+1}{2x-3} - \dfrac{x-1}{2x+3} - \dfrac{2x^2-7x}{9-4x^2}$

32. $\dfrac{1}{x-y} - \dfrac{1}{x+y} + \dfrac{2x}{y^2-x^2}$

33. $\dfrac{m^2-1}{m^3+1} - \dfrac{2}{m^2-m+1}$

34. $\dfrac{3x+y}{x^2-y^2} - \dfrac{2y}{x^2-xy} - \dfrac{1}{x+y}$

35. $\dfrac{2a^2-5b^2}{a^2+ab-2b^2} + \dfrac{a-2b}{a+2b} - \dfrac{a-2b}{a-b}$

36. $\dfrac{2}{y^2+7y+12} + \dfrac{5y+10}{y^2+3y-4} + \dfrac{y+4}{y^2+2y-3}$

37. $\dfrac{1}{2a^2+3a+1} + \dfrac{5a}{2a^2-a-3} + \dfrac{a+2}{4a^2-4a-3}$

4-8 COMPLEX FRACTIONS

Thus far in the chapter we have considered simple fractions—fractions whose numerators and denominators are nonfractional. We now introduce fractions which contain a fraction in the numerator or denominator, or both. A fraction of this kind is called a *complex fraction*.

To simplify a complex fraction means to express it as an equivalent fraction that has no fraction in either the numerator or denominator. The simplification may be accomplished by either of the following methods. Sometimes one of the methods is more convenient than the other.

1. Reduce the numerator and denominator to simple fractions and then divide.
2. Multiply the numerator and denominator of the main fraction by an expression such that the numerator and denominator of the new fraction will be nonfractional.

EXAMPLE 1. Simplify the complex fraction $\dfrac{\dfrac{1}{a} + \dfrac{1}{b}}{\dfrac{1}{a} - \dfrac{1}{b}}$.

Solution. Combining the fractions of the numerator and the denominator, we have

$$\frac{\dfrac{1}{a} + \dfrac{1}{b}}{\dfrac{1}{a} - \dfrac{1}{b}} = \frac{\dfrac{b+a}{ab}}{\dfrac{b-a}{ab}} = \frac{b+a}{ab} \cdot \frac{ab}{b-a} = \frac{b+a}{b-a}$$

EXAMPLE 2. Simplify the complex fraction $\dfrac{\dfrac{x+1}{x-2} - \dfrac{x-1}{x+2}}{\dfrac{8}{x-2} + 4}$.

Solution.

$$\frac{\dfrac{x+1}{x-2} - \dfrac{x-1}{x+2}}{\dfrac{8}{x-2} + 4} = \frac{\dfrac{x^2 + 3x + 2 - (x^2 - 3x + 2)}{(x-2)(x+2)}}{\dfrac{8 + 4x - 8}{x-2}}$$

$$= \frac{6x}{(x-2)(x+2)} \cdot \frac{x-2}{4x} = \frac{3}{2(x+2)}$$

We also simplify the given fraction by multiplying the numerator and denominator by the L.C.M. of the denominators of the three simple fractions which occur. Thus we have

$$\frac{(x-2)(x+2)\left[\dfrac{x+1}{x-2} - \dfrac{x-1}{x+2}\right]}{(x-2)(x+2)\left[\dfrac{8}{x-2} + 4\right]} = \frac{(x+2)(x+1) - (x-2)(x-1)}{8(x+2) + 4(x-2)(x+2)}$$

$$= \frac{x^2 + 3x + 2 - (x^2 - 3x + 2)}{8x + 16 + 4(x^2 - 4)}$$

$$= \frac{6x}{4x^2 + 8x} = \frac{3}{2(x+2)}$$

⊘ odd #

EXERCISE 4-7

Simplify the complex fractions.

1. $\dfrac{\dfrac{1}{2} - \dfrac{2}{5}}{\dfrac{1}{2} + \dfrac{3}{10}}$

2. $\dfrac{\dfrac{2}{3} + \dfrac{4}{7}}{1 - \dfrac{2}{21}}$

3. $\dfrac{\dfrac{2}{x} + \dfrac{1}{2}}{\dfrac{2}{x} - \dfrac{1}{2}}$

4. $\dfrac{\dfrac{a}{b} - \dfrac{c}{d}}{1 + \dfrac{ac}{bd}}$

5. $\dfrac{\dfrac{3}{x} - 1}{\dfrac{3}{x} + 1}$

6. $\dfrac{x - 1 + \dfrac{2}{x}}{x + 1 - \dfrac{2}{x}}$

7. $\dfrac{4 - \dfrac{4x - 3}{x}}{2 - \dfrac{x - 1}{x}}$

8. $\dfrac{1 + \dfrac{3}{x} - \dfrac{4}{x^2}}{1 + \dfrac{1}{x} - \dfrac{2}{x^2}}$

9. $\dfrac{\dfrac{3}{x - 1} - \dfrac{2}{x + 1}}{\dfrac{1}{x - 1} - \dfrac{5}{x + 1}}$

10. $\dfrac{\dfrac{x + 1}{x - 1} - \dfrac{x - 1}{x + 1}}{\dfrac{x + 1}{x - 1} + \dfrac{x - 1}{x + 1}}$

11. $\dfrac{\dfrac{1}{2x - 1} - \dfrac{2}{3x - 1}}{\dfrac{1}{2x - 1} + \dfrac{1}{3x - 1}}$

12. $\dfrac{\dfrac{x - y}{x + y} + \dfrac{1}{x^2 - y^2}}{\dfrac{x + y}{x - y} - \dfrac{x - y}{x + y}}$

13. $\dfrac{3 - \dfrac{x - 6}{x^2 - 6x + 8}}{2 + \dfrac{x + 7}{x^2 - 2x - 8}}$

14. $\dfrac{1 - \dfrac{x - 6}{x^2 - x - 6}}{2 + \dfrac{x + 6}{x^2 - x - 6}}$

15. $\dfrac{1 + \dfrac{9}{x^2 + 2x - 8}}{1 + \dfrac{6}{x^2 + x - 6}}$

16. $\dfrac{\dfrac{x^2}{x^2 - y^2} - 1}{\dfrac{2x}{y - x} + 2}$

17. $\dfrac{1}{1 - \dfrac{1}{1 - \dfrac{1}{x}}}$

18. $\dfrac{x}{x - \dfrac{x}{2 - \dfrac{1}{x}}}$

REVIEW EXERCISE

Factor the following expressions (Secs. 4-1, 4-2, and 4-3).

1. $x^2 - 3xy + x$ **2.** $2ab^2 - 4ab + 2a^2b$
3. $9a^2 - 1$ **4.** $4x^2 - 9y^2$ **5.** $8m^2 - 2n^2$
6. $3xy^2 - 12x^3$ **7.** $8 + a^3$ **8.** $y^3 - 64$
9. $8m^3 + 27n^3$ **10.** $3ap^3 + 24aq^3$
11. $1 - (x + y)^3$ **12.** $(x + y)^3 + 1$
13. $x^2 - 9x + 20$ **14.** $x^2 - 2x - 24$
15. $4 - 4a + a^2$ **16.** $m^2n^2 + 13mn + 42$
17. $2x^2 + 7x + 6$ **18.** $6a^2 + a - 12$
19. $x^4 - 3x^2 - 4$ **20.** $x^6 + x^3 - 2$
21. $9x^2 + 6xy + y^2$ **22.** $4a^2 - 20ab + 25b^2$

23. $4(a - b) - a(a - b)$
24. $x(y + 2) - y(y + 2)$
25. $a^3(x - y) - 8b^3(x - y)$
26. $8a + 2ab - 4ay - aby$
27. $x^2 - y^2 + 3x - 3y$
28. $x^2 + 4xy + 4y^2 + x + 2y$
29. $x^2 - x - 2 + xy - 2y$
30. $m^3 - n^3 - m^2 + 2mn - n^2$
31. $25 - a^2 + 6a - 9$
32. $9x^2 - 6xy + y^2 - 16a^2$
33. $2x^2 - xy + xz - 2x + y - z$
34. $x^3 - 3x^2 + 3x - 1$
35. $8a^3 + 36a^2 + 54a + 27$
36. $1 - 12x + 48x^2 - 64x^3$
37. $m^2 + n^2 + 9 + 2mn - 6m - 6n$
38. $x^2 - 4xy + 4y^2 - a^2 - 2ab - b^2$
39. $16x^4 - y^4$
40. $25a^4 - 16$
41. $8m^6 + n^6$
42. $m^6 - y^3$
43. $x^4 + x^2 + 1$
44. $m^4 + 3m^2 + 4$
45. $4x^4 - 21x^2y^2 + 9y^4$
46. $9x^4 - 31x^2y^2 + y^4$
47. $(x + y)^2 + 3(x + y) - 4$
48. $(2x + y)^2 - (2x + y) - 6$
49. $9(m - n)^2 - 9(m - n) + 2$

By making only sign changes, find three more fractions that are equal to each of the given fractions (Sec. 4-4).

50. $\dfrac{x - 7}{4 - x}$

51. $\dfrac{x^2 - 3}{4x + 2}$

52. $\dfrac{x^2 - x - 5}{x^2 + x + 3}$

53. $\dfrac{2 - x - x^2}{1 + x - x^2}$

Perform the indicated operations and express each result in lowest terms (Secs. 4-5 and 4-6).

54. $\dfrac{2x^2 + x - 3}{x^2 - 1} \cdot \dfrac{x^2 + 2x + 1}{3x^2 + 5x + 2}$

55. $(x^2 - y^2) \cdot \dfrac{3x^2 + xy - 4y^2}{2x^2 + 3xy + y^2} \cdot \dfrac{2x + y}{(x - y)^2}$

56. $\dfrac{6x^2 - xy - y^2}{3x^2 + 7xy + 2y^2} \cdot \dfrac{x^2 + 5xy + 6y^2}{2x^2 + xy - y^2} \div \dfrac{x^2 - 9y^2}{x^2 - y^2}$

57. $\dfrac{6x^2 + 13xy + 5y^2}{2xy - x^2} \cdot \dfrac{2x^2 - 7xy + 3y^2}{y^2 - 4x^2} \div \dfrac{3x^2 - 4xy - 15y^2}{2x^2 - 4xy}$

58. $\dfrac{y^2 - 5xy + 6x^2}{y^2 + 7xy + 10x^2} \cdot \dfrac{y^2 - 25x^2}{y^2 + xy - 6x^2} \div \dfrac{y^2 - 3xy - 10x^2}{y^2 + 5xy + 6x^2}$

59. $\dfrac{3x^2 + 13xy - 10y^2}{x^2 - 4y^2} \cdot \dfrac{3x^2 + 4xy - 4y^2}{4x^2 - 4xy + y^2} \div \dfrac{9x^2 - 12xy + 4y^2}{2x^2 - 5xy + 2y^2}$

60. $\left(\dfrac{x^2}{1 - x^2} \div \dfrac{2x}{1 - x} \right) \div \dfrac{1 - x}{1 + x}$

61. $\dfrac{x^2}{1 - x^2} \div \left(\dfrac{2x}{1 - x} \div \dfrac{1 - x}{1 + x} \right)$

Combine into a single fraction in lowest terms (Sec. 4-4).

62. $\dfrac{3x - 1}{(x - 1)(x + 1)} - \dfrac{x + 3}{(x + 1)(x + 2)} - \dfrac{1}{x + 2}$

63. $\dfrac{2a + 1}{a^2 + a - 2} + \dfrac{3a - 2}{a^2 - 4} - \dfrac{2a - 3}{a^2 - 3a + 2}$

64. $\dfrac{2x^2 - 5y^2}{x^2 + xy - 2y^2} - \dfrac{x - 2y}{x - y} + \dfrac{x - 2y}{x + 2y}$

65. $\dfrac{2m + 7}{2m^2 + 5m - 7} - \dfrac{m - 2}{2m^2 + 3m - 14} - \dfrac{m - 1}{m^2 - 3m + 2}$

66. $\dfrac{3}{x^3 + 3x^2 - x - 3} - \dfrac{2}{x^3 - x^2 - 3x + 3} + \dfrac{4}{x^3 + x^2 - 3x - 3}$

Simplify the complex fractions (Sec. 4-8).

67. $\dfrac{x - \dfrac{x^2 + 1}{2x}}{\dfrac{2}{x} - \dfrac{2}{x^2}}$

68. $\dfrac{\dfrac{1}{x} - \dfrac{5}{x + 1}}{\dfrac{3}{x} - \dfrac{2}{x + 1}}$

69. $\dfrac{x + \dfrac{x - 1}{x + 1}}{x - \dfrac{x - 1}{x + 1}}$

70. $\dfrac{\dfrac{1}{x + 3} + \dfrac{1}{x - 3}}{\dfrac{4}{x + 3} - \dfrac{2}{x - 3}}$

71. $\dfrac{\dfrac{x + y}{x - y} - \dfrac{x - y}{x + 1}}{\dfrac{x}{x - y} - \dfrac{x}{x + 1}}$

72. $\dfrac{1 + \dfrac{6}{x^2 + x - 6}}{1 + \dfrac{9}{x^2 + 2x - 8}}$

73. $\dfrac{1 - \dfrac{6d}{2c + 3d}}{2 + \dfrac{d}{c - 2d}}$

74. $\dfrac{\dfrac{2x}{y - x} + 2}{\dfrac{x^2}{x^2 - y^2} - 1}$

75. $\dfrac{1}{1 - \dfrac{2}{1 + (2/x)}}$

76. For what values of x is the fraction $\dfrac{x(2 - x)}{x^2 - 3x + 2}$ not equal to $\dfrac{x}{1 - x}$?

77. Is the fraction $\dfrac{x - 3}{2 + x^2}$ defined for all real values of x?

78. Is the fraction $\dfrac{x^2 - 5x + 6}{x - 2} = x - 3$ for all real values of x?

Chapter 5

FUNCTIONS AND RELATIONS

5-1 RECTANGULAR COORDINATES

In Sec. 2-5 we discussed a scheme for putting the system of real numbers into a one-to-one correspondence with the points of a line. We now develop a plan for putting the points of a plane into a one-to-one correspondence with a set of pairs of real numbers. The device for this association of points and number pairs is called a *rectangular Cartesian coordinate system*. This coordinate system was introduced in 1637 by René Descartes, a French mathematician and philosopher. The Cartesian and other coordinate systems open the way for reducing geometric problems to arithmetic ones and consequently have contributed greatly to the advancement of mathematics and its applications in various areas of learning.

We draw a horizontal line and a vertical line meeting at O (Fig. 5-1). The point O is called the *origin*; the horizontal line OX, the x *axis*; and the vertical line OY, the y *axis*. With a convenient unit of length we make a number scale on each axis, letting the origin be the zero point. The positive numbers are to the right on the x axis and above the origin on the y axis. Arrows are placed on the axes to indicate their positive directions.

Each point P of the plane determined by the axes has associated with it a pair of coordinates. The coordinates are defined in terms of the perpendicular distances from the axes to the point.

Definition 5-1   The x *coordinate*, or *abscissa*, of a point P is the directed distance from the y axis to the point.

The y *coordinate*, or *ordinate*, of a point P is the directed distance from the x axis to the point.

The abscissa of a point P is positive if P is to the right of the y axis and negative if P is to the left of the y axis; the ordinate is positive if P is above the x axis and negative if P is below the x axis. A point whose abscissa is x and whose ordinate is y is designated by (x, y), the abscissa always coming first. Hence the coordinates of a point are referred to as an *ordered pair* of numbers.

A point of given coordinates is plotted by measuring the proper distances from the axes and marking the point thus located. For example, if the co-

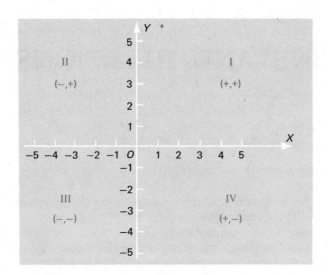

Fig. 5-1

ordinates of a point are $(-4, 3)$, the abscissa -4 means the point is 4 units to the left of the y axis and the ordinate 3 (plus sign understood) means the point is 3 units above the x axis. Consequently, we locate the point by going from the origin 4 units to the left along the x axis and then 3 units upward parallel to the y axis (Fig. 5-2).

The coordinate axes divide their plane into four parts, called *quadrants*, which are numbered I to IV in Fig. 5-1. The coordinates of a point in the first

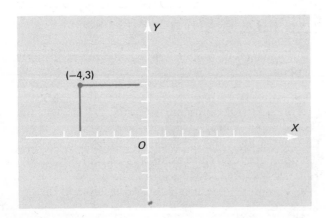

Fig. 5-2

quadrant are both positive, which is indicated in the figure by $(+, +)$. The signs of the coordinates in each of the other quadrants are similarly indicated.

We assume that to any pair of real numbers (coordinates) there corresponds one definite point. Conversely, we assume that to each point of the plane there corresponds one definite pair of coordinates. This relation of points on a plane and pairs of real numbers is called a *one-to-one correspondence*.

The cross product of two sets of real numbers A and B is defined to be the set of all ordered pairs (x, y) such that x is a member of A and y is a member of B. Using this definition and letting $A = B = R$, we may denote the set of number pairs corresponding to the points of a coordinate plane by the notation

$$R \times R = \{(x, y) | x \in R, y \in R\}$$

5-2 RELATIONS AND FUNCTIONS

The concepts of relations and functions, which we shall discuss in this chapter, are fundamental in many areas of mathematics. We begin our discussion of these topics by defining a relation.

Definition 5-2 A *relation* is a set of ordered pairs of numbers. The set of first elements of the ordered pairs is called the *domain* of the relation and the set of second elements is called the *range* of the relation.

EXAMPLE 1. The set of number pairs

$$\{(-1, -1), (0, 1), (1, 1), (1, 2), (3, 0)\}$$

defines a relation. The set $\{-1, 0, 1, 3\}$ is the domain and the set $\{-1, 1, 2, 0\}$ is the range. The graph of the relation is plotted in Fig. 5-3.

EXAMPLE 2. Relations are frequently determined by a given domain and an equation which provides a rule for computing the second element of each number pair from the first element. To illustrate, let the second element of each number pair (x, y) be given by the equation $y = 3x + 1$ and let the domain be the set of real numbers R. These conditions definitely define the relation. The set of number pairs comprising the relation can, with the specified domain, be expressed in the form

$$\{(x, y) | y = 3x + 1\}$$

This relation consists of infinitely many number pairs, and the elements therefore cannot be listed. We can of course find y for any particular value of x. Thus the number pairs for $x = -2, 0$, and 3 are

$$(-2, -5), (0, 1), (3, 10)$$

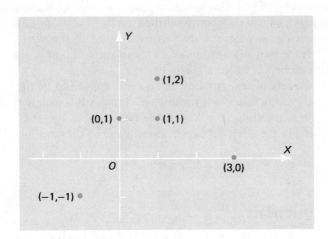

Fig. 5-3

In this chapter we shall be mainly concerned with a special kind of relation which is defined as follows.

Definition 5-3 Let X and Y denote sets of real numbers and let (x, y) denote an ordered pair of numbers such that to each $x \in X$ there is a unique $y \in Y$. The totality of number pairs thus determined is called a *function*. The set X is called the *domain* of the function and the set Y is called the *range* of the function.

We emphasize that this definition requires that for any value x of the domain there is one and only one corresponding value of y. A relation does not have this stipulation. Consequently a relation may have more than one value of the range corresponding to a value of the domain. Thus we see that a function is a relation but a relation is not necessarily a function. Consider, for example, the sets

$$\{(x, y)| y = \sqrt{x}\} \quad \text{and} \quad \{(x, y)| y = \pm\sqrt{x}\}$$

with the domain of each set given by $X = \{1, 2, 3\}$. The first set has a unique value of y for each value of x, and therefore is both a function and a relation. The second set, having two values of y for each value of x, is a relation but not a function. We note too that the set of number pairs in Example 1 above is a relation but not a function while the set of number pairs in Example 2 is a relation and also a function.

The second element of a number pair of a function is called the *value of the function* corresponding to the value of the first element. The value of a function at x is sometimes denoted by $f(x)$ which is read "f of x." We warn that the symbol is not a product—not f times x. Suppose, for example, that we define a function by a given domain and an equation such as $y = x^2 - 2x$. Since $y = f(x)$, we may write the equation as

$$f(x) = x^2 - 2x$$

The equation yields the value of the function at any chosen value of x. For example, denoting the value of the function at $x = 3$ by $f(3)$ and substituting for x, we find

$$f(3) = 3^2 - 2(3) = 3$$

Similarly,

$$f(-2) = (-2)^2 - 2(-2) = 8$$
$$f(5) = 5^2 - 2(5) = 15$$

Hence $(3, 3)$, $(-2, 8)$, $(5, 15)$ are three elements of the function.

Although a function or a relation may be completely defined by an equation and a specified domain, often an equation is given with no mention of the domain. In all such cases, it is to be understood that the domain consists of all real numbers for which the equation yields real numbers for the values of the function or relation.

EXAMPLE 3. Find the domain and range of the function defined by $y = \sqrt{9 - x^2}$.

Solution. A square root of a negative number is not a real number. This is true because the square of either a positive or a negative number is positive. So the domain of this function consists of the set of all numbers from -3 to 3, inclusive. Any value of $x < -3$ or of $x > 3$ makes $9 - x^2$ negative and does not belong to the domain. The range consists of the set of all numbers from 0 to 3, inclusive. We use the symbol $\leq$ (read "is less than or equal") and express these descriptions of the domain and range by

$$X = \{x | -3 \leq x \leq 3\} \qquad Y = \{y | 0 \leq y \leq 3\}$$

EXERCISE 5-1

1. Plot the points whose coordinates are $(4, 3)$, $(4, -3)$, $(-4, 3)$, $(-4, -3)$.
2. In which quadrant is a point located if (a) both coordinates are positive, (b) both coordinates are negative, (c) the abscissa is positive and the ordinate negative?
3. Where does a point lie if (a) its ordinate is zero, (b) its abscissa is zero?
4. What points have their abscissas equal to 3? What points have their ordinates equal to -4?
5. Where may a point lie if (a) the abscissa is equal to the ordinate, (b) the abscissa is equal to the negative of the ordinate?
6. Three vertices of a rectangle are at the points $A(2, 3)$, $B(6, 0)$, and $C(8, 11)$. Find the coordinates of the fourth vertex and draw the rectangle.

7. The points $A(0,0)$, $B(5,1)$, and $C(1,3)$ are vertices of a parallelogram. Find the coordinates of the fourth vertex (a) if AB is a diagonal, (b) if AC is a diagonal, (c) if BC is a diagonal.

8. The points $(-2,-1)$, $(4,0)$, and $(3,3)$ are vertices of a parallelogram. Find the three possible positions of the fourth vertex.

Determine which of the following sets of number pairs determine a function. List the elements of the domain and the elements of the range of each function.

9. $\{(2,1),(3,-1),(4,-3)\}$ 10. $\{(-2,4),(0,5),(1,-3)\}$

11. $\{(2,0),(2,3),(5,0)\}$ 12. $\{(3,a),(4,b),(5,c)\}$

13. $\{(5,1),(1,5),(5,2)\}$ 14. $\{(x,y),(y,x),(x,y)\}$

15. If $f(x) = 1 - x$, find (a) $f(0)$, (b) $f(-1)$, (c) $f(11)$.
16. If $g(t) = t^2 - 2t - 3$, find (a) $g(3)$, (b) $g(-1)$, (c) $g(4)$.
17. If $h(s) = (s-1)/(s+2)$, find (a) $2h(-1)$, (b) $[h(-1)]^2$, (c) $h(2s)$.
18. If $F(t) = (2t-1)/t$, find (a) $F(2)$, (b) $F(\frac{1}{2})$, (c) $F(t^2)$.
19. If $f(x) = x^2$, find (a) $f(x+h)$, (b) $f(x+h) - f(x)$.

Determine the domain and range of the function defined by each of the following equations.

20. $f(x) = x^2 - 4$ 21. $f(x) = \sqrt{3x}$ 22. $f(x) = 1/(x-2)$

23. $f(x) = \sqrt{16 - x^2}$ 24. $f(x) = \sqrt{x^2 - 16}$ 25. $f(x) = \sqrt{x^2 + 1}$

5-3 GRAPHS OF FUNCTIONS AND RELATIONS

Consider the function defined by the equation

$$f(x) = x^2 - 3x - 2$$

If a value is assigned to x, the corresponding value of $f(x)$ may be computed. Several values of x and the corresponding values of $y = f(x)$ are shown in the table. These number pairs furnish a picture of the relation of x and y. A better representation is had, however, by plotting each value of x and the corresponding value of y as the abscissa and ordinate of a point, and then drawing a smooth curve through the points thus determined. This process is called *graphing the function*, and the curve is called the *graph* or *locus* of the function.

x	-1.5	-1	0	1	2	3	4	4.5
$y = f(x)$	4.75	2	-2	-4	-4	-2	2	4.75

The plotted points (Fig. 5-4) extend from $x = -1.5$ to $x = 4.5$. Points farther to the left and right, as well as any number of intermediate points, could be located. But the plotted points show about where the intermediate points would lie. Hence we can use a few points to draw a curve which is reasonably accurate. The exact graph satisfies the following definition.

$$f(x+h)^2 - f(x) =$$
$$(x+h)^2 - x^2 =$$
$$2hx + h^2$$

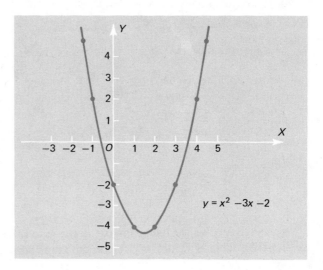

Fig. 5-4

Definition 5-4 The graph of a function (or relation) consists of the set of all points whose coordinates are the ordered pairs of the function (or relation). If the function or relation is defined by an equation, the graph of the equation is the same as the graph of the function (or relation).

The point or points of the graph of a function which are on the x axis are of special interest. The abscissa of such a point makes the value of y, or the function, equal to zero and is called a *zero of the function*. We estimate the abscissas of the points where the graph in Fig. 5-4 cuts the x axis to be $x = -0.6$ and $x = 3.6$. These values are approximations to the zeros of the function represented by the graph. In Chapter 8 we shall discuss the zeros of functions in considerable detail.

EXAMPLE 1. Draw the graph of the function defined by $y = 1 - 2x$.

Solution. We assign certain values to x and obtain the corresponding values of y, as shown in the table. The tabulated values enable us to plot some points of the graph and draw a curve through them (Fig. 5-5). The graph is a straight line and shows the zero of the function to be $x = \frac{1}{2}$.

x	-2	-1	0	1	2	3
y	5	3	1	-1	-3	-5

EXAMPLE 2. Draw the graph of $y = |x - 2|$.

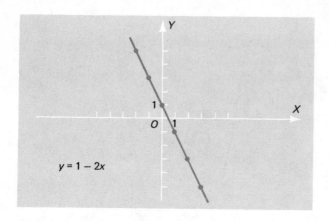

Fig. 5-5

Solution. Recalling the definition of the absolute value of a number (Sec. 3-1), we prepare a table of values and sketch the graph shown in Fig. 5-6.

x	-2	0	2	4	6
y	4	2	0	2	4

EXAMPLE 3. Construct the graph of the relation defined by $y^2 = 4x$.

Solution. Solving for y gives $y = \pm 2\sqrt{x}$. We draw the graph (Fig. 5-7) by use of the following table.

x	0	1	4	9
y	0	± 2	± 4	± 6

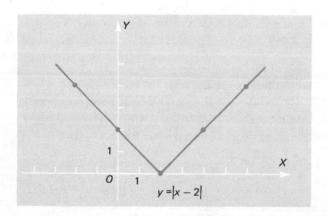

Fig. 5-6

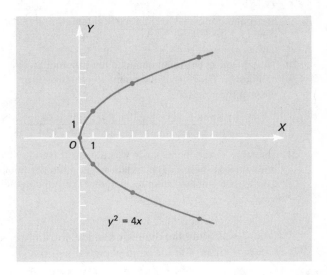

Fig. 5-7

EXERCISE 5-2

Draw the graph of the function defined by each of the following equations. Give the zeros of the functions.

1. $y = x - 3$
2. $y = 2x + 5$
3. $y = -x - 1$
4. $3x - y = 6$
5. $x + 3y - 3 = 0$
6. $2x - y = 0$
7. $y = |x|$
8. $y = 1 - |x|$
9. $y = |x + 2|$

Draw the graph of the function defined by each of the following equations and give (or estimate) its zeros. Use all the integral values of x in the indicated domain and enough fractional values to determine the shape of the curve.

10. $y = x^2$ from $x = -3$ to $x = 3$
11. $y = x^2 - 6x$ from $x = -2$ to $x = 7$
12. $y = 2x^2 - 5$ from $x = -3$ to $x = 3$
13. $y = -x^2 + 3x + 4$ from $x = -2$ to $x = 5$
14. $y = 3x^2 - 6x - 4$ from $x = -1$ to $x = 3$
15. $y = x^3$ from $x = -3$ to $x = 3$
16. $y = -x^3$ from $x = -3$ to $x = 3$

17. Draw the graph of the relation defined by $y = \pm 3\sqrt{x - 1}$ from $x = 1$ to $x = 10$.
18. Draw the graph of the relation defined by $y = \pm\sqrt{x(5 - x)}$ from $x = 0$ to $x = 5$.

19. Hourly Fahrenheit temperature readings from 8:00 A.M. to 5:00 P.M. are as follows:
20°, 33°, 46°, 51°, 53°, 55°, 58°, 57°, 52°, 45°. Select convenient scales along the
horizontal (time) axis and the vertical (temperature) axis and plot the points determined by these data. Make a graph by connecting the points consecutively with
line segments.

20. The table gives the miles obtained per gallon of gasoline by a car at various speeds
in miles per hour. Plot the points to represent these data. Draw a smooth curve
through the points.

x (speed)	10	20	30	40	50	60	70
y (miles per gal.)	15	17.5	19	19.4	18.5	16	14

21. The table shows the distance which a car traveled while being brought to a stop
from various speeds. The speeds are in miles per hour and the distances in feet.
Construct a smooth curve to represent the stopping distance as a function of the
speed.

x (speed)	10	20	30	40	50	60	70
y (distance)	3	12	27	48	75	108	147

5-4 THE DISTANCE FORMULA AND THE CIRCLE

The distance between two points of the coordinate plane, or the length of the
line segment connecting them, can be determined from the coordinates of the
points. Let $P_1(x_1, y_1)$ and $P_2(x_2, y_2)$ be the coordinates of any two points
of the plane. Then a line through P_1 parallel to the x axis and a line through
P_2 parallel to the y axis intersect at a point P_3 whose abscissa is x_2 and whose
ordinate is y_1. Referring to Fig. 5-8, or Fig. 5-9, we express the distances
between P_1 and P_3 and between P_2 and P_3, respectively, by

$$P_1P_3 = |x_2 - x_1| \quad \text{and} \quad P_2P_3 = |y_2 - y_1|$$

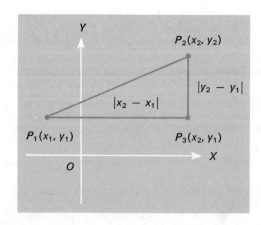

Fig. 5-8

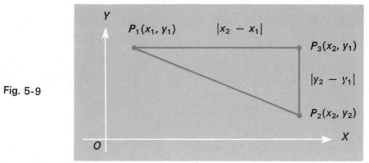

Fig. 5-9

Designating the distance between P_1 and P_2 by d and applying the Pythagorean theorem (see problem 28, Exercise 5-3), we have

$$d = \sqrt{|x_2 - x_1|^2 + |y_2 - y_1|^2}$$

Finally, noting that $|x_2 - x_1|^2 = (x_2 - x_1)^2$ and $|y_2 - y_1|^2 = (y_2 - y_1)^2$, we express the formula for the distance between two points in the form

$$d = \sqrt{(x_2 - x_1)^2 + (y_2 - y_1)^2} \tag{1}$$

In applying this formula, either point may be designated by (x_1, y_1) and the other by (x_2, y_2). This results because the two differences involved are squared.

EXAMPLE 1. Find the lengths of the sides of the triangle (Fig. 5-10) with vertices at $A(-2, -3)$, $B(6, 1)$, and $C(-2, 5)$.

Solution.

$$AB = \sqrt{[(6 - (-2)]^2 + [1 - (-3)]^2}$$
$$= \sqrt{64 + 16} = \sqrt{80} = 4\sqrt{5}$$
$$BC = \sqrt{(-2 - 6)^2 + (5 - 1)^2}$$
$$= \sqrt{64 + 16} = \sqrt{80} = 4\sqrt{5}$$
$$CA = \sqrt{[-2 - (-2)]^2 + (5 + 3)^2}$$
$$= \sqrt{0 + 64} = 8$$

Two of the sides have equal lengths and the triangle is isosceles.

The distance formula (1) has numerous important uses in mathematical situations. We shall now apply the formula to find the equation of a circle. We recall, of course, that a circle is the locus of all points in a plane that are equally distant from a fixed point of the plane. The fixed point is called the *center*, and the distance from the center to a point of the circle is called the *radius*.

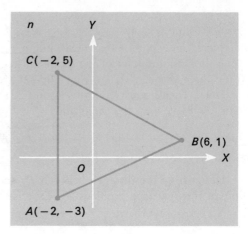

Fig. 5-10

If we let $C(h, k)$ be the coordinates of the center of a circle of radius r (Fig. 5-11), then any point $P(x, y)$ of the circle must satisfy the condition

$$\sqrt{(x - h)^2 + (y - k)^2} = r$$

or, squaring,

$$(x - h)^2 + (y - k)^2 = r^2 \tag{2}$$

Equation (2) is the general equation of a circle. The circle consists of the set of all points whose coordinates satisfy the equation. For the special case in which the center is at the origin and the radius is equal to 1, the equation becomes

$$x^2 + y^2 = 1$$

This equation has important applications in trigonometry.

Fig. 5-11

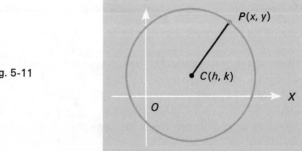

EXAMPLE 2. Find the equation of the circle which passes through the points $(0, 0)$, $(4, 0)$, and $(0, -4)$.

Solution. These points are vertices of a right triangle (Fig. 5-12). Then, recalling that the midpoint of the hypotenuse of a right triangle is equidistant from the three vertices, we see that $(2, -2)$ are the coordinates of the center of the circle. The distance from this point to each of the vertices is $\sqrt{8}$. Hence, taking $r^2 = 8$, we have the equation

$$(x - 2)^2 + (y + 2)^2 = 8$$

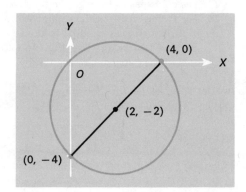

Fig. 5-12

EXERCISE 5-3

Find the distance between the pair of points in each problem 1 through 6.

1. $(3, 1)$, $(7, 4)$ 2. $(4, -3)$, $(-8, 2)$
3. $(-3, -2)$, $(0, 1)$ 4. $(5, -12)$, $(0, 0)$
5. $(-4, 0)$, $(0, 3)$ 6. $(7, 2)$, $(4, -1)$

In each problem 7 through 10 draw the triangle having the given vertices and find the lengths of the sides.

7. $A(-1, 1)$, $B(-1, 4)$, $C(3, 4)$ 8. $A(2, -1)$, $B(4, 2)$, $C(5, 0)$
9. $A(0, 0)$, $B(5, -2)$, $C(-3, 2)$ 10. $A(0, -3)$, $B(3, 0)$, $C(0, -4)$

Determine, by the distance formula, if the three points in each problem 11 through 16 are the vertices of a scalene triangle (no equal sides), an isosceles triangle (two equal sides), or a right triangle (see problem 28), or if the three points are on a straight line (the distance between two of the points equal to the sum of the other distances).

11. $A(6, 1)$, $B(2, -4)$, $C(-2, 1)$ 12. $A(1, 4)$, $B(10, 6)$, $C(2, 2)$
13. $A(-2, 1)$, $B(5, -2)$, $C(3, 3)$ 14. $A(1, -1)$, $B(1, -7)$, $C(8, 3)$
15. $A(2, 1)$, $B(-1, 2)$, $C(5, 0)$ 16. $A(3, 2)$, $B(0, 0)$, $C(9, 6)$

In each problem 17 through 27 write the equation of the circle which satisfies the given conditions.

17. Center $(-6, 2)$, radius 5   **18.** Center $(4, 3)$, radius 5

19. Center $(0, 0)$, radius 6   **20.** Center $(-4, -3)$, radius 7

21. The line segment joining $A(0, 0)$ and $B(6, 0)$ is a diameter.

22. The line segment joining $A(2, 3)$ and $B(2, -3)$ is a diameter.

23. The center is at $(2, 4)$, and the circle passes through $(-1, -1)$.

24. The center is at $(1, -3)$, and the circle passes through $(-3, 5)$.

25. The circle is tangent to the x axis, and the center is at $(-4, 1)$.

26. The circle passes through the points $(0, 0)$, $(2, 0)$, and $(0, 2)$.

27. The circle passes through the points $(1, 1)$, $(5, 1)$, and $(1, 5)$.

28. The Pythagorean theorem states that the square of the length of the hypotenuse of a right triangle is equal to the sum of the squares of the lengths of the remaining two sides. Use Fig. 5-13 to prove that $c^2 = a^2 + b^2$. Hint: The area of the outer square is given by

$$A = (a + b)^2 \quad \text{and also by} \quad A = c^2 + 4(\tfrac{1}{2}ab)$$

Fig. 5-13

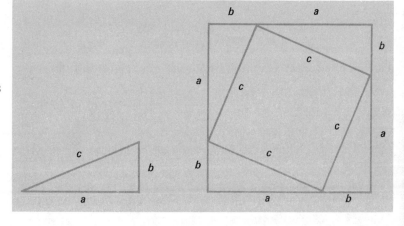

Chapter 6

LINEAR EQUATIONS

6-I CONDITIONAL EQUATIONS AND IDENTITIES

A statement that two quantities are equal is called an *equation*. The customary way of writing an equation is to place the symbol = (read "is equal to") between the equal quantities. An equation then has two members, the left member and the right member. Equations usually have one or more letters which are regarded as *variables* or *unknowns*. Numbers which, when substituted for the variables, make the two members of the equation equal are said to *satisfy* or be a *solution* of the equation. The totality of solutions is called the *solution set*.

EXAMPLES.

$$x - 3 = 4 \tag{1}$$

$$x^2 - 2x - 15 = 0 \tag{2}$$

$$(x - 2)^2 = x^2 - 4x + 4 \tag{3}$$

$$\frac{1}{x - 1} - \frac{1}{x} = \frac{1}{x(x - 1)} \tag{4}$$

The first equation is satisfied when $x = 7$ and for no other number. The second equation is satisfied for $x = -3$ and $x = 5$; any other value for x makes the two members of the equation unequal. The right member of equation (3) is the result of the operation indicated on the left side. Clearly, then, the equation is satisfied when x is replaced by any real number. The right member of equation (4) is the sum of the fractions of the left member, and the equation is true for all values of x except $x = 0$ and $x = 1$. Either of these values produces a zero in a denominator on the left and the denominator on the right. Consequently, we say that each member of the equation is *undefined* for these values of x.

The two types of equations which we have illustrated are named in accordance with the following definitions.

Definition 6-1 An equation which is satisfied by some, but not all, of the values of the variables for which the members of the equation are defined is called a *conditional equation*.

Definition 6-2   An equation which is satisfied by all the values of the variables for which the members of the equation are defined is called an *identity*.

According to these definitions, equations (1) and (2) are conditional equations, and equations (3) and (4) are identities. The solution sets of the four equations are respectively

$$\{7\}, \qquad \{-3, 5\}, \qquad \{x | x \in R\}, \qquad \{x | x \in R, x \neq 0 \text{ or } 1\}$$

We dealt largely with identities in Chapter 4. Now, however, we wish to consider conditional equations with only one variable. A solution of an equation in one variable is called a *root* of the equation. To solve the equation means to find all its roots. Before solving equations we discuss certain operations on equations.

6-2 OPERATIONS ON EQUATIONS

Two equations are said to be *equivalent* if their solution sets are the same. The equations $x - 3 = 2$ and $x + 1 = 6$, for example, are equivalent; each has the one root 5. Neither of these equations, however, is equivalent to $(x - 5)(x - 1) = 0$. The values $x = 1$ and $x = 5$ satisfy the last equation. But $x = 1$ does not satisfy the other equations.

In the process of finding the solutions of an equation it is usually necessary to perform operations, starting with the given equation, which lead to other equations. It is important, therefore, to know if the new equations are equivalent to the original equation. Pertaining to this question, we have the following theorems.

Theorem 6-1   If the same number or polynomial is added to, or subtracted from, both members of a conditional equation, an equivalent equation is obtained.

Proof. Suppose we replace the variables of an equation by numbers which constitute a solution of the equation. Then the members of the equation become equal to some number a. The same solution will make the polynomial which is added to both members of the equation equal to some number c. Thus we reduce the first equation to $a = a$ and the second to $a + c = a + c$. Hence a solution of the original equation is also a solution of the new equation. Now starting with a solution of the new equation and making substitutions for the variables, we obtain an equation of the form $d + e = f + e$, where e is the value which the polynomial takes and d and f are the values which the members of the original equation take. From this equation we see that $d = f$ (Theorem 2-20), and consequently a solution of the new equation is a solution of the original equation. Since a solution of the original equation is a solution of the new equation and a solution of the new equation is a solution of the original equation, we conclude that the equations are equivalent. ∎

Theorem 6-2   If both members of an equation are multiplied by or divided by the same nonzero constant, an equivalent equation is obtained.

We leave the proof of this theorem to the student.

If both members of an equation are multiplied by or divided by an expression which contains a variable, the resulting equation will in some cases not be equivalent to the original one. Thus, for example, multiplying both members of $4x - 8 = 12$ by x gives $4x^2 - 8x = 12x$. The new equation is satisfied by $x = 5$ and also by $x = 0$; whereas the first equation is satisfied by $x = 5$ but not by $x = 0$. Conversely, if we start with $4x^2 - 5x = 12x$ and divide by x, a root is lost rather than gained. Notice, however, that each of these operations, yielding nonequivalent equations, involves the variable.

A root of a derived equation which is not a root of the original equation is called an *extraneous root*. An extraneous root can be detected by substituting solutions of the derived equation in the given equation. An operation which leads to an equation with fewer roots than the original equation should be avoided if all the roots are desired.

6-3   LINEAR EQUATIONS IN ONE VARIABLE

In this section we shall solve equations of the form, or reducible to the form

$$ax + b = 0$$

where b stands for any number and a for any number except zero. This equation is of the first degree in x and is called a *linear equation*.

EXAMPLE 1.   Solve the equation $4x + 10 = 1 - 2x$.

Solution.

$$
\begin{aligned}
4x + 10 &= 1 - 2x && \text{given equation} \\
4x + 10 + 2x - 10 &= 1 - 2x + 2x - 10 && \text{adding } 2x - 10 \\
6x &= -9 && \text{collecting terms} \\
x &= -1.5 && \text{dividing by 6}
\end{aligned}
$$

Check.   We substitute -1.5 for x in each member of the given equation and find

$$
\begin{aligned}
4(-1.5) + 10 &= -6 + 10 = 4 && \text{for the left member} \\
1 - 2(-1.5) &= 1 + 3 = 4 && \text{for the right member}
\end{aligned}
$$

The value $x = -1.5$ is a root because it makes the two members of the equation equal.

EXAMPLE 2.   Solve the equation $3x + 4 = 2ax + 5c$ for x.

Solution.

$$
\begin{aligned}
3x + 4 - 2ax - 4 &= 2ax + 5c - 2ax - 4 && \text{adding } -2ax - 4 \\
3x - 2ax &= 5c - 4 && \text{collecting terms} \\
(3 - 2a)x &= 5c - 4 && \text{factoring}
\end{aligned}
$$

$$x = \frac{5c - 4}{3 - 2a} \qquad \text{dividing by } 3 - 2a$$

This result may be checked by substituting for x in the given equation.

An equation having simple fractions may be reduced to nonfractional form by multiplying both members by the L.C.M. of the denominators. This process is called *clearing of fractions*. If none of the denominators contains the variable the new equation is equivalent to the original equation. If, however, the variable appears in any denominator, the new equation may have roots in addition to those of the original equation. It is essential therefore that all roots of the new equation be checked in the given equation.

EXAMPLE 3. Solve the equation $1/(x - 3) + 1 = 1/(x - 3)$.
Solution. Multiplying both members by $x - 3$, we get

$$1 + x - 3 = 1$$

$$x = 3$$

But 3 is not a root of the given equation because this value makes a denominator equal to zero. In fact, the equation has no solution since the left member exceeds the right member by 1 for all permissible values of x.*

EXAMPLE 4. Solve the equation

$$\frac{x - 1}{2x - 1} - \frac{x}{x + 1} = \frac{2 - x^2}{2x^2 + x - 1}$$

Solution. Multiplying both members of the equation by $(2x - 1)(x + 1)$, the L.C.M. of the denominators, we have

$$(x - 1)(x + 1) - x(2x - 1) = 2 - x^2$$

$$x^2 - 1 - 2x^2 + x = 2 - x^2 \qquad \text{simplifying}$$

$$x = 3 \qquad \text{adding } x^2 + 1$$

Check.

$$\frac{3 - 1}{6 - 1} - \frac{3}{3 + 1} = \frac{-7}{20} \qquad \text{for the left member}$$

$$\frac{2 - 9}{18 + 3 - 1} = \frac{-7}{20} \qquad \text{for the right member}$$

* The permissible values of the variables in an equation are all values of the variables for which the members of the equation are defined.

EXERCISE 6-1

Do odd

Solve each equation for x, and reject as a root any number which makes a denominator equal to zero.

1. $5x - 4 = 3x - 6$
2. $x - 5 = 5x + 11$
3. $3(4 - 5x) - 2(x + 4) = 4$
4. $2(x - 7) + 5(3 + 2x) + 6 = 0$
5. $4x + 13 = 3(5x + 20) + 8$
6. $(x + 1)^2 - (x - 1)^2 = x + 9$
7. $(3x - 1)(x + 1) = 3x^2$
8. $(x + 4)(x - 1) = (x + 2)(x - 3)$
9. $x(10 - 3a) = 4 + 2b$
10. $3 - ax + 4x - 36 = 0$
11. $2(x - a) - 5 = 3(ax - 4)$
12. $5x - a = 2ax + 7$

13. $\dfrac{3x + 4}{4} - \dfrac{5x}{8} = \dfrac{2x + 4}{6}$

14. $\dfrac{x}{7} + \dfrac{x - 1}{3} = \dfrac{2x + 1}{21} - \dfrac{1}{3}$

15. $\dfrac{2x + 7}{4} + \dfrac{x - 5}{3} = \dfrac{x + 1}{6}$

16. $\dfrac{3 - 4x}{5} = \dfrac{4 - 3x}{4} - \dfrac{5 - 2x}{2}$

17. $\dfrac{2}{x} - \dfrac{2}{5} = \dfrac{1}{x} + \dfrac{11}{10}$

18. $\dfrac{5}{7x} - \dfrac{7}{2x} - 3 = \dfrac{3}{14}$

19. $\dfrac{x}{x - 2} + \dfrac{2}{x + 2} = \dfrac{x^2 - 1}{x^2 - 4}$

20. $\dfrac{2}{x - 1} + \dfrac{3}{x + 1} = \dfrac{4}{x^2 - 1}$

21. $\dfrac{2}{x - 1} + \dfrac{3}{x + 1} = \dfrac{4}{x^2 - 1}$

22. $\dfrac{1}{x} + \dfrac{1}{x^2 - x} = \dfrac{2}{x - 1}$

23. $\dfrac{x + 1}{x} - \dfrac{x + 4}{x + 5} = \dfrac{3x + 5}{x^2 + 5x}$

24. $\dfrac{1}{2x + a} + \dfrac{2}{3x} = \dfrac{7}{6x + 3a}$

6-4 WORD PROBLEMS

The equations in Exercise 6-1 are either linear or expressible linearly by clearing of fractions. We next consider word problems which lead to equations of the same type. The equation in each problem is to be found from the described relations between known numbers and unknown numbers. We make the following suggestions for attacking such problems.

1. Read and reread the problem until it is clear what is stated, getting well in mind the given numbers and the unknown numbers (numbers to be found).
2. Express each unknown number in terms of a single letter.
3. Find the quantities, involving the given numbers and the unknown numbers, which are equal. Then form an equation.
4. Solve the equation and check the result.

EXAMPLE 1. The length of a rectangle exceeds its width by 2 feet. If each dimension were increased by 3 feet, the area would be increased by 51 square feet. Find the original dimensions.

Solution. If x feet by $x + 2$ feet are the original dimensions, then $x + 3$ feet by $x + 5$ feet are the new dimensions. Since the increase in the area is 51 square feet, we have

$$(x + 3)(x + 5) - x(x + 2) = 51$$
$$x^2 + 8x + 15 - x^2 - 2x = 51$$
$$6x = 36$$
$$x = 6$$

The original dimensions are 6 by 8 feet and the new dimensions are 9 by 11 feet. The areas, then, are 48 square feet and 99 square feet, which yield a difference of 51 square feet.

EXAMPLE 2. A can do a certain task in 8 hours, B in 10 hours, and C in 12 hours. How long will it take to do the task if A and B work 1 hour and then A and C finish the task?

Solution. The fractional parts of the task done in 1 hour by A, B, and C are $\frac{1}{8}, \frac{1}{10}$, and $\frac{1}{12}$, respectively. The contribution of each worker is the part he does in 1 hour multiplied by the number of hours he works. If we let x be the total number of hours required to do the task, then A works x hours, B works 1 hour, and C works $x - 1$ hours. Hence

$$\frac{x}{8} = \text{part of task done by A}$$

$$\frac{1}{10} = \text{part of task done by B}$$

$$\frac{x - 1}{12} = \text{part of task done by C}$$

The whole task is completed by the sum of these parts and therefore

$$\frac{x}{8} + \frac{1}{10} + \frac{x - 1}{12} = 1$$

$$15x + 12 + 10x - 10 = 120 \qquad \text{multiplying by 120, the L.C.D.}$$
$$25x = 118 \qquad \text{combining terms}$$
$$x = 4\tfrac{18}{25}$$

Hence the time required is $4\tfrac{18}{25}$ hours.

EXAMPLE 3. How many ounces of pure silver must be added to 100 ounces, 40% pure, to make a mixture which is 65% pure?

Solution. The equation may be set up by noticing that the amount of silver in the two separate parts is just the same as the amount of silver in the mixture.

Using x to stand for the number of ounces of silver to be added, we arrange the information thus:

x oz	and	100 oz	give	$(100 + x)$ oz
100% pure		40% pure		65% pure
x oz silver		40 oz silver		$.65(100 + x)$ oz silver

Since the original 40 ounces of silver and the additional x ounces constitute the silver in the mixture, we have

$$x + 40 = .65(100 + x)$$
$$.35x = 25$$
$$x = 71\tfrac{3}{7} \text{ oz}$$

Check.

$$71\tfrac{3}{7} + 40 = 111\tfrac{3}{7} \quad \text{and} \quad .65(171\tfrac{3}{7}) = 111\tfrac{3}{7}$$

EXERCISE 6-2

1. The sum of three numbers is 138. The second is 5 more than the smallest and the third is 10 more than the smallest. Find the numbers.
2. Three numbers are so related that the second number is 2 more than the first number and the third is 4 more than the first. Find the numbers if the sum of their squares is 56 more than three times the square of the smallest number.
3. The length of a rectangle exceeds its width by 6 units. If each dimension were increased by 3 units, the area would be increased by 57 square units. Find the dimensions of the rectangle.
4. The altitude of a triangle is $\tfrac{3}{4}$ the length of its base. If the altitude were increased by 3 feet and the base decreased by 3 feet, the area would be unchanged. Find the length of the base and altitude.
5. A can paint a certain house in 10 days and B can paint the house in 12 days. How long will it take to paint the house if both men work?
6. A man can do a certain task in 21 hours, another man can do the task in 28 hours, and a boy can do the task in 48 hours. Find how long it will take to do the task if all three work.
7. A can do a certain job in 4 hours, B can do the job in 6 hours, and C can do the job in 8 hours. How long will it take to do the job if A and B work 1 hour and B and C finish the job?
8. A tank can be filled by one pipe in 9 hours and by another pipe in 12 hours. Starting empty, how long will it take to fill the tank if water is being taken out by a third pipe at a rate per hour equal to one-sixth the capacity of the tank?
9. How many ounces of pure silver must be added to 18 ounces, 60% pure, to make an alloy which is 76% pure silver?
10. A perfumer wishes to blend perfume valued at $4.10 an ounce with perfume worth $2.50 an ounce to obtain a mixture of 40 ounces worth $3.00 an ounce. How much of the $4.10 perfume should he use?

11. An alloy of silver and gold weighs 15 ounces in air and 14 ounces in water. Assuming that silver loses $\frac{1}{10}$ of its weight in water and that gold loses $\frac{1}{19}$ of its weight, how many ounces of each metal are in the alloy?

12. The denominator of a fraction exceeds the numerator by 2. If 1 is added to the numerator and 2 subtracted from the denominator, the sum of the original fraction and the new fraction is equal to 2. Find the original fraction.

13. The numerator of a fraction is 4 less than the denominator. If the numerator is doubled and the denominator diminished by 2, the sum of the original fraction and the new one is 3. Find the original fraction.

14. A man loans $10,000, part at 6% annual interest and the rest at 11%. Find the amount of each loan if the total annual income is $624.

15. A man loans $4000 at one interest rate and $5000 at a 1% greater rate. The $5000 loan earns $110 more per year than the $4000 loan earns. Find the rates.

16. A man invests $4000 at a certain interest rate and $6000 at a rate $\frac{2}{3}$ as high. His annual income from the two investments is $720. Find the interest rates.

17. In college algebra a student earns an average grade of 86 on his homework and 75 on his hour tests. What final examination grade will give him a total average of 80 if homework counts $\frac{1}{10}$, hour tests $\frac{6}{10}$, and final examination $\frac{3}{10}$?

18. Mr. Jones has two health insurance policies. For the first policy he pays $120 a year which entitles him to 80% of his covered medical expenses after the insured pays the first $100. The premium for the second policy is $90 for which the company will pay 25% of his medical bill plus 20% of the amount which the first company pays. Find the amount of a medical bill which the insured may incur during a year so that he will recover exactly the medical bill plus the cost of the two premiums.

If a body travels at a constant speed for a given time, the relation among the distance d, the speed s, and the time t is expressed by each of the formulas

$$d = st, \qquad s = d/t, \qquad t = d/s$$

These formulas will be helpful in Problems 19 through 23.

19. A man drove 220 miles in $3\frac{1}{2}$ hours. Part of the trip was at 60 miles per hour and the rest at 65 miles per hour. Find the time spent at the lower speed.

20. Town A is 11 miles west of town B. A man walks from A to B at the rate of 3 miles per hour, and another man, starting at the same time, walks from B to A at the rate of 4 miles per hour. Find the times after starting that the men are 2 miles apart.

21. Two men starting at a point on a 1-mile circular race track walk in opposite directions and meet in 6 minutes. But if they walk in the same direction, the faster walker gains one lap in 30 minutes. Find the speed of each man in miles per hour.

22. The speed of a motorboat in still water is 16 miles per hour. Find the speed of a river's current if the motorboat goes 5 miles downstream in the same time required to go 3 miles upstream.

23. A motorboat goes 3 miles upstream in the same time required to go 5 miles downstream. If the rate of flow of the river is 3 miles per hour, find the speed of the motorboat in still water.

6-5 LINEAR EQUATIONS IN TWO VARIABLES

An equation of the form

$$ax + by + c = 0 \qquad (1)$$

where a, b, and c are constants with a and b not both zero, is called a *linear equation* in the variables x and y. A pair of values, one for x and one for y, which satisfy the equation is called a *solution* of the equation. We have already solved equations of the form (1) with $b = 0$. Now we assume that $b \neq 0$ and express the equation in the equivalent form (Sec. 6-2)

$$y = -\frac{a}{b}x - \frac{c}{b}$$

This equation reveals that there is a unique value of y corresponding to any arbitrary value of x. Hence we say there are infinitely many solutions of the equation. The totality of solutions is called the *solution set* of the equation. If we let A stand for the solution set, we may express the set as

$$A = \left\{ (x, y) \middle| y = -\frac{a}{b}x - \frac{c}{b} \right\}$$

or, equivalently,

$$A = \{(x, y) | ax + by + c = 0\}$$

Referring to Definition 5-4, we note that the graph of a linear equation in x and y is the set of all points (x, y) in the plane whose coordinates satisfy the equation. And it is proved in analytic geometry that the graph is a straight line. The name linear equation comes from this fact. The function defined by the equation is quite appropriately called a *linear function*. Clearly, the graph of this kind of equation (or function) can be drawn from two of its points.

Two linear equations in the same two variables are said to constitute a *system* of equations. We have seen how to solve a linear equation in one variable, and now we consider the problem of solving a system of linear equations in two variables. We represent such a system by the equations

$$\begin{aligned} a_1 x + b_1 y &= c_1 \\ a_2 x + b_2 y &= c_2 \end{aligned} \qquad (2)$$

If we let

$$A = \{(x, y) | a_1 x + b_1 y = c_1\} \quad \text{and} \quad B = \{(x, y) | a_2 x + b_2 y = c_2\}$$

be the solution sets of the equations, then we define the solution set of the system of equations to be the intersection $A \cap B$ of the sets A and B.

If the graph of two linear equations in x and y consists of two intersecting lines, the solution set of the system has only one member, namely, the pair of coordinates of the intersection point of the lines. Because of this restriction on x and y, we shall frequently refer to these letters in a system of the form (2) as *unknowns* instead of variables.

We next show how to solve graphically a system of two linear equations in two unknowns.

EXAMPLE 1. Solve graphically the system

$$2x + 3y = 6$$

$$3x - 10y = 15$$

Solution. We prepare a table of values for each equation. This is done by assigning values to one of the unknowns and solving for the other. Although two number pairs are sufficient for drawing the line, we suggest a third pair which serves as a check against errors.

$$2x + 3y = 6:\quad \begin{array}{c|c|c|c} x & 0 & 3 & 1.5 \\ \hline y & 2 & 0 & 1 \end{array}$$

$$3x - 10y = 15:\quad \begin{array}{c|c|c|c} x & 0 & 5 & 2 \\ \hline y & -1.5 & 0 & -0.9 \end{array}$$

The lines intersect as shown in Fig. 6-1. We estimate the coordinates of the intersection point as $x = 3.6$ and $y = -0.4$. These values are approximately correct. They satisfy the first equation but make the left side of the second equation 14.8 and the right side 15.

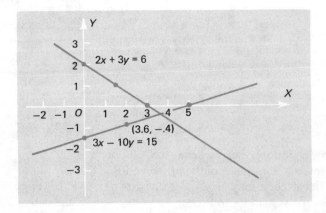

Fig. 6-1

EXAMPLE 2. Construct the graph of the system

$$3x - 4y = 5$$

$$6x - 8y = -2$$

Solution. We make a table of values for each equation.

$3x - 4y = 5$:

x	-1	3
y	-2	1

$6x - 8y = -2$:

x	-3	-1
y	-2	1

The graph (Fig. 6-2) consists of two lines which appear to be parallel. If the lines are parallel, there is no pair of numbers for x and y which satisfy both equations. We can show independently of the graph that there is no solution. Thus multiplying both members of the first equation by 2, which yields an equivalent equation, we may write the system as

$$6x - 8y = 10$$

$$6x - 8y = -2$$

A pair of numbers which makes $6x - 8y$ equal to 10 would certainly not make $6x - 8y$ equal to -2.

Had the two equations been given as

$$3x - 4y = 5$$

$$6x - 8y = 10$$

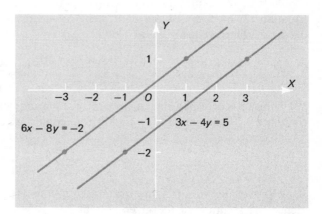

Fig. 6-2

then the members of the first equation multiplied by 2 yield the second equation. Any solution of one equation would obviously be a solution of the other. Hence the equations of the system are equivalent, and there is an unlimited number of solutions.

We have illustrated the following kinds of solution sets of a system of the form (2).

1. The solution set is a single pair of numbers. In this case the graphs of the equations intersect in a single point, and the equations are said to be *consistent*.
2. The solution set has no element, or is the null set $\emptyset$. The graphs of the equations are distinct parallel lines, and the equations are said to be *inconsistent*.
3. The solution set of one equation is the same as the solution set of the other, and the equations are said to be *dependent*. The graphs of the equations coincide.

6-6 SOLUTION BY ALGEBRAIC METHODS

We next discuss two different algebraic methods for solving a system of linear equations in two unknowns. These methods yield exact solutions of consistent equations, whereas the graphical method, due to imperfect graphing and possible imperfect estimation of coordinates, often gives only approximate results.

A. Elimination by Addition or Subtraction. We apply this method in solving the system

$$5x + 3y = -26 \tag{1}$$

$$4x - 9y = 2 \tag{2}$$

Solution. If we multiply the members of equation (1) by 4 and those of equation (2) by 5, we obtain equivalent equations with the coefficients of x the same. Thus we get

$$20x + 12y = -104 \tag{3}$$

$$20x - 45y = 10 \tag{4}$$

We subtract corresponding members of equation (4) from equation (3) to eliminate x. Thus, subtracting equals from equals, we get

$$57y = -114$$

$$y = -2$$

If we substitute -2 for y in either of equations (1) or (2), we find $x = -4$. Hence the solution of the given system is $x = -4$, $y = -2$. We check by

substituting these values for x and y in the left members of the given equations.
  Check.

$$5(-4) + 3(-2) = -20 - 6 = -26$$

$$4(-4) - 9(-2) = -16 + 18 = 2$$

B. Elimination by Substitution. We illustrate this method by solving the system of the previous example:

$$5x + 3y = -26 \tag{1}$$

$$4x - 9y = 2 \tag{2}$$

Solving equation (2) for x, we get

$$x = \frac{9y + 2}{4}$$

By substituting this expression for x in equation (1), we obtain a linear equation in y. Thus

$$5\left(\frac{9y + 2}{4}\right) + 3y = -26$$

Multiplying by 4, we have

$$45y + 10 + 12y = -104$$

$$57y = -114$$

$$y = -2$$

Then

$$x = \frac{9y + 2}{4} = \frac{-18 + 2}{4} = -4$$

Hence we obtain, as before, the solution $x = -4$, $y = -2$.

If either of the above methods of solution is applied to an inconsistent or a dependent system of equations, both x and y will be eliminated simultaneously. If the elimination process leads to an equation whose members are different constants, the equations are inconsistent. If the resulting members are the same constant, the equations are dependent.

EXERCISE 6-3

Draw the graphs of the equations in each of the following systems. Estimate the co-ordinates of the intersection point of each consistent system.

1. $x + y = 5$	**2.** $2x + 3y = 2$	**3.** $3x - 2y = 6$
$x - y = 1$	$x - 2y = -6$	$2x + 3y = 6$
4. $y - 4x = 3$	**5.** $7x + 3y = 4$	**6.** $4x + 6y = 12$
$2y - 8x = 6$	$14x + 6y = 8$	$2x + 3y = -6$
7. $5y + 5x = 18$	**8.** $5x + y = 1$	**9.** $4x - 3y = 5$
$y - x = 0$	$3x + 2y = -2$	$3x - 2y = 3$

Solve each pair of equations, if found consistent, by the substitution method. Then check your solution by the addition or subtraction method.

10. $x + y = 1$
$3x - y = 3$

11. $2x - 3y = 2$
$4x + 3y = 22$

12. $4x + 3y = 1$
$2x + 5y = 11$

13. $3x - 4y = 3$
$6x - 8y = 6$

14. $15x + 3y = 2$
$10x + 2y = 3$

15. $2x - 4y = -3$
$4x + 2y = 9$

Solve each system for x and y by either elimination process and check your results.

16. $3x + 2y = -3$
$2x - 5y = 17$

17. $2x - 7y = 17$
$4x - 5y = 25$

18. $3x + ay = 4$
$2x + ay = 2$

19. $x - y = 2a$
$x + y = 2b$

20. $ax + by = a^2 + b^2$
$bx - ay = a^2 + b^2$

21. $a_1x + b_1y = c_1$
$a_2x + b_2y = c_2$

22. $\dfrac{2x + 2}{2y - 3} = \dfrac{x - 2}{y + 3}$

$\dfrac{x + 2}{3y - 2} = \dfrac{x - 1}{3y + 8}$

23. $\dfrac{2x - 3}{2x - 1} = \dfrac{y + 7}{y + 5}$

$\dfrac{5x - 4}{x + 5} = \dfrac{5y - 14}{y + 3}$

Solve each system, first eliminating one unknown without clearing of fractions.

24. $\dfrac{1}{x} - \dfrac{2}{y} = 3$

$\dfrac{1}{x} + \dfrac{3}{y} = 2$

25. $\dfrac{2}{x} + \dfrac{3}{y} = 1$

$\dfrac{4}{x} + \dfrac{3}{y} = 4$

26. $\dfrac{3}{5x} + \dfrac{1}{2y} = \dfrac{1}{10}$

$\dfrac{5}{3x} + \dfrac{1}{y} = \dfrac{2}{3}$

27. $\dfrac{1}{3x} + \dfrac{3}{4y} = \dfrac{5}{2}$

$\dfrac{5}{9x} + \dfrac{2}{3y} = 3$

28. Show that the equations below are dependent if $a_1/a_2 = b_1/b_2 = c_1/c_2$

$$a_1x + b_1y = c_1$$
$$a_2x + b_2y = c_2$$

Hint: Set each of the three fractions equal to a constant k. Show that the system is inconsistent if

$$a_1/a_2 = b_1/b_2 \neq c_1/c_2$$

6-7 LINEAR EQUATIONS IN THREE UNKNOWNS

We next consider the problem of solving a system of three linear equations in three unknowns. The algebraic methods which were used for solving a system of equations in two unknowns will serve in an analogous fashion to yield the solution for the case of three unknowns.

A system of three linear equations in three unknowns has one solution, no solution, or infinitely many solutions. This same situation was discussed in the case of two unknowns. But we shall consider systems of three unknowns which have one and only one solution.

EXAMPLE. Solve the system of equations

$$2x + 3y + 4z = 8 \tag{1}$$

$$3x + 2y + 3z = 8 \tag{2}$$

$$5x - 4y - 2z = 7 \tag{3}$$

Solution. Selecting y as the first unknown to be eliminated, we multiply the members of equation (1) by 2 and the members of equation (2) by 3 and obtain

$$4x + 6y + 8z = 16$$

$$9x + 6y + 9z = 24$$

Subtracting, we have

$$5x + z = 8$$

Next we multiply the members of equation (2) by 2 and use equation (3) unchanged. This gives

$$6x + 4y + 6z = 16$$

$$5x - 4y - 2z = 7$$

Adding, we have

$$11x + 4z = 23$$

We now have the equations $5x + z = 8$ and $11x + 4z = 23$. The solution of these equations is $x = 1$, $z = 3$. Substituting these values for x and z in any of the given equations yields $y = -2$. Hence the solution of the given system is $x = 1$, $y = -2$, $z = 3$. These values satisfy each of the given equations.

The following steps may be employed to solve a system of linear equations of three unknowns.

1. Choose a pair of equations from the three given equations and eliminate one of the unknowns.
2. Select one of the same pair and the unused equation and eliminate the same unknown.
3. Solve the equations resulting from steps 1 and 2 for the two unknowns.
4. Substitute these values in one of the given equations and obtain the remaining unknown.

In solving a system of linear equations in two unknowns we performed operations which led to a new system in which each unknown is equal to a constant. And in the example above we obtained equations with each of the three unknowns equal to a constant. In each case the key operation is the replacement of one equation by a new equation. The resulting new system of equations is equivalent to the original system, as we shall now show.

Consider the system

$$a_1x + b_1y = c_1 \qquad a_1 \neq 0, \quad b_1 \neq 0$$
$$a_2x + b_2y = c_2 \qquad a_2 \neq 0, \quad b_2 \neq 0 \tag{1}$$

and the system

$$a_1x + b_1y = c_1$$
$$k_1(a_1x + b_1y) + k_2(a_2x + b_2y) = k_1c_1 + k_2c_2 \tag{2}$$

where k_1 and k_2 are nonzero numbers. Clearly any solution of system (1) is a solution of system (2). Suppose next that an ordered pair of numbers (x', y') is a solution of system (2). Then the second equation may be expressed as

$$k_1c_1 + k_2(a_2x' + b_2y') = k_1c_1 + k_2c_2$$

which may be reduced to

$$a_2x' + b_2y' = c_2$$

Hence (x', y') is a solution of system (1). Since a solution of system (1) is a solution of system (2) and a solution of system (2) is a solution of system (1), we conclude that the two systems are equivalent.

In solving a system of the form (1), we select values for k_1 and k_2 so that either $k_1a_1 + k_2a_2 = 0$ or $k_1b_1 + k_2b_2 = 0$, thus eliminating one of the unknowns. We note that the reasoning which we have used in establishing the equivalence of systems (1) and (2) is applicable for the case of three linear equations in three unknowns.

EXERCISE 6-4

Solve the following systems of equations for x, y, and z or u, v, and w. Check your results.

1. $3x - 4y - z = 1$	**2.** $2x - y + 3z = 9$	**3.** $4u + v + 2w = 10$
$x - y + 3z = 3$	$3x + y + 2z = 11$	$3u + 2v + w = 5$
$3x - 2y + 2z = 0$	$x - y + z = 2$	$2u + 3v + 2w = 10$
4. $3u + v + 3w = 9$	**5.** $4x + 2y + 3z = 3$	**6.** $2x - 3y - 4z = 8$
$4u + v + 2w = 13$	$3x - 6y + 4z = 3$	$3x - 2y - 3z = 8$
$7u + 2v - w = 24$	$x + 2y + z = 1$	$4x + 4y + 2z = 6$
7. $3x - 2y + 5z = 28$	**8.** $5x - 7y - 9z = 14$	**9.** $5x + 4y - 3z = 3$
$4x + 5y + 4z = 9$	$7x + 3y - 4z = 11$	$4x - 2y + 2z = -1$
$5x - 4y - 3z = 26$	$3x + 4y + 8z = -5$	$6x - 4y + 2z = -1$

10. $u + v + w = 2a$
$u - v + w = 2b$
$u + v - w = 2c$

11. $u + 2v - w = a - b$
$2u - v + w = 3a + 3b$
$-u + v + 2w = 4a - 2b$

12. $ax + by + cz = 6$
$ax - by + cz = 8$
$ax - by - cz = 0$

13. $u - v = 1$
$u + w = -1$
$v - w = 6$

14. $2u + 3v = 2$
$u + 2w = 1$
$3v + 8w = 3$

15. $ax - by = 2$
$ax - 2cz = 3$
$by + 3cz = -2$

16. $4u - 9v = 4$
$u - 6w = 4$
$3v - 4w = 1$

17. $\dfrac{1}{x} - \dfrac{2}{y} + \dfrac{1}{z} = \dfrac{4}{3}$

$\dfrac{2}{x} - \dfrac{3}{y} + \dfrac{2}{z} = \dfrac{5}{3}$

$\dfrac{3}{x} - \dfrac{4}{y} + \dfrac{6}{z} = 3$

18. $\dfrac{2}{x} + \dfrac{2}{y} + \dfrac{3}{z} = 4$

$\dfrac{1}{x} + \dfrac{4}{y} - \dfrac{4}{z} = 0$

$\dfrac{3}{x} + \dfrac{3}{y} + \dfrac{2}{z} = 1$

6-8 WORD PROBLEMS LEADING TO SYSTEMS OF EQUATIONS

There are many kinds of verbal problems involving two or more unknowns which can be solved by setting up a system of equations from the known data. The reasoning by which the equations are obtained is much the same as for the equations of one unknown in Sec. 6-4.

EXAMPLE. A and B can do a certain job in 9 days; A and C can do the job in 8 days; and B and C can do the job in 12 days. Find how long it would take each person working alone to do the job.

Solution. We let x, y, and z stand for the days required for A, B, and C, respectively, working alone to do the job. Then $1/x$, $1/y$, and $1/z$ give the fractional parts of the job completed in 1 day by A, B, and C. Therefore, we may write the equations

$$\frac{1}{x} + \frac{1}{y} = \frac{1}{9}$$

$$\frac{1}{x} + \frac{1}{z} = \frac{1}{8}$$

$$\frac{1}{y} + \frac{1}{z} = \frac{1}{12}$$

These equations, though not linear in x, y, and z, are linear in $1/x$, $1/y$, and $1/z$. The equations yield

$$x = 13\tfrac{1}{11} \text{ days}, \qquad y = 28\tfrac{4}{5} \text{ days}, \qquad z = 20\tfrac{4}{7} \text{ days}$$

do checked

Find the solution of each of the following problems by first setting up a system of two equations in two unknowns.

1. The difference of two numbers is 14 and twice the smaller number is 5 less than the larger number. Find the numbers.

2. A cattleman sold 60 calves and 240 sheep for $17,160; and, pricing the animals in the same way, he sold 40 calves and 180 sheep to another buyer for $12,240. Find the price per head for each kind of animal.

3. A merchant has $36 worth of dimes and quarters. How many coins of each kind are there if the total number of coins is 192?

4. If the numerator and denominator of a certain fraction are each decreased by 2, the value of the new fraction is $\frac{1}{2}$. But if the numerator of the original fraction is increased by 2 and the denominator decreased by 2, the resulting fraction is equal to $\frac{3}{4}$. Find the original fraction.

5. A man has two investments, one paying 3% annual interest and the other 4%. The total annual income from the investments is $170. If the interest rates were interchanged, the total annual income would be $180. Find the amount of each investment.

6. If the length of a certain rectangle were increased by 2 feet and the width decreased by 1 foot, the area would be decreased by 2 square feet. If the length were decreased by 2 feet and the width increased by 2 feet, the area would be increased by 16 square feet. Find the dimensions of the rectangle.

7. A and B can paint a certain house if A works 6 days and B works 12 days, or they can paint the house if A works 9 days and B works 8 days. How long would it take each working alone to paint the house?

8. A man and a boy perform a task in 2 days. They completed one-half of the task the first day with the man working 8 hours and the boy working 10 hours. On the next day they finished the task with the man working 4 hours and the boy working 15 hours. Find how many hours it would take for each working alone to do the task.

9. The sum of the reciprocals of two numbers is 11. Three times the reciprocal of one of the numbers is 3 more than twice the reciprocal of the other number. Find the numbers.

10. A certain two-digit number is equal to 9 times the sum of its digits. If 63 were subtracted from the number the digits would be reversed. Find the number. [Hint: If x is the tens digit and y the units digit, the value of the number is $10x + y$.]

11. An airplane travels 360 miles with the wind in 1 hour and 20 minutes and returns against the wind in 2 hours and 15 minutes. Find the speed of the airplane in still air and the speed of the wind.

12. A boat goes 6 miles downstream in a river in 30 minutes and returns upstream in 45 minutes. Find the speed of the boat in still water and the speed of the current.

13. A man rows 12 miles upstream in a river and back in $7\frac{1}{2}$ hours. He can row 1 mile upstream in the same time that he can row 4 miles downstream. Find the speed of the boat in still water and the speed of the current.

14. A laborer was employed to work for a daily wage of $15 and pay $2 per day for board. He was idle on certain days and received no wage for that time, but was

charged for food every day. His net pay at the end of the period was $382. Had his daily wage been $18 and his board $3, the net pay would have been $438. How many days did he work and how many days was he idle?

Find the solution of each of the following problems by first setting up a system of three equations with three unknowns.

15. The sum of the three angles of a triangle is 180°. The sum of two of the angles is equal to the third angle and the difference of the two angles is equal to $\frac{2}{3}$ the third angle. Find the angles.

16. Find three numbers such that the sum of the first and second is 67, the sum of the first and third is 80, and the sum of the second and third is 91.

17. To provide change for a day's business, a merchant buys $75 worth of half dollars, quarters, dimes, and nickels. Altogether there are 360 coins, and the number of half dollars is the same as the number of quarters, and the number of dimes is the same as the number of nickels. Find the number of coins of each kind.

18. A and B working together can do a job in 6 days. A and C can do the job in 8 days. B and C can do the job in 9 days. Find how long it would take each man working alone to do the job.

19. A, B, and C working together can do a certain task in 6 hours. A and B can do the task in 9 hours, and B and C can do the task in 12 hours. How long would it take each working alone to do the task?

20. A man makes three investments totaling $24,000 with interest rates at 6%, 7%, and 8%. The total annual income is $1720, and the income from the 7% investment is $40 less than the combined income from the other investments. Find the amount of each investment.

21. The sum of the second and third digits of a three-digit number is equal to the first digit. The sum of the first digit and the second digit is 2 more than the third digit. If the second and third digits were interchanged, the new number would be 54 more than the original number. Find the number.

22. The sum of the digits of a three-digit number is 18. If the second and third digits were interchanged, the number would be increased by 36. If the first and third digits were interchanged, the new number would be 99 less than the original number. Find the number.

23. A, B, and C complete a task in 3 days. On the first day they do $\frac{1}{3}$ of the task with A working 10 hours, B working 4 hours, and C working 10 hours. On the second day they do $\frac{1}{4}$ of the task with A working 4 hours, B working 6 hours, and C working 9 hours. They finish the task on the third day with A working 7 hours, B working 12 hours, and C working 12 hours. Find the number of hours it would take each man working alone to do the task.

6-9 RATIO AND PROPORTION

Two numbers may be compared in magnitude by dividing one by the other. The quotient is called the *ratio* of one number to the other. If x and y ($\neq 0$) stand for two numbers, the ratio of x to y may be expressed in any of the forms

$$x/y, \qquad x \div y, \qquad x:y$$

The fractional form is usually preferable when algebraic operations are to be performed. The form $x:y$ (read "x is to y") employs the colon to indicate division.

Two like quantities may be compared in magnitude by first expressing them in terms of the same unit of measure. For example, the ratio of 16 hours to 3 days is

$$\frac{16}{72} = \frac{2}{9}$$

Hence the ratio of 16 hours to 3 days is 2 to 9. The same information is given by saying the ratio of 3 days to 16 hours is 9 to 2.

A statement of equality of two ratios is called a *proportion*. A proportion may be expressed in any of the forms

$$\frac{a}{b} = \frac{c}{d}, \qquad a \div b = c \div d, \qquad a:b = c:d$$

The form $a:b = c:d$ is read "a is to b as c is to d." The numbers a and d are called the *extremes* and b and c the *means* of the proportion. The number d is called the *fourth* proportional to a, b, and c. If the means are equal ($b = c$), the proportion becomes

$$\frac{a}{b} = \frac{b}{d}$$

In this special case b is called the *mean proportional* to a and d, and d is the *third proportional* to a and b.

The following theorems are met in elementary plane geometry. The student should make the proofs. The values of a, b, c, and d are assumed to be such that no denominator is equal to zero.

Theorem 6-3   If $\dfrac{a}{b} = \dfrac{c}{d}$, then $ad = bc$.

Theorem 6-4   If $\dfrac{a}{b} = \dfrac{c}{d}$, then $\dfrac{a}{c} = \dfrac{b}{d}$.

Theorem 6-5   If $\dfrac{a}{b} = \dfrac{c}{d}$, then $\dfrac{b}{a} = \dfrac{d}{c}$.

Theorem 6-6   If $\dfrac{a}{b} = \dfrac{c}{d}$, then $\dfrac{a+b}{b} = \dfrac{c+d}{d}$.

Theorem 6-7   If $\dfrac{a}{b} = \dfrac{c}{d}$, then $\dfrac{a-b}{b} = \dfrac{c-d}{d}$.

Theorem 6-8   If $\dfrac{a}{b} = \dfrac{c}{d}$, then $\dfrac{a+b}{a-b} = \dfrac{c+d}{c-d}$.

Hints: To prove Theorem 6-4, divide both members of $ad = bc$ by cd. To prove Theorem 6-6, add 1 to both members of the given proportion. Theorem 6-8 may be had readily from Theorems 6-6 and 6-7.

EXAMPLE 1. Find the third proportional to 8 and 14.

Solution. To form the proportion, we observe that each of the means is 14. Then letting x stand for the third proportional, we have

$$8:14 = 14:x \quad \text{or} \quad \frac{8}{14} = \frac{14}{x}$$

whence, clearing of fractions, $8x = 196$ and $x = 24.5$.

EXAMPLE 2. A triangle has sides of lengths 12, 17, and 22 inches. If the length of the shortest side of a similar triangle is 8 inches, find the length of the longest side.

Solution. We recall that the corresponding sides of similar figures are proportional. Then if x denotes the length of the longest side, we have

$$\frac{x}{22} = \frac{8}{12}$$

$$12x = 176$$

$$x = 14\tfrac{2}{3} \text{ inches}$$

Sometimes we express the relative magnitudes of three numbers in the form of a ratio. Suppose, for example, that the ratio of x to y is 2 to 5 and that the ratio of y to z is 5 to 7. Then we say that x is to y is to z as 2 is to 5 is to 7, and write

$$x:y:z = 2:5:7$$

EXERCISE 6-6

Express each ratio as a fraction in lowest terms.

1. 3 feet to 9 inches
2. 15¢ to $1.05
3. 9 hours to 3 days
4. 72 sq in. to 2 sq ft

Find the fourth proportional to each set of numbers, and the third proportional to the first two numbers.

5. 4, 3, 6
6. $-2, 6, 5$
7. $-3, 11, -9$

Find the unknown (or unknowns) in each problem.

8. $4:b = 2:5$
9. $9:7 = 6:x$
10. $x:4 = 9:x$
11. $(2a - 1):(4a + 1) = (a - 4):(2a + 3)$
12. $(3x + 2):(6x - 2) = (x - 2):(2x - 3)$
13. $x:y:2 = 2:3:5$
14. $a:3:b = -2:3:-4$

15. Two school buildings are 7 inches apart on a city map. If the map scale is 4 inches to 3 miles, find the distance between the buildings.
16. Divide 72 into two parts which have the ratio 4:5.
17. The sides of a quadrilateral are 6, 7, 9, and 10 inches long. The longest side of a similar quadrilateral is 8 inches. Find the lengths of the other sides of the quadrilateral.
18. Three men do a piece of work for $246. If this sum is to be divided among the men in the ratio 3:4:5, find the three resulting amounts.

6-10 VARIATION

In our discussion of ratios thus far we have dealt with fixed quantities. We next consider the ratio of two variables which are so related that their ratio is constant, or, equivalently, one variable is a constant times the other variable. Variables of this kind are plentiful in science, business, and many everyday situations.

If a variable y is equal to a constant times another variable x, then y is said to *vary directly as* x. The relation may be written as

$$y = kx$$

The fixed number k is called the *constant of proportionality*, or the *constant of variation*. The expressions

y varies directly as x
y varies as x
y is directly proportional to x
y is proportional to x

are all used with the same meaning.

The formula for the circumference of a circle $C = 2\pi r$ shows that the circumference varies directly as the radius and that 2π is the constant of variation. The formula for the area of a circle $A = \pi r^2$ shows that the area varies directly as the square of the radius and that π is the constant of proportionality.

Suppose that z is always equal to a constant times the product of two variables x and y; that is

$$z = kxy$$

We describe this relation by saying that z *varies jointly as* x *and* y.

The area of a triangle varies jointly as the base and altitude. The formula $A = \frac{1}{2}bh$ shows the constant of variation to be $\frac{1}{2}$.

EXAMPLE 1. The weight of a sphere is proportional to the cube of its radius. A certain sphere of radius 4 inches weighs 24 pounds. Find the weight of a sphere of radius 6 inches which is made of the same kind of material.

Solution. We let W pounds stand for the weight and r inches for the radius and write

$$W = kr^3$$

To determine k, we substitute 24 for W and 4 for r. Thus we find

$$24 = 64k \quad \text{and} \quad k = \tfrac{3}{8}$$

Then for all spheres of this material, we have

$$W = \frac{3r^3}{8}$$

Hence the weight of the sphere of radius 6 inches is

$$W = \frac{3(6)^3}{8} = 81 \text{ pounds}$$

If a variable y is always equal to a constant times the reciprocal of another variable x, then y is said to *vary inversely as x*, or *y is inversely proportional to x*. This relation is expressed by the equation

$$y = \frac{k}{x}$$

Boyle's law is an example of inverse variation. According to this law, the volume of a confined gas varies inversely with the pressure if the temperature remains constant. Thus if V denotes the volume and p the pressure, then $V = k/p$.

The different kinds of variation may occur simultaneously in problems. In the relation

$$z = \frac{kx}{y}$$

z varies directly as x and inversely as y. The relation

$$W = \frac{kxy^3}{z^2}$$

means W varies jointly as xy^3 and inversely as z^2.

EXAMPLE 2. The maximum safe load for a beam varies jointly as the width and the square of the depth and inversely as the length. If a beam 3 inches wide, 8 inches deep, and 12 feet long can safely bear a load up to 2000 pounds, find the maximum safe load for a beam of the same material which is 2 inches wide, 6 inches deep, and 10 feet long.

Solution. Using L, w, d, and l for the load, width, depth, and length, respectively, we write the law of the beam as

$$L = \frac{kwd^2}{l}$$

When we substitute $L = 2000$, $w = 3$, $d = 8$, and $l = 12$, we find

$$2000 = \frac{k(3)(8)^2}{12} \quad \text{and} \quad k = 125$$

With this value for k, the formula becomes

$$L = \frac{125wd^2}{l}$$

where L is in pounds, w and d in inches, and l in feet. Then the maximum safe load for the beam which has $w = 2$, $d = 6$, and $l = 10$ is given by

$$L = \frac{125(2)(6)^2}{10} = 900 \text{ pounds}$$

EXERCISE 6-7

Express the statements 1 to 5 as equations, using k for the constant of variation.

1. The surface area S of a sphere varies as the square of the radius r.
2. The area A of a rectangle varies jointly as the base b and altitude h.
3. The volume V of a circular cylinder is proportional to its altitude h and the square of its radius r.
4. The attraction F between two bodies varies inversely as the square of the distance s between them.
5. Neglecting air resistance, the distance s which a body falls from rest varies directly as the square of the time t.

6. The force required to stretch a spring is proportional to the elongation. If 24 pounds stretches a spring 3 inches, find the force required to stretch the spring (a) 2 inches, (b) $5\frac{1}{2}$ inches.
7. The amount of paint required to paint a circular floor varies as the square of the radius. If it takes 3 quarts for a floor of radius 10 feet, find the amount required for a floor of radius 16 feet.
8. A constant force moves a body in a line on a smooth surface. The distance which the body moves varies jointly as the force and the square of the time the force acts. If the distance is 48 feet when the force is 8 pounds and the time 2 seconds, find the distance when the force is 6 pounds and the time 3 seconds.
9. The volume of a confined gas varies inversely as the pressure if the temperature remains unchanged. If the volume is 720 cubic inches when the pressure is 9 pounds per square inch, find the volume when the pressure is (a) 6 pounds, (b) 12 pounds, (c) 15 pounds.

10. The weight of a body above the earth's surface varies inversely as the square of the distance from the center of the earth. A body weighing 64 pounds at the surface (4000 miles from the center) would have what weight at a distance above the surface of (a) 1000 miles, (b) 2000 miles, (c) 4000 miles?

11. The time required for a crew of men to do a job is inversely proportioned to the product of the number of men and the number of hours they work per day. If a crew of 9 men working 10 hours a day do a job in 14 days, find how long it would take 12 men working 7 hours per day to do the job.

12. The volume of gas in a container varies directly as its absolute temperature and inversely as the pressure. If the volume is 600 cubic inches when the absolute temperature is 300° and the pressure 15 pounds per square inch, find the volume when the absolute temperature is 150° and the pressure is 60 pounds per square inch.

13. The electrical resistance of a wire varies directly as the length and inversely as the square of its radius. The resistance of a certain piece of wire of length 100 feet and radius 0.02 inches is 300 ohms. Find the resistance of a wire of the same material 2500 feet long and radius 0.05 inches.

14. The maximum safe load for a beam 16 feet long, 4 inches wide, and 5 inches deep is 4000 pounds. Find the maximum safe load for a beam of the same material which is 12 feet long, 3 inches wide, and 9 inches deep. [See Example 2, Sec. 6-10.]

15. The kinetic energy of a moving body is proportional to the square of its velocity. Compare the kinetic energy of a car going 30 miles per hour with the same car going 40 miles per hour.

16. The illumination on a page is proportional to the wattage of the lamp and inversely proportional to the square of the distance from the lamp. Compare the illumination furnished by a 100-watt lamp at a distance of 3 feet with that of a 75-watt lamp at a distance of 2 feet.

17. Find the wattage of a lamp which will give the same amount of light at a distance of 15 feet as a 75-watt lamp will give at a distance of 5 feet. (See Problem 16.)

18. Compare the maximum safe load for a beam 18 feet long, 4 inches wide and 8 inches deep with that of a beam 12 feet long, 3 inches wide, and 12 inches deep.

19. Compare the maximum safe load for a beam 15 feet long, 3 inches wide, and 5 inches deep with that of the same beam when it is turned so that the 3-inch dimension is the depth and the 5-inch dimension is the width.

REVIEW EXERCISE

Solve each equation for x. Reject as a root any number which makes a denominator equal to zero (Secs. 6-1, 6-2, and 6-3).

1. $3x^2 + 12 = (x - 1)(3x - 1)$

2. $(x - 4)(x + 3) = (x - 1)(x + 2)$

3. $\dfrac{5x}{8} + \dfrac{x + 2}{2} = \dfrac{3x + 4}{4}$

4. $\dfrac{x - 1}{3} + \dfrac{2x + 1}{21} = \dfrac{x}{7}$

5. $\dfrac{7}{2x} + \dfrac{3}{14} = \dfrac{5}{7x} - 3$

6. $\dfrac{x^2 - 1}{x^2 - 4} - \dfrac{x}{x - 2} = \dfrac{2}{x + 2}$

7. $\dfrac{4x}{x-1} - \dfrac{3}{x+1} = \dfrac{4x^2}{x^2-1}$

8. $\dfrac{4}{x^2-1} - \dfrac{2}{x-1} = \dfrac{3}{x+1}$

9. $\dfrac{3}{x+2} + \dfrac{2}{x-2} = \dfrac{4x-4}{x^2-4}$

10. $\dfrac{11x+10}{x(5-x)} + \dfrac{x+4}{x-5} = \dfrac{x-3}{x}$

Solve each system for x and y by either elimination process (Secs. 6-5 and 6-6).

11. $5x + y = 6$
$\quad\ 3x + 2y = -2$

12. $7x + 3y = 4$
$\quad\ 4x + 2y = 3$

13. $3x - 4y = 7$
$\quad\ 5x - 5y = 8$

14. $x - y = 3a$
$\quad\ x + y = 3b$

15. $ax + 3y = 6$
$\quad\ ax - 2y = 2$

16. $ax + 2by = 2c$
$\quad\ 2ax - by = c$

Solve each system for $1/x$ and $1/y$ and, from the results, obtain x and y (Secs. 6-4 and 6-5).

17. $\dfrac{3}{x} + \dfrac{2}{y} = 3$
$\quad \dfrac{2}{x} - \dfrac{3}{y} = 4$

18. $\dfrac{4}{x} - \dfrac{3}{y} = 2$
$\quad \dfrac{3}{x} + \dfrac{4}{y} = 3$

19. $\dfrac{5}{2x} + \dfrac{3}{y} = 2$
$\quad \dfrac{3}{x} - \dfrac{2}{y} = 3$

20. A man can do a certain task in 6 days, another man can do the task in 8 days, and a boy can do the task in 10 days. How long will it take to do the task if all three work?

21. The denominator of a fraction exceeds the numerator by 2. If 3 is subtracted from the numerator and 6 is subtracted from the denominator, the sum of the original fraction and the new fraction is 2. Find the original fraction.

22. A man loans \$20,000, part at 8.5% and the rest at 9%. Find the amount of each loan if the total annual income is \$1725.00.

23. A man loans \$6250 at one annual interest rate and \$7500 at a 1% greater rate. The \$7500 loan earns \$25 more than the other loan. Find the interest rates.

24. A man invests \$5600 at a certain interest rate and \$7400 at a $\frac{1}{2}$% greater rate. His annual income from the investments is \$914.50. Find the interest rates.

25. In college algebra a student earned an average of 83 on his homework and 73 on his hour tests. What final examination grade will give him a total average of 80 if homework counts one-tenth, hour test six-tenths, and the final examination three-tenths?

26. Suppose a person pays \$70 for health insurance during a year. The insurance company will pay 80% of covered medical expenses which remain after the insured pays the first \$100. Now suppose that the insured has another policy which will pay 20% of the medical bill plus an amount equal to 20% of the amount which the first company pays. If the premium for the second policy is \$90 per year, find the amount of a medical bill which the insured may incur during a year so that he will recover exactly the medical expense plus the cost of the premiums.

27. Prove Theorems 6-3 through 6-8, Sec. 6-9. The theorems may be worded, in order, as follows. If four quantities are in proportion, then:

(a) The product of the *extremes* is equal to the product of the *means*.

(b) They are in proportion by *alteration*.

(c) They are in proportion by *inversion*.

(d) They are in proportion by *addition*.

(e) They are in proportion by *subtraction*.

(f) They are in proportion by *addition* and *subtraction*.

Chapter 7

EXPONENTS AND RADICALS

7-1 LAWS OF EXPONENTS

In Sec. 3-3 we defined a positive integral exponent. Accordingly a^n means the product formed by taking a as a factor n times. That is,

$$a^n = a \cdot a \cdot a \cdots a \quad (n \text{ factors})$$

The quantity a^n is called *the nth power of the base a* and n is called the *exponent*.

In this chapter we shall extend the definition of exponents to include all rational numbers. Before taking up new exponents, however, we state five laws of exponents and prove that the laws hold for positive integral exponents. The bases a and b in the statement of the laws may be any real numbers which do not make a denominator equal to zero. The proofs of Law I and Law II are given in Secs. 3-3 and 3-4.

Law I $\quad a^m a^n = a^{m+n}$

Law II

$$\frac{a^m}{a^n} = a^{m-n} \quad \text{if } m > n$$

$$= 1 \quad \text{if } m = n$$

$$= \frac{1}{a^{n-m}} \quad \text{if } m < n$$

Law III $\quad (a^m)^n = a^{mn}$

Proof. By definition, $(a^m)^n$ means a^m is to be taken as a factor n times. But each a^m has a as a factor m times. Hence, in all, a appears as a factor mn times, giving a^{mn}. ∎

Law IV $\quad (ab)^m = a^m b^m$

Proof. By definition $(ab)^m$ means the product obtained by taking ab as a factor m times. Hence the factor a occurs m times and the factor b occurs m times. By the commutative and associative axioms we may rearrange the factors so that all the a factors come first and all the b factors follow. Thus we may write $(ab)^m = a^m b^m$. ∎

Law V $(a/b)^m = a^m/b^m$ if $b \neq 0$

Proof. By definition $(a/b)^m$ means the product of m factors each of which is the fraction a/b. Recalling the definition of the products of fractions, we have a^m for the numerator of the product of the fractions and b^m for the denominator. ∎

EXAMPLES.

$$(a^2)^3 = a^{2 \cdot 3} = a^6 \qquad\qquad (4a)^3 = 4^3 a^3 = 64a^3$$

$$\left(\frac{x^2}{2y}\right)^4 = \frac{(x^2)^4}{(2y)^4} = \frac{x^8}{16y^4} \qquad \frac{2^{n+2}x^{2m}}{2^n x^m} = 2^2 x^m = 4x^m$$

7-2 ZERO AND NEGATIVE INTEGRAL EXPONENTS

We have defined positive integral exponents and have established five laws of exponents which apply to them. Our next step is to extend the idea of exponents to include zero and the negative integers. The new exponents are to be defined in such a way that they obey the five laws of exponents.

First let us determine what meaning should be given to the symbol a^0. If Law II is to hold when $m = n$, we have

$$\frac{a^n}{a^n} = a^{n-n} = a^0 \qquad (a \neq 0)$$

This division yields a zero exponent. But any nonzero number divided by itself has 1 as a quotient. This leads us to define the zero exponent as follows.

Definition 7-1 If a is a nonzero number, then

$$a^0 = 1$$

Next, in a similar way, we determine what meaning is to be given to a^{-n} when $-n$ is a negative integer. If Law I holds when $m = -n$, then

$$a^{-n}a^n = a^0 = 1$$

Dividing both members of this equation by a^n, we have

$$a^{-n} = \frac{1}{a^n} \qquad (a \neq 0)$$

Hence we make the following definition.

Definition 7-2 If n is a positive integer and $a \neq 0$, then

$$a^{-n} = \frac{1}{a^n}$$

The definitions of a^0 and a^{-n} were reached, in each case, by applying one of the laws of exponents. However, it would be easy to show that all the laws are satisfied by the definitions.

EXAMPLES.

$$(-3)^2(-3)^{-2} = (-3)^0 = 1 \qquad\qquad a^5 a^{-3} = a^{5-3} = a^2$$

$$\left(\frac{a}{b}\right)^{-3} = \left(\frac{b}{a}\right)^3 \qquad\qquad \frac{ax^{-2}}{by^{-5}} = \frac{ay^5}{bx^2}$$

As illustrated by the last equation, a factor of the numerator of a fraction may be moved to the denominator, or vice versa, if the sign of the exponent is changed. Terms of the numerator or denominator, however, cannot be handled in this manner. Thus,

$$\frac{1}{x^{-2} + y^{-2}} = \frac{1}{\dfrac{1}{x^2} + \dfrac{1}{y^2}} = \frac{x^2 y^2}{x^2 + y^2}$$

and not equal to $x^2 + y^2$.

EXERCISE 7-1

Find the value of each of the following expressions, using the laws of exponents.

1. $2^2 \cdot 2^3$ 2. $(2^3)^2$ 3. 4^{-1}
4. $(-4)^0$ 5. $(2a)^0$ 6. 3^{-3}
7. $5^2 \cdot 5^{-3}$ 8. $(5^2)^{-2}$ 9. $(2 \cdot 3)^{-2}$
10. $(4^{-2})^{-2}$ 11. $(-3)^{-2}$ 12. $(3 \cdot 8)^0$
13. $(\frac{1}{7})^{-1}$ 14. $(\frac{2}{3})^{-1}$ 15. $(2 \cdot 7^0)^{-4}$

Simplify each expression by performing the indicated operations and leaving the result without zero or negative exponents.

16. $(2xy)^{-2}(3xy^3)$ 17. $(x^2 y^{-2})^{-1}(x^3 y^0)^2$ 18. $(ab^{-3})(a^{-1}b^{-1})^{-1}$

19. $\dfrac{a^{-2}b^{-1}}{a^{-2}b^{-2}}$ 20. $\dfrac{3^{-1}a^{-2}b^2}{2^{-1}a^{-4}b}$ 21. $\dfrac{5^{-2}a^2b^3}{10^{-1}a^0b^{-2}}$

22. $\left(\dfrac{7x^3}{3yz}\right)^2 \left(\dfrac{z}{7x^2}\right)^2$ 23. $\dfrac{(8x^4y^5)^3}{(4x^3y^6)^3}$ 24. $\dfrac{(10xy^3)^2}{(5x^2y)^3}$

25. $\left(\dfrac{p^{-3}q^{-2}}{2^0 r^{-1}}\right)^{-2}$ 26. $\left(\dfrac{p^2 q^4}{pq^{-1}}\right)^{-3}$ 27. $\left(\dfrac{p^3 q^0}{r^2}\right)^{-1}$

28. $\dfrac{2^{-1} + 3^{-1}}{2^{-1}}$ 29. $\dfrac{2^{-1} + 1^{-1}}{2^{-1} - 1^{-1}}$ 30. $\dfrac{3^{-1} - 5^{-1}}{3^{-1}}$

31. $\dfrac{x^{-2}y^{-2}}{x^{-1} + y^{-1}}$ 32. $\dfrac{y^{-2}}{x^{-2} - y^{-2}}$ 33. $\dfrac{x^{-2} + y^{-2}}{(xy)^{-2}}$

34. $\dfrac{x^{-3} + y^{-3}}{x^{-3} - y^{-3}}$ 35. $\dfrac{x^{-1}y^2 + x^2 y^{-1}}{x^{-1} + y^{-1}}$ 36. $\dfrac{x^{-1} + y^{-1}}{x^{-2} - y^{-2}}$

7-3 FRACTIONAL EXPONENTS

In this section we shall extend the idea of exponents to include all rational numbers. Before introducing fractional exponents, however, we need to consider the following definition.

Definition 7-3   If a and b are two numbers such that the nth power of a (n a positive integer) is equal to b, then a is called an nth root of b. That is, if $a^n = b$, then a is an nth root of b.

According to this definition, the equations

$$2^2 = 4, \quad (-2)^2 = 4, \quad 3^3 = 27, \quad (-3)^3 = -27$$

show that $+2$ and -2 are square roots of 4, that 3 is a cube root of 27, and -3 is a cube root of -27. Since 4 has two square roots, we might raise the question as to how many nth roots a number has. Although the proof comes later (Sec. 15-5), we state now that any nonzero number has two square roots, three cube roots, four fourth roots, and so on. But some of these roots involve a new kind of number. This new number, which we shall introduce later, is not a real number. We see at once that a negative number has no real even root (square, fourth, sixth, and so on). This is true because an even power of a positive or a negative number is positive. We wish now, however, to give attention to real numbers only. Hence we shall defer consideration of numbers and roots of numbers which are not real.

In connection with roots of numbers, we make the following statements but do not prove them.

1. A positive number has exactly two real even roots, one positive and the other negative.
2. A positive or negative number has exactly one real odd root, the sign of the root being the same as the sign of the number.
3. A negative number has no real even root.

If n is a positive even integer, the positive nth root of a positive number a is called the *principal nth root of a*. When n is odd, the real nth root of a positive or negative number a is called the principal nth root. The principal nth root of a number is denoted by $\sqrt[n]{a}$. The symbol $\sqrt[n]{a}$ is called a *radical*, a is called the *radicand*, and n is called the *index*, or *order*, of the radical. We are excluding from consideration the case in which the radicand is negative and the index is an even number.

The following equations are illustrations of principal roots of numbers

$$\sqrt{36} = 6, \quad \sqrt[3]{8} = 2, \quad \sqrt[4]{81} = 3, \quad \sqrt[5]{-32} = -2$$

We note that -6 is a square root of 36 and -3 is a fourth root of 81, but neither is a principal root. The negative of the principal nth root of a number a is denoted by $-\sqrt[n]{a}$. Thus $-\sqrt[4]{81} = -3$.

We are now ready to consider exponents of the form m/n, where m is either a positive or negative integer and n is a positive integer. First we take $m = 1$ and seek an interpretation of $a^{1/n}$. If Law III is to hold, we have

$$(a^{1/n})^n = a^{n/n} = a$$

This equation shows that the nth power of $a^{1/n}$ is equal to a, or that $a^{1/n}$ is an nth root of a. Specifying this root as the principal nth of a, we have, by definition,

$$a^{1/n} = \sqrt[n]{a}$$

In this definition a may be any real number when n is odd, but we exclude negative values for a when n is even.

Applying Law III again, with the integer $m \neq 1$, we have

$$a^{m/n} = (a^m)^{1/n} = \sqrt[n]{a^m}$$

and also

$$a^{m/n} = (a^{1/n})^m = (\sqrt[n]{a})^m$$

Summarizing, we have the following definition.

Definition 7-4   If m/n is a rational number with n positive, then

$$a^{m/n} = \sqrt[n]{a^m} = (\sqrt[n]{a})^m$$

The form $\sqrt[n]{a^m}$ means the principal nth root of a^m, and the form $(\sqrt[n]{a})^m$ means the mth power of the principal nth root of a. In each form the denominator n of the exponent indicates a root and the numerator m indicates a power. We note again, however, that n here stands for any positive integer and m for any positive or negative integer.

We began our study of exponents by defining positive integral exponents and establishing the five laws of operations. Then we extended the definitions to include all rational numbers. Although complete proofs of the laws of exponents were made only for the positive integral exponents, it can be shown that the laws hold for all rational exponents.

Assuming that Law III holds for rational exponents, we can show that a fractional exponent can be reduced to lowest terms. Thus, if m, n, and c are integers, n and c not zero, we have

$$a^{cm/cn} = (a^{m/n})^{c/c} = a^{m/n}$$

EXAMPLE 1. $8^{2/3} = (\sqrt[3]{8})^2 = 2^2 = 4$

$$8^{2/3} = \sqrt[3]{8^2} = \sqrt[3]{64} = 4$$

Example 2.

$$81^{-3/4} = \frac{1}{(\sqrt[4]{81})^3} = \frac{1}{3^3} = \frac{1}{27}$$

Example 3. $(-32)^{3/5} = (\sqrt[5]{-32})^3 = (-2)^3 = -8$

Example 4. $(x^{5/3}y^{3/4})(x^{1/3}y^{5/4}) = x^{5/3+1/3}y^{3/4+5/4} = x^2y^2$

Example 5.

$$\left(\frac{4a^{-4/3}}{b^{2/3}c^{-2}}\right)^{-1/2} = \frac{4^{-1/2}a^{2/3}}{b^{-1/3}c} = \frac{a^{2/3}b^{1/3}}{2c}$$

EXERCISE 7-2

Find the value of each expression.

1. $16^{-1/2}$ 2. $4^{5/2}$ 3. $64^{-1/3}$

4. $(\frac{8}{27})^{2/3}$ 5. $(\frac{4}{9})^{-3/2}$ 6. $(\frac{16}{81})^{1/4}$

7. $(-32)^{4/5}$ 8. $(5^{1/4})^{-4}$ 9. $(5^{-1/4})^{-4}$

Simplify each expression, leaving the results without zero or negative exponents.

10. $x^{2/3}x^{4/3}$ 11. $x^{4/5}x^{-3/5}$ 12. $x^{-2/3}x^{-4/3}$

13. $5^{3/2} \div 5^{2/3}$ 14. $x^{1/4} \div x^{1/6}$ 15. $(x^2y^{-1})^{-1}$

16. $(x^{1/2} + 2)(x^{1/2} - 2)$ 17. $(x^{1/2} + y^{1/2})^2$ 18. $(x^{3/2} + 3)(x^{3/2} - 3)$

19. $(3x^{-2/3})^3$ 20. $(2x^{-1/4})^4$ 21. $(5^{1/3}x^{2/3})^3$

22. $(2y^{-5/8})^{-4}$ 23. $(x^{-1}y^{-3/4})^{-4}$ 24. $(5x^{-5/3})^{-3}$

25. $\dfrac{4^{3/2}x^{1/2}y^{2/3}}{8^{1/3}x^{-1/3}y^{5/3}}$ 26. $\dfrac{9^{3/2}a^{-1}b^{3/2}}{4^{5/2}a^{-5/2}b^{1/2}}$ 27. $\dfrac{27^{1/3}a^{3/4}b^{2/3}}{16^{1/2}a^{-1/4}b^{1/2}}$

28. $\left(\dfrac{x^0y^{-3/5}}{x^2y^{-4/5}}\right)^{-10}$ 29. $\left(\dfrac{2^{-1/3}x^{1/2}y^{3/2}}{3^{1/3}x^{-3/2}y^{1/2}}\right)^6$ 30. $\left(\dfrac{x^{1/2}y^{-3/5}}{x^0y^{-2/5}}\right)^{-5}$

7-4 LAWS OF RADICALS

From the laws of exponents certain useful laws of radicals may be had. We list here six laws of radicals which are immediate consequences of the corresponding laws of exponents appearing in the column to the right. In these formulas we restrict c, m, and n to positive integers, and restrict a and b to the extent that a zero must not appear in a denominator and radicals of even order must have positive radicands.

I. $\sqrt[n]{a^n} = (\sqrt[n]{a})^n = a$ $(a^n)^{1/n} = (a^{1/n})^n = a$

II. $\sqrt[n]{ab} = \sqrt[n]{a}\sqrt[n]{b}$ $(ab)^{1/n} = a^{1/n}b^{1/n}$

III. $\qquad \sqrt[n]{\dfrac{a}{b}} = \dfrac{\sqrt[n]{a}}{\sqrt[n]{b}} \qquad\qquad\qquad \left(\dfrac{a}{b}\right)^{1/n} = \dfrac{a^{1/n}}{b^{1/n}}$

IV. $\qquad \sqrt[cn]{a^{cm}} = \sqrt[n]{a^m} \qquad\qquad\qquad a^{cm/cn} = a^{m/n}$

V. $\qquad \sqrt[n]{\sqrt[m]{a}} = \sqrt[nm]{a} \qquad\qquad\qquad (a^{1/m})^{1/n} = a^{1/mn}$

VI. $\qquad \sqrt[n]{a^m}\,\sqrt[q]{a^p} = \sqrt[nq]{a^{mq+np}} \qquad\qquad a^{m/n}a^{p/q} = a^{(mq+np)/nq}$

These laws may be employed in making some of the needed changes in radicals, the most common of which are:

1. Removing factors from the radicand.
2. Making the radicand nonfractional.
3. Expressing a radical as one of lower order.
4. Bringing an outside factor inside the radical sign.

A radical is said to be in *simplest form* when operations 1, 2, and 3 are fully carried out. Operation 2 is called *rationalizing the denominator*.

In the illustrative examples and the exercises to follow, we shall assume that all literal numbers are positive.

EXAMPLE 1. Simplify the radicals $\sqrt{75a^3b^2}$ and $\sqrt[3]{8(x+y)^7}$.
Solution.

$$\sqrt{75a^3b^2} = \sqrt{25a^2b^2(3a)} = \sqrt{(5ab)^2(3a)} = 5ab\sqrt{3a}$$

$$\sqrt[3]{8(x+y)^7} = \sqrt[3]{2^3(x+y)^6(x+y)} = 2(x+y)^2\sqrt[3]{x+y}$$

EXAMPLE 2. Rationalize the denominators of

$$\sqrt{\dfrac{2}{5}} \quad \text{and} \quad \sqrt[3]{\dfrac{b}{2x^2}}$$

Solution.

$$\sqrt{\dfrac{2}{5}} = \sqrt{\dfrac{10}{25}} = \dfrac{\sqrt{10}}{\sqrt{25}} = \dfrac{\sqrt{10}}{5} = \dfrac{1}{5}\sqrt{10}$$

$$\sqrt[3]{\dfrac{b}{2x^2}} = \sqrt[3]{\dfrac{4bx}{8x^3}} = \dfrac{\sqrt[3]{4bx}}{\sqrt[3]{8x^3}} = \dfrac{\sqrt[3]{4bx}}{2x} = \dfrac{1}{2x}\sqrt[3]{4bx}$$

EXAMPLE 3. Reduce the order of the radicals $\sqrt[4]{25a^2}$ and $\sqrt[6]{8x^3y^9}$.
Solution.

$$\sqrt[4]{25a^2} = \sqrt[4]{(5a)^2} = \sqrt{5a}$$

$$\sqrt[6]{8x^3y^9} = \sqrt[6]{(2xy^3)^3} = \sqrt{2xy^3} = y\sqrt{2xy}$$

EXAMPLE 4. Bring the coefficient of

$$2x\sqrt{1 - \frac{1}{4x^2}}$$

to the proper power, inside the radical sign.
   Solution.

$$2x\sqrt{1 - \frac{1}{4x^2}} = \sqrt{4x^2\left(1 - \frac{1}{4x^2}\right)} = \sqrt{4x^2 - 1}$$

EXERCISE 7-3

Express each of the following radicals in simplest form.

1. $\sqrt{12}$ 2. $\sqrt[3]{-16}$ 3. $\sqrt{20a^4b^2}$
4. $\sqrt[3]{48x^2y^4}$ 5. $\sqrt[3]{64x^4y^5}$ 6. $\sqrt[4]{32x^5y^4}$

7. $\sqrt{\dfrac{2}{3}}$ 8. $\sqrt{\dfrac{3}{5a}}$ 9. $\sqrt[3]{\dfrac{8}{9}}$

10. $\sqrt[3]{-\dfrac{3}{4}}$ 11. $\sqrt[4]{\dfrac{2}{27}}$ 12. $\sqrt{\dfrac{2}{3x^3}}$

13. $\sqrt[3]{\dfrac{2x}{3y}}$ 14. $\sqrt[3]{\dfrac{-2x^4}{9y^4}}$ 15. $\sqrt[4]{\dfrac{b}{4c^3}}$

16. $\sqrt[4]{\dfrac{3}{8a^3}}$ 17. $\sqrt[4]{9}$ 18. $\sqrt[4]{81x^4}$

19. $\sqrt[6]{8x^6y^9}$ 20. $\sqrt[6]{81x^8y^4}$ 21. $\sqrt[4]{\dfrac{9}{x^2}}$

22. $\sqrt[4]{\dfrac{25x^2}{y^2}}$ 23. $\sqrt[4]{\dfrac{x^2 + 4x + 4}{4x^2}}$ 24. $\sqrt[4]{\dfrac{9x^6}{a^2 - 2a + 1}}$

Bring the coefficient, to the proper power, inside the radical sign.

25. $2\sqrt{3}$ 26. $2x\sqrt{y}$ 27. $2x\sqrt{\dfrac{4 - x}{4x^2}}$

28. $2a\sqrt{\dfrac{1}{4} - \dfrac{1}{a^2}}$ 29. $\dfrac{a}{b}\sqrt{\dfrac{3b^3}{a}}$ 30. $\dfrac{3a}{x^2}\sqrt{\dfrac{2x^5}{9a}}$

Employ appropriate radical laws and express each of the repeated radicals as single
radical.

31. $\sqrt[3]{\sqrt{3}}$ $6\sqrt{3}$ 32. $\sqrt[3]{\sqrt{a^3}}$ 33. $\sqrt{2\sqrt[3]{16}}$
34. $\sqrt[3]{3\sqrt{5}}$ 35. $\sqrt[4]{2\sqrt[3]{16}}$ 36. $\sqrt[4]{8\sqrt{8}}$

7-5 ADDITION AND SUBTRACTION OF RADICALS

Radicals of the same order and the same radicand are called *like radicals*. An algebraic sum of like radicals may be expressed as a single radical by use of the distributive law. Unlike radicals often become like radicals when they are simplified. Radicals not expressible as like radicals may have addition and subtraction indicated by the proper signs; they cannot be written as a single radical.

EXAMPLE 1.

$$2\sqrt{18} - 6\sqrt{\tfrac{1}{2}} + \sqrt[4]{4} = 2\sqrt{9\cdot 2} - 6\sqrt{\tfrac{2}{4}} + \sqrt[4]{2^2}$$

$$= 6\sqrt{2} - 3\sqrt{2} + \sqrt{2}$$

$$= 4\sqrt{2}$$

EXAMPLE 2.

$$\sqrt[3]{2a^4} - 3\sqrt[3]{16a} - \sqrt{2a} = a\sqrt[3]{2a} - 6\sqrt[3]{2a} - \sqrt{2a}$$

$$= (a - 6)\sqrt[3]{2a} - \sqrt{2a}$$

The unlike radicals $\sqrt[3]{2a}$ and $\sqrt{2a}$ cannot be combined into a single radical.

EXAMPLE 3.

$$\sqrt{\frac{a}{b}} - \sqrt{\frac{b}{a}} = \frac{1}{b}\sqrt{ab} - \frac{1}{a}\sqrt{ab} = \left(\frac{1}{b} - \frac{1}{a}\right)\sqrt{ab}$$

EXERCISE 7-4

Simplify the radicals in each of the following problems and then combine all the like terms.

1. $\sqrt{50} - \sqrt{32} + \sqrt{18}$ **2.** $\sqrt{75} - \sqrt{27} - \sqrt{12}$ **3.** $\sqrt{28} + 3\sqrt{63} - \sqrt{112}$

4. $\sqrt{20} + 2\sqrt{75} - 4\sqrt{12}$ **5.** $\sqrt{50} + \sqrt{63} + \sqrt{28}$ **6.** $\sqrt{\tfrac{1}{2}} + 2\sqrt{2} - \sqrt{\tfrac{1}{8}}$

7. $\sqrt{\tfrac{1}{3}} - 3\sqrt{\tfrac{1}{27}} + \sqrt{12}$ **8.** $3\sqrt{\tfrac{5}{3}} + 5\sqrt{\tfrac{3}{5}} - \sqrt{60}$ **9.** $2\sqrt[3]{16} + \sqrt[3]{54} + \sqrt[3]{50}$

10. $\sqrt[3]{16} + \sqrt[3]{81} + \sqrt[3]{54}$ **11.** $\sqrt{8x^3} - 3\sqrt{18x} + \sqrt{2x}$

12. $\sqrt{4xy^3} - \sqrt{16x^3y} - \sqrt{xy}$ **13.** $\sqrt{3x^2y} + \sqrt{12x^4y} - 3\sqrt{75x^6y^3}$

14. $\sqrt[3]{2ab} - \sqrt[3]{54ab^4} + \sqrt[3]{16a^4b^4}$ **15.** $\sqrt[3]{3a^2} + \sqrt[3]{24a^5b^3} - \sqrt[3]{81a^2b^6}$

16. $\sqrt[3]{16a} + \sqrt[3]{54a^4} + \sqrt[3]{24a}$ **17.** $\sqrt[3]{2ab^2} + \sqrt[3]{2a^4b^2} + \sqrt[3]{16ab^5}$

18. $\sqrt[4]{25a^2} + \sqrt{20a^3} + \sqrt{5a}$ **19.** $\sqrt{3a^2} - \sqrt[4]{9a^2} + \sqrt[6]{27a^3}$

20. $3\sqrt{2a} + 4\sqrt[4]{4a^6} + 6\sqrt[6]{8a^3}$

21. $\sqrt{\dfrac{x}{2}} - \sqrt{\dfrac{1}{2x}} - \sqrt{\dfrac{2}{x}}$

22. $\sqrt{\dfrac{x}{y}} + \sqrt{\dfrac{y}{x}} - \sqrt{\dfrac{1}{xy}} + \sqrt[4]{x^2y^2}$

7-6 MULTIPLICATION AND DIVISION OF RADICALS

Two radicals of the same order can be multiplied by the use of formula II, Sec. 7-4. To multiply radicals of different orders, it is first necessary to express them as radicals of the same order. The order of the new radicals should be the L.C.M. of the orders of the original radicals. The order of a radical can be raised (formula IV, Sec. 7-4) by multiplying the order of the radical and the exponent of the radicand by the same positive integer greater than 1.

EXAMPLE 1. Multiply $2\sqrt[3]{2a}$ by $5\sqrt[3]{3a^2b}$.
Solution.

$$2\sqrt[3]{2a} \cdot 5\sqrt[3]{3a^2b} = 10\sqrt[3]{6a^3b} = 10a\sqrt[3]{6b}$$

EXAMPLE 2. Find the product $(2\sqrt{3} + 3\sqrt{5})(4\sqrt{3} - \sqrt{5})$.
Solution. The binomials have like radicals, and we multiply in the usual way.

$$(2\sqrt{3} + 3\sqrt{5})(4\sqrt{3} - \sqrt{5}) = 8 \cdot 3 + 10\sqrt{15} - 3 \cdot 5$$

$$= 9 + 10\sqrt{15}$$

EXAMPLE 3. Find the product of $5\sqrt{3}$ and $6\sqrt[3]{2}$.
Solution. We first express each radical as a radical of order 6, the lowest common multiple of the orders of the given radicals. Thus

$$5\sqrt{3} \cdot 6\sqrt[3]{2} = 30\sqrt[6]{3^3} \cdot \sqrt[6]{2^2} = 30\sqrt[6]{3^3 \cdot 2^2} = 30\sqrt[6]{108}$$

Formula III, Sec. 7-4 shows how to express the quotient of two radicals of the same order as a single radical. As in multiplication, however, radicals of different orders must first be converted to radicals of the same order. We shall consider the division of two radicals as fully carried out when the quotient has no radical in the denominator and the radical in the numerator, if any, is expressed in simplest form. The process of ridding a denominator of a radical or radicals is called *rationalizing the denominator*.

EXAMPLE 4. Find the quotient of $\sqrt{6}$ divided by $\sqrt{5}$.
Solution.

$$\frac{\sqrt{6}}{\sqrt{5}} = \sqrt{\frac{6}{5}} = \sqrt{\frac{6 \cdot 5}{5 \cdot 5}} = \frac{1}{5}\sqrt{30}$$

or, alternatively,

$$\frac{\sqrt{6}}{\sqrt{5}} \cdot \frac{\sqrt{5}}{\sqrt{5}} = \frac{\sqrt{30}}{5} = \frac{1}{5}\sqrt{30}$$

EXAMPLE 5. Find the quotient of $6\sqrt[3]{5}$ divided by $2\sqrt{2}$.
Solution.

$$\frac{6\sqrt[3]{5}}{2\sqrt{2}} = \frac{6\sqrt[3]{5}}{2\sqrt{2}} \cdot \frac{\sqrt{2}}{\sqrt{2}} = \frac{6\sqrt[6]{5^2} \cdot \sqrt[6]{2^3}}{4} = \frac{3}{2}\sqrt[6]{200}$$

Alternatively, converting to fractional exponents, we have

$$\frac{6 \cdot 5^{1/3}}{2 \cdot 2^{1/2}} = \frac{6 \cdot 5^{1/3} \cdot 2^{1/2}}{2 \cdot 2^{1/2} \cdot 2^{1/2}} = \frac{6 \cdot 5^{2/6} \cdot 2^{3/6}}{4} = \frac{3}{2}\sqrt[6]{5^2 \cdot 2^3} = \frac{3}{2}\sqrt[6]{200}$$

When the divisor is a binomial which contains a second-order radical in one or both terms, we rationalize the denominator by multiplying the dividend and divisor by a properly chosen expression. For this purpose we observe that the product of $\sqrt{a} + \sqrt{b}$ and $\sqrt{a} - \sqrt{b}$ is the rational expression $a - b$. Hence a rationalizing factor of the kind in question is obtained by changing the sign of one term of the divisor.

EXAMPLE 6. Divide $(3\sqrt{2} - 2\sqrt{3})$ by $(4\sqrt{2} - 3\sqrt{3})$.
Solution.

$$\frac{3\sqrt{2} - 2\sqrt{3}}{4\sqrt{2} - 3\sqrt{3}} = \frac{3\sqrt{2} - 2\sqrt{3}}{4\sqrt{2} - 3\sqrt{3}} \cdot \frac{4\sqrt{2} + 3\sqrt{3}}{4\sqrt{2} + 3\sqrt{3}}$$

$$= \frac{24 + \sqrt{6} - 18}{32 - 27} = \frac{6 + \sqrt{6}}{5}$$

EXERCISE 7-5

Multiply as indicated and simplify the result.

1. $\sqrt{2} \cdot \sqrt{7}$ 2. $\sqrt{2} \cdot \sqrt{7} \cdot \sqrt{28}$ 3. $\sqrt{3} \cdot \sqrt{5} \cdot \sqrt{30}$
4. $\sqrt{18x^2y} \cdot \sqrt{2xy^3}$ 5. $\sqrt[3]{6} \cdot \sqrt[3]{9}$ 6. $\sqrt[3]{3a} \cdot \sqrt[3]{4a^2}$
7. $\sqrt[3]{16a^2} \cdot \sqrt[3]{4ab}$ 8. $\sqrt{3} \cdot \sqrt[3]{2}$ 9. $\sqrt{2} \cdot \sqrt[3]{3}$
10. $\sqrt{2} \cdot \sqrt[4]{8}$ 11. $\sqrt{x} \cdot \sqrt[3]{x} \cdot \sqrt[4]{x}$ 12. $\sqrt{2} \cdot \sqrt[3]{2} \cdot \sqrt[3]{3}$
13. $\sqrt{2} \cdot \sqrt[3]{3} \cdot \sqrt[4]{4}$ 14. $(2\sqrt{3} + 4)(2\sqrt{3} - 3)$ 15. $\sqrt{3} - \sqrt{a} \cdot \sqrt{3} + \sqrt{a}$

16. Find the value of $x^2 - 6x + 7$ if $x = 3 + \sqrt{2}$.
17. Find the value of $2x^2 + x + 1$ if $x = \sqrt{2} - 1$.
18. Find the value of $x^2 + x - 5$ if $x = \sqrt{3} + \sqrt{2}$.

Perform the divisions and express each result in simplest form.

19. $\sqrt{63} \div \sqrt{7}$ **20.** $\sqrt{11} \div \sqrt{33}$ **21.** $\sqrt{7x^2} \div \sqrt{28x}$

22. $\sqrt{15a^4} \div \sqrt{3a}$ **23.** $\sqrt{20x} \div 2\sqrt{5x^3}$ **24.** $\sqrt[3]{108} \div \sqrt[3]{4}$

25. $2\sqrt[3]{7a} \div \sqrt[3]{2a^2}$ **26.** $\sqrt[3]{15x^4} \div \sqrt[3]{4x}$ **27.** $\sqrt{3} \div \sqrt[3]{3}$

28. $\sqrt[3]{9} \div \sqrt{3}$ **29.** $\sqrt{ab^2} \div \sqrt[3]{a^2b}$

30. $\dfrac{\sqrt{15} + \sqrt{35}}{\sqrt{5}}$ **31.** $\dfrac{1}{3 - \sqrt{5}}$ **32.** $\dfrac{1}{\sqrt{3} - \sqrt{2}}$ **33.** $\dfrac{\sqrt{2}}{5 + \sqrt{2}}$

34. $\dfrac{\sqrt{7} + \sqrt{5}}{\sqrt{7} - \sqrt{5}}$ **35.** $\dfrac{4\sqrt{5} - \sqrt{3}}{2\sqrt{3} + \sqrt{5}}$ **36.** $\dfrac{3\sqrt{7} + 2\sqrt{2}}{2\sqrt{7} - 3\sqrt{2}}$

7-7 COMPLEX NUMBERS

In our treatment of fractional exponents we excluded even roots of negative numbers. But even roots, especially square roots, of negative numbers are of much importance in mathematics. Their use has contributed to the development of a large body of mathematics, much of which has vital applications in the engineering and physical sciences.

Many equations cannot be solved and many problems cannot be investigated in the real number system. The equation $x^2 + 1 = 0$, for example, has a solution if and only if $x^2 = -1$. But the square of any real number is not negative, and consequently the equation has no real solution. In order to obtain numbers whose squares are negative, we enlarge our number system to include a new set of numbers. The basis of the new system is contained in the following definitions.

Definition 7-5 An ordered pair of real numbers (a, b) is called a *complex number.*

Definition 7-6 Two complex numbers (a, b) and (c, d) are equal if and only if $a = c$ and $b = d$.

Definition 7-7 The sum of the complex numbers (a, b) and (c, d) is $(a + c, b + d)$.

Definition 7-8 The product of the complex numbers (a, b) and (c, d) is $(ac - bd, ad + bc)$.

These definitions are purposely drawn up so that addition and multiplication obey the six field axioms (Sec. 2-2). We shall consider proofs of properties of complex numbers in a later chapter.

If in the complex numbers (a, b) and (c, d) we let $b = d = 0$, then the sum and product become

$$(a, 0) + (c, 0) = (a + c, 0 + 0) = (a + c, 0)$$

$$(a \cdot 0)(c, 0) = (ac - 0 \cdot 0, a \cdot 0 + 0 \cdot c) = (ac, 0)$$

Hence complex numbers of the form $(x, 0)$ behave like real numbers with respect to addition and multiplication. Because of this behavior a complex number $(x, 0)$ is often equated to the real number x.

The letter i is frequently used to stand for the complex number $(0, 1)$ and its negative $-i$ for $(0, -1)$. We next show that i^2 and $(-i)^2$ are both equal to -1 by applying Definition 7-8 for multiplication. Thus

$$i^2 = (0, 1)(0, 1) = (0 \cdot 0 - 1 \cdot 1, 0 \cdot 1 + 1 \cdot 0) = (-1, 0) = -1$$

$$(-i)^2 = (0, -1)(0, -1) = [0 \cdot 0 - (-1)(-1), 0(-1) + (-1)0] = (-1, 0) = -1$$

Since $i^2 = -1$ and $(-i)^2 = -1$, it follows that i and $-i$ are square roots of -1. To distinguish between these square roots, we use the notation $i = \sqrt{-1}$ and $-i = -\sqrt{-1}$.

Similarly, we can show that any negative number has two square roots. As examples, $\sqrt{-9} = \pm 3i$ and $\sqrt{-7} = \pm i\sqrt{7}$.

Now using $a = (a, 0)$, $b = (b, 0)$ and $i = (0, 1)$, we have, by Definitions 7-7 and 7-8,

$$a + bi = (a, 0) + (b, 0)(0, 1)$$

$$= (a, 0) + (b \cdot 0 - 0 \cdot 1, b \cdot 1 + 0 \cdot 0)$$

$$(a, 0) + (0, b) = (a, b)$$

In each of these notations $a + bi$ and (a, b) for a complex number, a is called the *real component* and b is called the *imaginary component* of the complex number. The number i in $a + bi$ is called the *imaginary unit*.

We may now express the sum and product of two complex numbers in the alternate forms

$$(a + bi) + (c + di) = (a + c) + (b + d)i$$

$$(a + bi)(c + di) = (ac - bd) + (ad + bc)i$$

We note that the right members of these equations may be obtained by carrying out the indicated operations in the left members as though i were a real number and, for the product, replacing i^2 by -1. Hence the sum and product of two complex numbers may be written without memorizing formulas.

EXAMPLE 1. $(2 + 3i) + (4 - 5i) = 6 - 2i$.

EXAMPLE 2.

$$(3 - 4i)(5 + 5i) = 15 - 5i - 20i^2$$

$$= 15 - 5i - 20(-1)$$

$$= 35 - 5i$$

To perform the operation of subtraction in the complex number system, we add the negative of the subtrahend. Thus

$$(a + bi) - (c + di) = (a + bi) + (-c - di) = (a - c) + (b - d)i$$

EXAMPLE 3. $(6 - 3i) - (4 + 5i) = 2 - 8i$.

Depending on the values of a and b, we have the following special cases of complex numbers.

Definition 7-9

$a + bi$ is called an *imaginary number* if $b \neq 0$.

$a + bi$ is called a *pure imaginary number* if $a = 0$ and $b \neq 0$.

$a + bi$ is called a *real number* if $b = 0$.

EXAMPLES.

$2i, 7 - 5i,$ and $-4 + 3i$ are imaginary numbers.

$2i, -6i,$ and $i\sqrt{5}$ are pure imaginary numbers.

EXERCISE 7-6

Express each number as a real number times i.

1. $\sqrt{-4}$ 2. $\sqrt{-16}$ 3. $\sqrt{-11}$

4. $2\sqrt{-6}$ 5. $-3\sqrt{-8}$ 6. $5\sqrt{-12}$

Perform the indicated operations and leave the results in the form $a + bi$.

7. $(3 + 4i) + (4 + 3i)$ 8. $(9 + 4i) - (-8 + 3i)$

9. $(-4 - 13i) - (7 - 8i)$ 10. $(15 + 3i) + (17 - 16i)$

11. $(2 + 3i)(2 - 3i)$ 12. $(4 - 3i)(4 + 3i)$

13. $(2 + i\sqrt{3})(2 - i\sqrt{3})$ 14. $(\sqrt{2} + 3i)(\sqrt{2} - 3i)$

15. $(c + di)(c - di)$ 16. $4i(6i - 7)$

17. $(5 + 8i) - (5 - 8i)$ 18. $(5 + 8i) + (5 - 8i)$

19. $(c + di) + (c - di)$ 20. $(c + di) - (c - di)$

21. Prove that $a + bi = 0$ if and only if $a = 0$ and $b = 0$. [Hint: Use Definition 7-6.]

REVIEW EXERCISE

Simplify each expression by performing the indicated operations and leaving the result without zero or negative exponents (Secs. 7-1 and 7-2).

1. $(xy^2)^3(6x^2y)^2$ 2. $(x^3y^0)^2(x^{-2}y^3)^2$

3. $(a^{-1}b^{-1})^{-1}(a^{-2}b^2)$ 4. $\dfrac{4^{-2}a^2b^3}{10^{-1}a^0b^{-2}}$

5. $\dfrac{14a^6b^{-3}}{7a^4b^{-3}}$

6. $\dfrac{3x^2y^{-3}z^2}{6x^{-4}y^{-2}z}$

7. $\dfrac{(4x^3y^3)^3}{(8x^5y^4)^3}$

8. $\dfrac{(5xy^3)^3}{(10x^2y^3)^2}$

9. $\left(\dfrac{25b^3q^{-3}}{15b^2q^3}\right)^{-2}$

10. $\dfrac{2^{-1} - 1^{-1}}{2^{-2} + 1^{-1}}$

11. $\dfrac{x^{-1} + y^{-1}}{x^{-3}y^{-3}}$

12. $\dfrac{y^{-3}}{x^{-3} + y^{-3}}$

13. $\dfrac{x^{-4} - y^{-4}}{x^{-4} + y^{-4}}$

14. $\dfrac{y^{-4}}{x^{-4} - y^{-4}}$

15. $\dfrac{3^{-1} + 4^{-1}}{3^{-1}}$

16. $\left(\dfrac{3xy}{y^4}\right)^2 \left(\dfrac{2x^3}{y}\right)^3$

17. $\dfrac{8x^2y^3z^{-3}}{5xy^{-2}z^{-4}}$

18. $\left(\dfrac{5x^{-1}y^3}{10x^2y^4}\right)^2 \left(\dfrac{6x^2y^2}{4x^{-1}y^{-1}}\right)^{-3}$

Find the rational value of each expression (Sec. 7-3).

19. $27^{2/3}$	**20.** $49^{-1/2}$	**21.** $9^{-3/2}$
22. $25^{3/2}$	**23.** $(-8)^{4/3}$	**24.** $(-64)^{2/3}$
25. $27^{-4/3}$	**26.** $(-243)^{3/5}$	**27.** $(-32)^{3/5}$
28. $(16^{-1/4})^{-2}$	**29.** $(-125)^{-4/3}$	**30.** $(16^{-1/3})^{3/4}$

Simplify each expression, leaving the result without zero or negative exponents (Sec. 7-3).

31. $a^{1/3}a^{3/5}$ **32.** $a^{-4/3}a^{3/2}$ **33.** $a^{-4/5}a^{3/5}$

34. $(2a^{-3/8})^{-4}$ **35.** $(a^{5/2} + 1)(a^{5/2} - 1)$

36. $(a^{-1}b^{-1/3})^{-3}$ **37.** $\dfrac{36^{1/2}x^{-2}y^{-1}}{6^{1/2}x^{-1/3}y^2}$

38. $\dfrac{81^{1/2}x^{-1/4}y^{-1/2}}{9^{5/2}x^{3/4}y^{5/4}}$ **39.** $\dfrac{4^{3/2}x^{1/2}y^{2/3}}{8^{1/3}x^{-1/3}y^{1/2}}$

40. $\dfrac{16^{1/4}x^{-1/4}y^{1/2}}{27^{1/3}x^{3/4}y^{2/3}}$ **41.** $\left(\dfrac{3^{-1/3}x^{-2/3}y^{1/2}}{2^{-1/3}x^{1/2}y^{3/2}}\right)^6$

42. $\dfrac{64^{1/2}x^{-1/4}y^{1/2}}{27^{1/3}x^{3/4}y^{2/3}}$

Express each of the following radicals in simplest form (Sec. 7-4).

43. $\sqrt{32}$	**44.** $\sqrt[3]{-81}$	**45.** $\sqrt{18a^4}$
46. $\sqrt{12a^2b^4}$	**47.** $\sqrt[3]{32x^4y^2}$	**48.** $\sqrt{100x^7y^4}$
49. $\sqrt[3]{27x^4y^5}$	**50.** $\sqrt[4]{32x^3y^6}$	**51.** $\sqrt[5]{32x^6}$

52. $\sqrt{\dfrac{5}{2}}$

53. $\sqrt{\dfrac{5a}{3}}$

54. $\sqrt{\dfrac{8}{7}}$

55. $\sqrt[3]{\dfrac{9}{64}}$

56. $\sqrt[3]{-\dfrac{3}{2}}$

57. $\sqrt[4]{\dfrac{3}{5}}$

58. $\sqrt[6]{\dfrac{x^2}{25}}$

59. $\sqrt[4]{\dfrac{4b^3}{a}}$

60. $\sqrt[3]{\dfrac{y^4 z^4}{x^2}}$

61. $\sqrt[6]{\dfrac{x^2 + 2x + 1}{9x^2}}$

62. $\sqrt[4]{\dfrac{4x^6}{x^2 - 2x + 1}}$

63. $\sqrt[6]{\dfrac{(a-1)^3}{b^3 c^3}}$

Bring the coefficient, to the proper power, inside the radical sign (Sec. 7-4).

64. $4\sqrt{2}$

65. $4x\sqrt{y}$

66. $2x\sqrt[3]{xy^2}$

67. $3a\sqrt{\dfrac{1}{9} - \dfrac{1}{a^2}}$

68. $\dfrac{b}{a}\sqrt{\dfrac{2a^3}{b}}$

69. $\dfrac{3a}{x}\sqrt[3]{\dfrac{4x^3}{9a^2}}$

Use appropriate radical laws and express each repeated radical as a single radical (Sec. 7-4).

70. $\sqrt[3]{\sqrt{2}}$

71. $\sqrt[3]{\sqrt{a^5}}$

72. $\sqrt{3\sqrt[3]{9}}$

73. $\sqrt[3]{5\sqrt{3}}$

74. $\sqrt[4]{2^3 \sqrt{5}}$

75. $\sqrt[4]{8\sqrt{8}}$

Simplify each radical and then combine all like terms (Sec. 7-5).

76. $\sqrt{28} - 3\sqrt{63} - \sqrt{112}$

77. $\sqrt{20} - 2\sqrt{75} - 4\sqrt{12}$

78. $3\sqrt{2} - \sqrt{\tfrac{1}{2}} - \sqrt{\tfrac{1}{8}}$

79. $3\sqrt{\tfrac{5}{3}} - \sqrt{60} - 5\sqrt{\tfrac{3}{5}}$

80. $2\sqrt[3]{54} + 2\sqrt[3]{40} - 6\sqrt[3]{16}$

81. $\sqrt[3]{54} + \sqrt[3]{81} - \sqrt[3]{16}$

82. $\sqrt{12x^4 y} + \sqrt{3x^2 y} - 3\sqrt{75y^3}$

83. $\sqrt[3]{24a^5 b^3} + \sqrt[3]{3a^2} - \sqrt[3]{81a^2 b^6}$

84. $\sqrt[4]{25a^2} + \sqrt{20a} + \sqrt{45a}$

85. $\sqrt{3a} + \sqrt[4]{9a^2} - \sqrt[6]{27a^3}$

86. $\sqrt{\dfrac{a+b}{a-b}} + \sqrt{\dfrac{a-b}{a+b}} - \sqrt{\dfrac{4a^2}{a^2 - b^2}}$

87. $\sqrt[4]{\dfrac{a^2}{b^2}} - \dfrac{1}{3}\sqrt{\dfrac{9a}{b}} + \sqrt{\dfrac{4a^3}{b}}$

88. $\sqrt{\dfrac{2a}{b}} + \sqrt[4]{\dfrac{324a^2}{b^2}} + \sqrt[8]{\dfrac{16a^4}{b^4}}$

89. $\sqrt[9]{\dfrac{x^3 y^3}{512z^3}} - \sqrt[6]{\dfrac{x^2 y^2}{64z^2}} - 15\sqrt{\dfrac{x^5 y^5}{z^5}}$

Multiply as indicated and simplify the results (Sec. 7-6).

90. $\sqrt{5} \cdot \sqrt{7}$

91. $3\sqrt{13} \cdot 2\sqrt{26}$

92. $\sqrt{2} \cdot \sqrt{5} \cdot \sqrt{45}$

93. $\sqrt{3xy} \cdot \sqrt{18x^2 y^4}$

94. $\sqrt[3]{3} \cdot \sqrt[3]{18}$

95. $\sqrt[3]{32a^2} \cdot \sqrt[3]{2ab}$

96. $\sqrt[3]{4} \cdot \sqrt[3]{3}$

97. $\sqrt[3]{6} \cdot \sqrt{2}$

98. $\sqrt[4]{2} \cdot \sqrt[3]{2} \cdot \sqrt{2}$

99. $\sqrt{3x} \cdot \sqrt[3]{2x} \cdot \sqrt[3]{4x}$

100. $\sqrt[3]{6} \cdot \sqrt{2}$

101. $\sqrt[4]{4} \cdot \sqrt[3]{2} \cdot \sqrt{3}$

102. $(\sqrt{3} + 3\sqrt{2})(4\sqrt{3} - 2\sqrt{2})$

103. $\sqrt{4 - \sqrt{a}} \cdot \sqrt{4 + \sqrt{a}}$

104. $\sqrt{7 + 4\sqrt{2}} \cdot \sqrt{7 - 4\sqrt{2}}$

105. $\sqrt[4]{8} \cdot \sqrt{2}$

Perform the divisions and simplify each result (Sec. 7-6).

106. $\sqrt{72} \div \sqrt{8}$

107. $\sqrt{68} \div \sqrt{17}$

108. $\sqrt{20x} \div \sqrt{5x^2}$

109. $\sqrt[3]{21a} \div \sqrt[3]{7a^2}$

110. $\sqrt[3]{15x} \div \sqrt[3]{4x^2}$

111. $\sqrt[3]{ab^2} \div \sqrt{a^2b}$

112. $\sqrt[4]{24x^4y} \div \sqrt[4]{8x^2y^3}$

113. $\sqrt[4]{xy^2} \div \sqrt{x^2y}$

114. $\sqrt{5} \div \sqrt[4]{5}$

115. $\dfrac{\sqrt{5}}{\sqrt{15} - \sqrt{35}}$

116. $\dfrac{1}{\sqrt{5} - \sqrt{3}}$

117. $\dfrac{\sqrt{6} - \sqrt{7}}{\sqrt{6} + \sqrt{7}}$

118. $\dfrac{\sqrt{11} - \sqrt{3}}{\sqrt{11} + \sqrt{3}}$

119. $\dfrac{\sqrt{13} + \sqrt{7}}{\sqrt{13} - \sqrt{7}}$

120. $\dfrac{2\sqrt{5} - \sqrt{3}}{\sqrt{5} + \sqrt{3}}$

121. $\dfrac{2\sqrt{7} - 3\sqrt{2}}{\sqrt{7} + \sqrt{2}}$

122. $\dfrac{4\sqrt{2} + 3\sqrt{3}}{2\sqrt{2} + 4\sqrt{3}}$

123. $\dfrac{2\sqrt{7} - 3\sqrt{5}}{3\sqrt{7} - 5\sqrt{5}}$

Chapter 8

QUADRATIC EQUATIONS

$$x = \frac{-\mathcal{B} \pm \sqrt{\mathcal{B}^2 - 4AC}}{2A}$$

8-1 SOLUTION BY FACTORING

An equation which can be expressed in the form

$$ax^2 + bx + c = 0 \qquad (1)$$

where a, b, and c are constants with $a \neq 0$, is called a *quadratic equation in x.* An important operation on a quadratic equation is that of finding the *solution set*, or *roots*, of the equation. If the first-degree term is missing, the roots can be readily obtained. For example, the roots of $x^2 - 16 = 0$ are $x = \pm 4$, and the roots of $x^2 + 9 = 0$ are $x = \pm\sqrt{-9} = \pm 3i$.

If the left member of an equation of the form (1) can be separated into two linear factors, the roots can be had at once from the factors. The key principle in this method is the fact that the product of two factors is equal to zero if either of the factors is equal to zero (Theorem 2-6, Sec. 2-3). Hence the two linear equations obtained by setting each factor equal to zero yield the roots.

EXAMPLE. Solve the equation $6x^2 + 5x - 4 = 0$.

Solution. Factoring the left member, we have

$$(2x - 1)(3x + 4) = 0$$

We now have the two linear equations

$$2x - 1 = 0 \quad \text{and} \quad 3x + 4 = 0$$

From these equations we find $x = \frac{1}{2}$ and $x = -\frac{4}{3}$. Either of these values will satisfy the given equation. Checking $x = \frac{1}{2}$, we find

$$6(\tfrac{1}{2})^2 + 5(\tfrac{1}{2}) - 4 = \tfrac{6}{4} + \tfrac{5}{2} - 4 = 0$$

EXERCISE 8-1

Solve the following quadratic equations.

1. $x^2 - 3 = 0$	2. $x^2 - a^2 = 0$	3. $x^2 + 25 = 0$
4. $9x^2 + 2 = 0$	5. $a^2x^2 + b^2 = 0$	6. $4x^2 + 1 = 0$

Solve each of the following equations for x by the factoring method. Check your solutions.

7. $x^2 - x - 2 = 0$ **8.** $x^2 + x - 6 = 0$ **9.** $x^2 + 9x + 8 = 0$
10. $2x^2 - x = 10$ **11.** $2x^2 + 3x + 1 = 0$ **12.** $7x^2 - 11x = 0$
13. $81x^2 - 1 = 0$ **14.** $9x^2 - 25 = 0$ **15.** $4x^2 - 49 = 0$
16. $6x^2 + 11x = -3$ **17.** $2x^2 - 5x + 3 = 0$ **18.** $6x^2 - 5x + 1 = 0$
19. $18x^2 + 3 = 29x$ **20.** $8x^2 = 14x - 5$
21. $12x^2 + 5x = 2$ **22.** $x^2 + (a + 1)x + a = 0$
23. $x^2 - ax + bx - ab = 0$ **24.** $a^2x^2 + abx - 2b^2 = 0$
25. $ax^2 - abx + x - b = 0$

8-2 SOLUTION BY FORMULA

The roots of a quadratic equation are best obtained by factoring if the factoring can be done readily. For other cases we need to find a practicable procedure. If by some means we solve the equation $ax^2 + bx + c = 0$, we obtain the roots in terms of the literal coefficients a, b, and c. Then the roots of this equation can be used as a formula for solving other quadratic equations. We derive the desired formula by the following steps.

$$ax^2 + bx + c = 0$$

$$ax^2 + bx = -c \qquad\qquad \text{adding } -c$$

$$x^2 + \frac{b}{a}x = -\frac{c}{a} \qquad\qquad \text{dividing by } a$$

$$x^2 + \frac{b}{a}x + \left(\frac{b}{2a}\right)^2 = \left(\frac{b}{2a}\right)^2 - \frac{c}{a} \qquad\qquad \text{adding } \left(\frac{b}{2a}\right)^2$$

$$\left(x + \frac{b}{2a}\right)^2 = \frac{b^2 - 4ac}{4a^2}$$

$$x + \frac{b}{2a} = \frac{\pm\sqrt{b^2 - 4ac}}{2a} \qquad\qquad \text{taking square roots}$$

$$x = \frac{-b \pm \sqrt{b^2 - 4ac}}{2a} \qquad\qquad \text{adding } -\frac{b}{2a}$$

Thus we have established the following theorem.

Theorem 8-1   The roots x_1 and x_2 of the equation

$$ax^2 + bx + c = 0 \qquad a \neq 0$$

are

$$x_1 = \frac{-b + \sqrt{b^2 - 4ac}}{2a}, \qquad x_2 = \frac{-b - \sqrt{b^2 - 4ac}}{2a}$$

or, expressed in set form, the solution is

$$\left\{ \frac{-b + \sqrt{b^2 - 4ac}}{2a}, \frac{-b - \sqrt{b^2 - 4ac}}{2a} \right\}$$

The solution set of the equation of this theorem is called the *quadratic formula*. The correctness of the results which we have found for x may be verified by direct substitution in the given equation.

The steps by which we derived the quadratic formula is called the *completing the square* method. That is, we manipulated the equation to make the left member a perfect square.

EXAMPLE 1. Solve the equation $3x^2 + 5x - 2 = 0$ by completing the square.
Solution.

$$3x^2 + 5x - 2 = 0 \qquad \text{given equation}$$

$$3x^2 + 5x = 2 \qquad \text{adding 2}$$

$$x^2 + \tfrac{5}{3}x = \tfrac{2}{3} \qquad \text{dividing by 3}$$

$$x^2 + \tfrac{5}{3}x + \tfrac{25}{36} = \tfrac{2}{3} + \tfrac{25}{36} \qquad \text{adding } (\tfrac{5}{6})^2$$

$$(x + \tfrac{5}{6})^2 = \tfrac{49}{36}$$

$$x + \tfrac{5}{6} = \pm\tfrac{7}{6} \qquad \text{taking square roots}$$

$$x = \frac{-5 \pm 7}{6} \qquad \text{adding } -\tfrac{5}{6}$$

$$x = \tfrac{1}{3} \quad \text{and} \quad -2$$

EXAMPLE 2. Use the quadratic formula to solve the equation

$$3x^2 - 4x + 5 = 0$$

Solution. Comparing coefficients of this equation with the corresponding coefficients of the equation $ax^2 + bx + c = 0$, we see that $a = 3$, $b = -4$, and $c = 5$. Therefore we substitute these values for a, b, and c in the quadratic formula. Thus we get

$$x = \frac{4 \pm \sqrt{(-4)^2 - 4(3)(5)}}{2(3)} = \frac{4 \pm \sqrt{-44}}{6}$$

$$= \frac{4 \pm 2\sqrt{-11}}{6} = \frac{2 \pm i\sqrt{11}}{3}$$

Check. We show that these values of x reduce the left member of the given equation to zero, and therefore are the roots.

$$3\left(\frac{2 \pm i\sqrt{11}}{3}\right)^2 - 4\left(\frac{2 \pm i\sqrt{11}}{3}\right) + 5 = 3\left(\frac{4 \pm 4i\sqrt{11} + 11i^2}{9}\right) - \frac{8 \pm 4i\sqrt{11}}{3} + 5$$

$$= \frac{4 \pm 4i\sqrt{11} - 11 - 8 \mp 4i\sqrt{11} + 15}{3} = 0$$

EXERCISE 8-2

Solve the following equations by completing the square, and check each result.

1. $x^2 - 4x + 3 = 0$ 2. $x^2 + 6x + 8 = 0$
3. $2x^2 - 5x + 2 = 0$ 4. $2x^2 - x - 3 = 0$
5. $x^2 - 4x + 1 = 0$ 6. $2x^2 + 5x + 3 = 0$
7. $x^2 - 8x + 25 = 0$ 8. $9x^2 + 10 = 12x$
9. $2x^2 - 6x + 9 = 0$

Solve for x by the quadratic formula, and check each result.

10. $x^2 - 6x + 8 = 0$ 11. $x^2 + 8x + 16 = 0$
12. $x^2 - 10x + 25 = 0$ 13. $4x^2 - 12x + 9 = 0$
14. $42x^2 - x - 1 = 0$ 15. $x^2 - x - 3 = 0$
16. $3x^2 + 10x + 4 = 0$ 17. $x^2 + x + 1 = 0$
18. $x^2 - 6x + 10 = 0$ 19. $4x^2 - 9x + 3 = 0$
20. $3x^2 + 11x + 7 = 0$ 21. $2x^2 - 2x + 5 = 0$
22. $x^2 + (b - a)x - ab = 0$ 23. $mx^2 + (1 + m)x + 1 = 0$
24. $abx^2 + (2b - 3a)x - 6 = 0$ 25. $(a + b)^2x^2 - (a + b)x - 2 = 0$

26. A rectangular pasture is to be fenced along four sides and divided into three parts by two fences parallel to one of the sides. Find the dimensions of the pasture if the total amount of fencing is 800 yards and the area of the pasture is (a) 19,200 square yards, (b) 20,000 square yards.

   Solution. Referring to Fig. 8-1, we have

$$400x - 2x^2 = 19,200$$
$$x^2 - 200x + 9600 = 0$$
$$(x - 80)(x - 120) = 0$$
$$x = 80 \text{ or } 120$$

   If the width is 80 yards, the length is 240 yards; if the width is 120 yards, the length is 160 yards. Each of these solutions satisfies the conditions of the problem. Part (b), however, has only one solution, as the student may verify.

27. The product of two consecutive odd integers is 143. Find the integers.

28. Two numbers differ by 1; their cubes differ by 91. Find the numbers.

29. A gardener wishes to enclose a rectangular piece of ground with 300 yards of fencing. A river runs along one side and no fence is needed there. Find the dimensions of the rectangle if (a) the area is 10,000 square yards, (b) 11,250 square yards.

30. A page, with length 3 inches more than its width, has 80 square inches of printed area. The margins at the top and bottom are each 1.5 inches wide and the side margins are each 1 inch wide. Find the dimensions of the page.

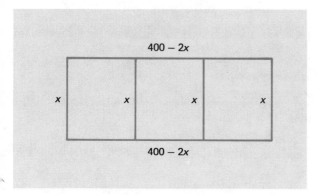

Fig. 8-1

31. A jet plane flying 500 miles per hour faster than a propeller plane travels a distance of 1400 miles in 5 hours less time than is required for the propeller plane to travel the same distance. Find the speeds of the planes.

32. A man can do a job in 9 hours less time than is required by a boy. Working together, they can do the job in 20 hours. How long would it take each alone to do the job?

33. A rectangular pasture is to be fenced along four sides and divided into two parts by a fence parallel to one of the sides. Find the dimensions of the pasture if the total amount of fencing is 600 yards and the area of the pasture is (a) 11,250 square yards, (b) 15,000 square yards.

34. A railroad and highway intersect at right angles. A train is 10 miles from the intersection at the same time a car is 6 miles from the intersection. If each is traveling at the rate of 0.5 mile per minute, at what times will they be 4 miles apart?

Do Problems 35 through 39, using the following formulas.

If an object is thrown vertically downward with a velocity v_0, the distance s which the object will fall in t seconds is given by the formula

$$s = v_0 t + \tfrac{1}{2} g t^2$$

where s is in feet, v_0 in feet per second, and g is 32 (approximately).

If the object is projected vertically upward, the formula is

$$s = v_0 t - \tfrac{1}{2} g t^2$$

35. If a ball is dropped ($v_0 = 0$) from a tower, find how long it takes to reach the ground if the height of the tower is (a) 256 feet; (b) 144 feet; (c) 64 feet.

36. From a balloon a marble is thrown downward with a velocity of 64 feet per second. Find (a) how far the ball falls in 3 seconds; (b) how long it takes to fall 512 feet.

37. A ball is thrown vertically upward with a starting velocity of 80 feet per second. In how many seconds will the ball be (a) 100 feet high; (b) 96 feet high? Explain the two answers for part (b).

38. A bullet is shot vertically upward at a balloon with a velocity of 1100 feet per second. If the balloon is 9400 feet high, find the time required for the bullet to reach the balloon.

39. If a ball is thrown upward with a velocity of 48 feet per second, find when the ball will be 32 feet above the starting point. Explain the two answers.

8-3 EQUATIONS IN QUADRATIC FORM

An equation is often met which is not quadratic but can be reduced to a quadratic by the substitution of a new unknown.

EXAMPLE 1. Solve the equation $y^4 - 3y^2 - 4 = 0$.

Solution. This fourth-degree equation when written as

$$(y^2)^2 - 3y^2 - 4 = 0$$

is seen to be quadratic in y^2. The substitution $y^2 = x$ gives the quadratic equation

$$x^2 - 3x - 4 = 0$$

whose roots are $x = -1$ and $x = 4$. Then replacing x by y^2, we have

$$y^2 = -1 \quad \text{and} \quad y^2 = 4$$

$$y = \pm i \quad \text{and} \quad y = \pm 2$$

These four values are the roots of the given equation. The check is left for the student.

Alternatively, we could find y^2 directly by use of the quadratic formula or by factoring. Thus, factoring, we find

$$(y^2 + 1)(y^2 - 4) = 0$$

Then equating each factor to zero yields the four roots.

EXAMPLE 2. Solve the equation $x + 2x^{1/2} - 8 = 0$.

Solution. This equation is quadratic in $x^{1/2}$. We apply the quadratic formula and obtain

$$x^{1/2} = \frac{-2 \pm \sqrt{4 + 32}}{2} = -1 \pm 3$$

$$x^{1/2} = -4 \quad \text{and} \quad x^{1/2} = 2$$

We recall that $x^{1/2} = +\sqrt{x}$, not $-\sqrt{x}$. Therefore $x^{1/2}$ cannot be negative, and consequently $x^{1/2} = -4$ has no solution. The equation $x^{1/2} = 2$ yields $x = 4$, which is the only root of the given equation.

EXAMPLE 3. Solve the equation

$$\frac{x^2 - 4}{x} - \frac{4x}{x^2 - 4} + 3 = 0$$

Solution. Substituting $y = (x^2 - 4)/x$ we get

$$y - \frac{4}{y} + 3 = 0$$

$$y^2 + 3y - 4 = 0$$

The roots of the last equation are $y = -4$ and $y = 1$. Hence we form the equations

$$\frac{x^2 - 4}{x} = -4 \quad \text{and} \quad \frac{x^2 - 4}{x} = 1$$

These equations yield the quadratic equations

$$x^2 + 4x - 4 = 0 \quad \text{and} \quad x^2 - x - 4 = 0$$

whose solutions are respectively $x = -2 \pm 2\sqrt{2}$ and $x = (1 \pm \sqrt{17})/2$. The student may check these values.

EXERCISE 8-3

Solve the following equations.

1. $x^4 - 10x^2 + 9 = 0$

2. $4x^4 - 5x^2 + 1 = 0$

3. $x^4 - x^2 - 2 = 0$

4. $x - x^{1/2} - 30 = 0$

5. $x - 4x^{1/2} + 3 = 0$

6. $x^{2/3} + 7x^{1/3} + 12 = 0$

7. $x^{4/3} - 5x^{2/3} + 4 = 0$

8. $x^{-2} + x^{-1} - 12 = 0$

9. $x^{-4} - 8x^{-2} - 9 = 0$

10. $x^{1/2} - 3x^{-1/2} + 2 = 0$

11. $6x^{1/2} + x^{-1/2} - 5 = 0$

12. $(2x^2 + x)^2 = 4(2x^2 + x) - 3$

13. $(x^2 - 5x)^2 + (x^2 - 5x) = 12$

14. $(x^2 - 3x) - 3\sqrt{x^2 - 3x} + 2 = 0$

15. $(x^2 + 2x) - \sqrt{x^2 + 2x} - 2 = 0$

16. $x^2 - 9x^{-2} - 8 = 0$

17. $\left(\frac{x - 1}{x}\right)^2 - \frac{x - 1}{x} - 2 = 0$

18. $\frac{x^2 - 3}{x - 1} = 2\left(\frac{x - 1}{x^2 - 3}\right) - 1$

19. $\left(\frac{x + 1}{x - 3}\right) - \left(\frac{x + 1}{x - 3}\right)^{1/2} - 2 = 0$

20. $4x^2 - 3x + \frac{5}{4x^2 - 3x} - 6 = 0$

8-4 EQUATIONS CONTAINING RADICALS

An equation in which the unknown appears under a radical sign is called a *radical*, or *irrational*, equation. The usual process for solving a radical equation employs the operation of raising both members of an equation to some positive integral power. The new equation thus obtained has all the roots of the original equation. This is true because the positive integral powers of two equal numbers are equal. The new equation, however, may have roots in addition to those of the original equation. The extra roots are called *extraneous roots*. Any extraneous root may be detected by checking in the given equation.

We list the steps for solving second-order radical equations.

1. Arrange the members of the equation so that a single radical constitutes one member.
2. Eliminate this radical by squaring both members of the equation.
3. Repeat steps 1 and 2, if necessary, until an equation free of radicals is obtained.
4. Solve the resulting equation and check all of its roots in the given equation. Reject any extraneous roots.

EXAMPLE 1. Solve the equation $\sqrt{x + 4} - \sqrt{2x + 1} + 1 = 0$.
Solution.

Transpose: $\qquad \sqrt{x + 4} = \sqrt{2x + 1} - 1$

Square: $\qquad x + 4 = 2x + 1 - 2\sqrt{2x + 1} + 1$

Transpose: $\quad 2\sqrt{2x + 1} = x - 2$

Square: $\qquad 4(2x + 1) = x^2 - 4x + 4$

Transpose: $\quad x^2 - 12x = 0$

$$x(x - 12) = 0$$

$$x = 0 \quad \text{and} \quad 12$$

Check. We substitute 0 and 12, in turn, for x in the left member of the given equation. Thus

$$\sqrt{0 + 4} - \sqrt{0 + 1} + 1 = 2 - 1 + 1 \neq 0$$

$$\sqrt{12 + 4} - \sqrt{24 + 1} + 1 = 4 - 5 + 1 = 0$$

Although $x = 0$ is a root of the quadratic equation, it does not satisfy the given equation. Hence 12 is the only root of the radical equation. We emphasize that $\sqrt{4} = +2$, the principal square root, and $-\sqrt{25} = -5$, the negative of the principal square root.

EXAMPLE 2. Solve the equation $\sqrt{2x - 3} - \sqrt{7x - 5} + \sqrt{x + 2} = 0$.
Solution.

Transpose: $\qquad \sqrt{2x - 3} = \sqrt{7x - 5} - \sqrt{x + 2}$

Square: $2x - 3 = 7x - 5 - 2\sqrt{(7x - 5)(x + 2)} + x + 2$

Transpose: $2\sqrt{(7x - 5)(x + 2)} = 6x$

$\sqrt{(7x - 5)(x + 2)} = 3x$

Square: $7x^2 + 9x - 10 = 9x^2$

Transpose: $2x^2 - 9x + 10 = 0$

The roots of this equation are $x = 2$ and $x = \frac{5}{2}$. These values are also roots of the given equation. We leave the verifications to the student.

EXERCISE 8-4   ed°dd-#5

Solve the following equations. Check your results.

1. $\sqrt{x^2 + 5} = 3$ 2. $\sqrt[3]{x^2 - 7x} = 2$ 3. $\sqrt{6x - 8} = x$

4. $\sqrt{5x + 6} = x + 2$ 5. $\sqrt{4 - x} = x - 4$ 6. $\sqrt{3x + 2} = -5$

7. $\sqrt{x + 3} - \sqrt{x + 1} = 3$ 8. $\sqrt{2x + 13} - \sqrt{x + 10} = 1$

9. $\sqrt{x - 2} - \sqrt{x - 5} = 1$ 10. $\sqrt{2x + 11} - \sqrt{x + 2} = 2$

11. $\sqrt{5 - 4x} + \sqrt{13 - 4x} = 4$ 12. $\sqrt{x + 6} = \sqrt{6x + 6} - \sqrt{x}$

13. $\sqrt{x} + \sqrt{3x + 4} = \sqrt{6x + 12}$ 14. $\sqrt{x + 9} - \sqrt{x + 3} = 2\sqrt{x + 5}$

15. $\sqrt{7 - 2x} - \sqrt{3 - x} - \sqrt{4 - x} = 0$ 16. $\sqrt{1 - x} + \sqrt{7 - x} - \sqrt{12 - 6x} = 0$

17. $\sqrt{x + 10} - \sqrt{x + 4} - 2\sqrt{x + 6} = 0$

18. $\sqrt{x^2 - x - 2} - \sqrt{x^2 + x + 7} + 1 = 0$

19. $\sqrt{x^2 - 3x + 4} - \sqrt{x^2 - x + 3} - 1 = 0$

20. $\sqrt{4x^2 - x - 8} - \sqrt{4x^2 + 3x + 4} + 2 = 0$

8-5 NATURE OF THE ROOTS

The roots of the quadratic equation $ax^2 + bx + c = 0$ are given by the formulas

$$x_1 = \frac{-b + \sqrt{b^2 - 4ac}}{2a} \quad \text{and} \quad x_2 = \frac{-b - \sqrt{b^2 - 4ac}}{2a}$$

In deriving these formulas we placed no restriction on the coefficients a, b, and c except $a \neq 0$. With this single exception, the formulas hold where each coefficient is any real number or any imaginary number. We now point out certain properties of the roots when the coefficients are real numbers and also when they are rational numbers. The properties depend on the radicand $b^2 - 4ac$, which is called the *discriminant*.

1. If $b^2 - 4ac = 0$, the roots are real and equal.
2. If $b^2 - 4ac > 0$, the roots are real and unequal.
3. If $b^2 - 4ac < 0$, the roots are imaginary and unequal.

If we further restrict the coefficients a, b, and c to be rational numbers and the discriminant to be positive, the following statements are true.

4. If $b^2 - 4ac$ is a perfect square, the roots are rational.
5. If $b^2 - 4ac$ is not a perfect square, the roots are irrational.

These conclusions can be readily verified. If the discriminant is zero, the roots are both equal to $-b/2a$. Clearly, the roots are unequal when the discriminant is not zero. Further, the roots are imaginary if and only if the discriminant is negative. If a, b, and c are rational, then $b^2 - 4ac$ is rational and, if positive, statements 4 and 5 follow at once.

EXAMPLE 1. Without solving the equation, determine the nature of the roots of $2x^2 + 8x + 3 = 0$.

Solution. Computing, we find $b^2 - 4ac = 8^2 - 4(2)(3) = 40$. The discriminant is positive and not a perfect square; hence the roots are real, irrational, and unequal.

EXAMPLE 2. Compute the value of the discriminant and determine the nature of the roots of the equation $\sqrt{2}x^2 - 2x + \sqrt{3} = 0$.

Solution. $b^2 - 4ac = (-2)^2 - 4(\sqrt{2})(\sqrt{3}) = 4 - 4\sqrt{6} < 0$. Hence the roots are imaginary and unequal.

8-6 SUM AND PRODUCT OF THE ROOTS

Denoting the roots of the equation $ax^2 + bx + c = 0$ by x_1 and x_2, we find their sum to be

$$x_1 + x_2 = \frac{-b + \sqrt{b^2 - 4ac}}{2a} + \frac{-b - \sqrt{b^2 - 4ac}}{2a} = \frac{-2b}{2a} = \frac{-b}{a}$$

and their product to be

$$x_1 x_2 = \frac{-b + \sqrt{b^2 - 4ac}}{2a} \cdot \frac{-b - \sqrt{b^2 - 4ac}}{2a}$$

$$= \frac{b^2 - (b^2 - 4ac)}{4a^2} = \frac{4ac}{4a^2} = \frac{c}{a}$$

These formulas, $-b/a$ and c/a, enable us to find the sum and the product of the roots of a quadratic equation at a glance. The formulas may also be used as a rapid check after solving a quadratic equation. The test consists simply in observing if the sum and the product of the two supposed roots satisfy the conditions $-b/a$ and c/a.

EXAMPLE 1. The sum and product of the roots of $8x^2 - 10x - 3 = 0$ are respectively $\frac{5}{4}$ and $-\frac{3}{8}$. The roots of the equation are $\frac{3}{2}$ and $-\frac{1}{4}$ because the sum of these values is $\frac{5}{4}$ and the product is $-\frac{3}{8}$.

We note that the equation $x^2 + bx + c = 0$ has $a = 1$, and consequently the sum of the roots is $-b$ and the product is c. This fact may be used to form a quadratic equation whose roots are two given numbers.

EXAMPLE 2. Form a quadratic equation whose roots are $-\frac{3}{2}$ and 5.

Solution. By making the coefficient of the second-degree term unity, we can use the negative of the sum of the roots for the coefficient of the first-degree term and the product of the roots for the constant term. Thus, $-\frac{3}{2} + 5 = \frac{7}{2}$ and $5(-\frac{3}{2}) = -\frac{15}{2}$; consequently the desired equation is

$$x^2 - \tfrac{7}{2}x - \tfrac{15}{2} = 0 \quad \text{or} \quad 2x^2 - 7x - 15 = 0$$

Alternatively, the equation may be formed by reversing the operation of solving a quadratic equation by factoring. Thus we write

$$(x + \tfrac{3}{2})(x - 5) = 0$$

We see at once that the roots of this equation are $-\frac{3}{2}$ and 5. Then multiplying the factors, we have, as before, the equation $x^2 - \frac{7}{2}x - \frac{15}{2} = 0$.

EXAMPLE 3. Find the value of k so that one root of the equation $x^2 + 9x + k = 0$ is twice the other root.

Solution. Let x_1 and $2x_1$ stand for the roots. Then $x_1 + 2x_1 = -9$ and $x_1 = -3$. To find k we write $k = x_1(2x_1) = (-3)(-6) = 18$. The given equation then becomes $x^2 + 9x + 18 = 0$. The roots of this equation are -3 and -6.

EXERCISE 8-5

Without solving, determine the nature of the roots of each equation. Find also the sum and product.

1. $x^2 + 6x + 5 = 0$ 2. $x^2 + 2x + 6 = 0$ 3. $2x^2 - x - 7 = 0$
4. $2x^2 + 5x + 5 = 0$ 5. $4x^2 - 12x + 9 = 0$ 6. $3x^2 + 7x - 1 = 0$
7. $5y^2 + 3 = 0$ 8. $10y^2 + 3y = 0$ 9. $7y^2 - 4 = 0$
10. $\sqrt{2}x^2 + 2x - 1 = 0$ 11. $\sqrt{3}x^2 - 4x + \sqrt{2} = 0$
12. $x^2 + \sqrt{5}x - \sqrt{2} = 0$ 13. $\sqrt{2}x^2 - \sqrt{5}x - 3\sqrt{2} = 0$

Find a quadratic equation with integral coefficients whose roots are the given numbers.

14. $2, -1$ 15. $3, 4$ 16. $-2, -6$
17. $\frac{2}{3}, -\frac{4}{3}$ 18. $\frac{2}{5}, \frac{5}{2}$ 19. $-\frac{3}{2}, \frac{4}{5}$
20. $\sqrt{2}, -\sqrt{2}$ 21. $2i, -2i$ 22. $3 - i, 3 + i$
23. $3 \pm \sqrt{3}$ 24. $\sqrt{2} \pm \sqrt{5}$ 25. $\sqrt{3} \pm i\sqrt{7}$

Determine the value of k so that the given condition is satisfied.

26. $x^2 + 3x + k = 0$, the roots are equal.
27. $kx^2 - 6x + 3 = 0$, the roots are equal.
28. $2x^2 - kx + k = 0$, the roots are equal.
29. $kx^2 + 6x + 32 = 0$, sum of roots is -2.
30. $7x^2 - 3kx + 41 = 0$, sum of roots is 0.
31. $3x^2 + 41x + k = 0$, product of roots is 7.
32. $x^2 - kx + 18 = 0$, one root is double the other.
33. $4x^2 - 3x + k = 0$, one root is three times the other.
34. $x^2 - 4x + k = 0$, the difference of the roots is 6.
35. $x^2 - 7x + k - 4 = 0$, the difference of the roots is 5.
36. $3kx^2 - 5x + 4k - 26 = 0$, the product of the roots is -3.
37. $4x^2 + 10x + k = 0$, one root is the reciprocal of the other.
38. $5x^2 - (k + 2)x + 7k - 6 = 0$, one root is 2.
39. $kx^2 + (3k - 4)x - 5 = 0$, one root is $\frac{1}{2}$.

8-7 THE GRAPH OF A QUADRATIC FUNCTION

The function defined by the equation

$$y = ax^2 + bx + c \qquad a \neq 0$$

is called a *quadratic function*. The graph of the function is called a *parabola*. (We assume that a, b, and c are real numbers.) If the graph has a point on the x axis, the abscissa of the point is the value of x which makes y equal to zero. This value of x is a zero of the function (Sec. 5-3), and is a root of the equation $y = ax^2 + bx + c = 0$. Hence the real roots, if any, of a quadratic equation may be obtained (at least approximately) from a graph. In finding the zeros of a quadratic function, we shall make use of a theorem which is proved in calculus.

Theorem 8-2 If $f(x)$ denotes a polynomial with real coefficients and if a and b are two real numbers such that $f(a)$ and $f(b)$ have opposite signs, then the equation $f(x) = 0$ has at least one real root between a and b.

Interpreted geometrically, this theorem means that the graph of $y = f(x)$ crosses the x axis at least once between any two of its points which are on opposite sides of the x axis.

EXAMPLE 1. Draw the graph of the function defined by the equation

$$y = 2 + 2x - x^2$$

Solution. Proceeding as in Sec. 5-3, we prepare a table of corresponding values of x and y, and draw a smooth curve through the points thus determined (Fig. 8-2).

x	-1	-0.5	0	1	2	2.5	3
y	-1	0.75	2	3	2	0.75	-1

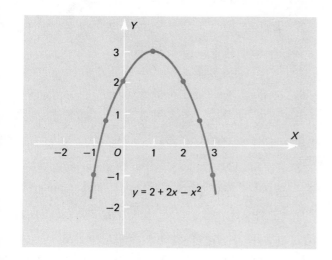

Fig. 8-2

The graph is a parabola which opens downward and crosses the x axis at two points. We estimate the abscissas of these points to be $x = -0.7$ and $x = 2.7$. These values are approximations to the zeros of the function and the roots of the equation $2 + 2x - x^2 = 0$.

It may be observed that the graph of $y = ax^2 + bx + c$ does not touch the x axis when $b^2 - 4ac < 0$. The negative discriminant means the function has no real zero and the corresponding quadratic equation has no real root. On the other hand, the graph crosses the x axis at two points if $b^2 - 4ac > 0$, and touches the x axis at one point only if $b^2 - 4ac = 0$. Hence the discriminant reveals if the graph of a quadratic function crosses the x axis at two points, touches the x axis at one point only, or has no point on the x axis. These three possibilities are illustrated by the graphs (Fig. 8-3) of the functions defined by $y = x^2 - 2x - 1$, $y = x^2 - 2x + 1$, and $y = x^2 - 2x + 3$, whose discriminants are positive, zero, and negative, respectively.

8-8 MAXIMUM AND MINIMUM VALUES

The parabola in Fig. 8-2 opens downward and the parabolas in Fig. 8-3 open upward. The direction in which the graph of the function $y = ax^2 + bx + c$ opens depends on the sign of the coefficient of x^2, as we shall prove.

Theorem 8-3   The graph of the function defined by the equation

$$y = ax^2 + bx + c$$

opens upward from a lowest point if a is positive and opens downward from a highest point if a is negative.

Proof. By steps easy to follow, we have

$$y = ax^2 + bx + c$$

$$= a\left[x^2 + \frac{b}{a}x + \frac{c}{a}\right]$$

$$= a\left[x^2 + \frac{b}{a}x + \frac{b^2}{4a^2} + \frac{c}{a} - \frac{b^2}{4a^2}\right]$$

$$= a\left(x + \frac{b}{2a}\right)^2 + \frac{4ac - b^2}{4a}$$

The quantity $(x + b/2a)^2$ is equal to zero when $x = -b/2a$ and is positive for all other values of x. Consequently this value of x yields the least value of y when a is positive and the greatest value when a is negative. Hence the point with coordinates

$$\left(-\frac{b}{2a}, \frac{4ac - b^2}{4a}\right)$$

is the lowest point or *minimum point* of the graph if $a > 0$ and the highest point or *maximum point* if $a < 0$. This point is called the *vertex* of the parabola. The ordinate of the vertex is the minimum value or maximum value of the function depending on the direction in which the graph opens.

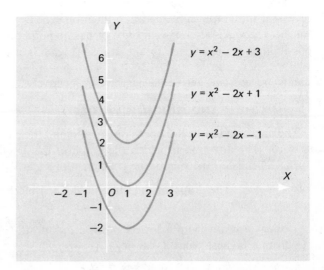

Fig. 8-3

EXAMPLE 1. Find the coordinates of the minimum point of the graph of the function defined by the equation $y = 2x^2 + 6x + 2$.

Solution. We change the form of the right member by adding and subtracting $\frac{9}{2}$. Thus

$$y = 2(x^2 + 3x) + 2$$
$$= 2(x^2 + 3x + \tfrac{9}{4}) + 2 - \tfrac{9}{2}$$
$$= 2(x + \tfrac{3}{2})^2 - \tfrac{5}{2}$$

Clearly, y takes its minimum value $-\frac{5}{2}$ when $x = -\frac{3}{2}$; any other value of x would require the addition of a positive number to $-\frac{5}{2}$. Hence the coordinates of the minimum point of the graph (Fig. 8-4) are $(-\frac{3}{2}, -\frac{5}{2})$.

EXAMPLE 2. Find the maximum value of the function defined by the equation $y = -3x^2 - 12x + 6$.

Solution. By subtracting and adding 12 in the right member, we have

$$y = -3(x^2 + 4x + 4) + 6 + 12$$
$$= -3(x + 2)^2 + 18$$

Since the expression $-3(x + 2)^2$ is never positive, the function has its greatest value when the expression is equal to zero. Hence we set $x = -2$ and obtain

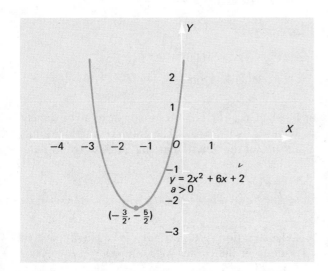

Fig. 8-4

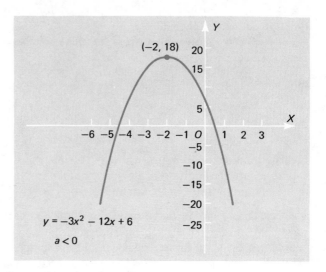

(-2, 18)

$y = -3x^2 - 12x + 6$

$a < 0$

Fig. 8-5

$y = 18$, which is the maximum value of the function. The coordinates of the maximum point of the graph (Fig. 8-5) are $(-2, 18)$.

EXAMPLE 3. Find the area of the largest rectangular plot which can be fenced on three of its sides with 160 yards of fencing.

Solution. If x yards is the width of the rectangle, then $160 - 2x$ yards is the length. Letting y square yards stand for the area, we have

$$y = 160x - 2x^2$$
$$= -2(x^2 - 80x + 1600) + 3200$$
$$= -2(x - 40)^2 + 3200$$

The maximum area, obtained when $x = 40$, is 3200 square yards. The dimensions of the largest plot are 40 yards by 80 yards. The graph (Fig. 8-6) exhibits the fact that the domain and range of the function arising in this problem are, respectively, the sets

$$\{x | 0 < x < 80\} \quad \text{and} \quad \{y | 0 < y \le 3200\}$$

To construct the graph of the equation of a parabola, it is usually best to locate the vertex, as we have done in the preceding examples. Then two or three additional points are sufficient for plotting a reasonably accurate graph.

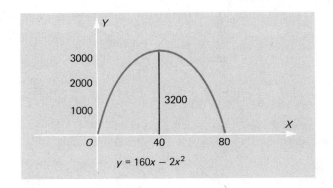

Fig. 8-6

EXERCISE 8-6

Find the coordinates of the maximum or minimum point of the graph of each of the following functions. From the graph give (or estimate) the real zeros, if any, of the function.

1. $y = x^2 - 8x + 10$ **2.** $y = x^2 + 4x + 3$
3. $y = 1 - 6x - x^2$ **4.** $y = 4 + 8x - x^2$
5. $y = 4x^2 + 4x + 1$ **6.** $y = 2x^2 - 3x + 2$
7. $y = 9 - 6x - x^2$ **8.** $y = -2x^2 + 2x - 5$
9. $y = 3x^2 - 3x - 4$ **10.** $y = -6x^2 + 4x + 3$

11. What are the dimensions of the largest rectangular field which can be enclosed with 440 rods of fencing?
12. Separate 18 into two parts such that the product of the parts is a maximum.
13. A rectangular pasture is to be fenced along four sides and divided into two equal parts by a fence parallel to two of the sides. Find the greatest possible area of the pasture if the total length of the fencing is 300 rods.
14. Find two numbers whose sum is 56 and the sum of their squares is a minimum.
15. If 20 children attend a skating party, the price per ticket will be 50 cents. For each child in excess of 20 the price of a ticket for each child will be decreased by 1 cent. Under this arrangement, what attendance will yield the greatest gross receipts?

REVIEW EXERCISE

Solve the following equations for x by the factoring method (Sec. 8-1).

1. $2x^2 - x - 3 = 0$ **2.** $3x^2 - x - 2 = 0$
3. $4x^2 - 11x - 3 = 0$ **4.** $3x^2 - 7x + 2 = 0$
5. $9x^2 - 3x - 2 = 0$ **6.** $5x^2 - 17x + 6 = 0$
7. $x^2 + (2b - 2a)x - 4ab = 0$ **8.** $a^2x^2 - abx - 2b^2 = 0$

Solve the equations by completing the square and check (Sec. 8-2).

9. $x^2 - x - 20 = 0$ **10.** $x^2 + 7x + 12 = 0$
11. $x^2 - x - 30 = 0$ **12.** $2x^2 + 5x + 2 = 0$
13. $6x^2 + 5x + 1 = 0$ **14.** $x^2 + 2x + 1 = 0$
15. $2x^2 - 4x + 1 = 0$ **16.** $9x^2 + 24x + 11 = 0$
17. $x^2 - 6x + 13 = 0$ **18.** $x^2 + 8x + 25 = 0$

Solve by the quadratic formula and check (Sec. 8-3).

19. $2x^2 + 3x - 2 = 0$ **20.** $x^2 + 8x + 15 = 0$
21. $3x^2 - x - 2 = 0$ **22.** $3x^2 - 2x - 5 = 0$
23. $20x^2 - 9x + 1 = 0$ **24.** $2x^2 - 5x + 1 = 0$
25. $8x^2 + 10x + 1 = 0$ **26.** $11x^2 - 20x + 8 = 0$
27. $9x^2 + 10x + 2 = 0$ **28.** $7x^2 + 13x + 7 = 0$

Solve the following equations (Sec. 8-3).

29. $x^4 + 13x^2 + 36 = 0$ **30.** $4x^4 + 9x^2 + 2 = 0$
31. $x^4 - 7x^2 + 12 = 0$ **32.** $3x^4 - 5x^2 - 12 = 0$
33. $2x^4 - 7x^2 - 4 = 0$ **34.** $x - 2x^{1/2} - 8 = 0$
35. $x - 3x^{1/2} - 10 = 0$ **36.** $x^{2/3} + 7x^{1/3} + 12 = 0$
37. $2x^{-2} - x^{-1} - 1 = 0$ **38.** $4x^{-2} + x^{-1} - 5 = 0$
39. $x^{4/3} - 10x^{2/3} + 9 = 0$ **40.** $x^{1/2} + 2x^{-1/2} + 3 = 0$
41. $(2x^2 - x)^2 - 4(2x^2 - x) + 3 = 0$ **42.** $(x^2 + 3x)^2 - 3(x^2 + 3x) + 2 = 0$

43. $\left(\dfrac{2}{x+1}\right)^2 + 7\left(\dfrac{2}{x+1}\right) - 30 = 0$ **44.** $5\left(\dfrac{x-1}{x+1}\right)^2 - 3\left(\dfrac{x-1}{x+1}\right) - 2 = 0$

45. $4x^2 - 3x + \dfrac{5}{4x^2 - 3x} + 6 = 0$ **46.** $\dfrac{x^2 - 3}{x+1} - 2\left(\dfrac{x+1}{x^2 - 3}\right) - 1 = 0$

Solve the following equations. Check your results (Sec. 8-4).

47. $\sqrt{x^2 - 8} = 1$ **48.** $\sqrt{x^2 + 9} = 5$
49. $\sqrt{6x + 7} = x$ **50.** $\sqrt{2x - 5} = x - 4$
51. $\sqrt{1 - x} = x + 1$ **52.** $\sqrt{3x + 11} = -2$
53. $\sqrt{1 - 5x} + \sqrt{3 - 3x} = 0$ **54.** $\sqrt{3x - 5} - \sqrt{2x + 2} = 0$
55. $\sqrt{2x + 11} - 1 = \sqrt{x + 9}$ **56.** $3 - \sqrt{x + 2} - \sqrt{3x - 5} = 0$
57. $\sqrt{4x + 5} = \sqrt{6x + 6} - 1$ **58.** $\sqrt{2x - 7} - 1 = \sqrt{x - 4}$
59. $\sqrt{3x + 3} = 2\sqrt{2x} - \sqrt{x - 1}$ **60.** $\sqrt{6x + 6} - \sqrt{x - 1} = \sqrt{3x + 1}$
61. $2\sqrt{2x - 2} - \sqrt{3x} = \sqrt{4 - x}$ **62.** $3\sqrt{x - 6} + \sqrt{x - 10} = 2\sqrt{x - 1}$
63. $\sqrt{2x - 5} + \sqrt{x + 1} = \sqrt{7x - 12}$ **64.** $\sqrt{3x - 2} + \sqrt{2x - 9} = \sqrt{7x + 1}$
65. $\sqrt{3x} + \sqrt{2x + 1} = \sqrt{10x + 1}$ **66.** $\sqrt{9 - 2x} - \sqrt{4 - x} = \sqrt{5 - x}$
67. $1 + \sqrt{x^2 - x + 2} = \sqrt{x^2 - 2x - 1}$
68. $\sqrt{x^2 - 6x + 4} = \sqrt{x^2 - 3x - 2}$

Show that the graph of the equation

$$x = ay^2 + by + c$$

is a parabola with vertex at the point

$$\left(\frac{4ac - b^2}{4a}, -\frac{b}{2a}\right)$$

Show also that the parabola extends to the right of the vertex if $a > 0$, and to the left if $a < 0$.

Find the coordinates of the vertex of the parabola defined by each equation. Plot two or three more points on the parabola and sketch the graph (Secs. 8-7 and 8-8).

69. $x = y^2 + 8y + 12$ **70.** $x = y^2 - 4y + 4$
71. $x = -y^2 + 6y - 7$ **72.** $x = -y^2 + 2y + 1$
73. $x = 4y^2 - 4y + 2$ **74.** $x = 2y^2 - 4y + 3$
75. $x = 3y^2 + 3x + 1$ **76.** $x = -4y^2 - 8y + 1$

Chapter 9

SYSTEMS INVOLVING QUADRATIC EQUATIONS

9-I THE GRAPH OF A QUADRATIC EQUATION IN x AND y

An equation of the form

$$Ax^2 + Bxy + Cy^2 + Dx + Ey + F = 0 \qquad (1)$$

where A, B, C, D, E, and F are constants with A, B, and C not all zero, is called a *quadratic equation* in the variables x and y. The graph of an equation of this kind is called a *conic section*. This name comes from the fact that an identical curve can be obtained by the intersection of a plane and a right circular cone. To construct the graph of an equation of the form (1) where all terms are present is a tedious task. We shall, however, consider certain special cases of the equation whose graphs can be drawn with little difficulty.

In Sec. 8-7, we constructed graphs, called parabolas, of equations of the form $y = ax^2 + bx + c$. An equation of this form may be obtained from equation (1) by assigning B and C the value zero and solving for y. Similarly, the equation

$$x = ay^2 + by + c$$

is a special case of equation (1). The graph of an equation of this form is a parabola which opens from a leftmost point (vertex) to the right if $a > 0$, and from a rightmost point to the left if $a < 0$ (see Fig. 9-5).

We shall next consider other special quadratic equations in x and y whose graphs, like the parabola, are conic sections with distinguishing names. We list here the steps involved in the graphing process.

1. Solve the equation for y in terms of x.
2. Assign values to x and compute the corresponding values of y.
3. Plot the points thus determined and draw a smooth curve through them.

If it is easier to solve for x in terms of y, interchange x and y in these directions.

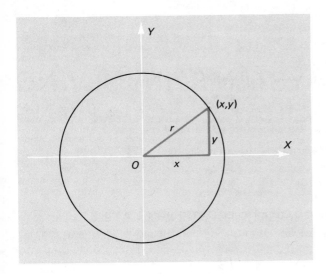

Fig. 9-1

The Circle. The graph of an equation of the form

$$x^2 + y^2 = r^2$$

is a circle of radius r with the center at the origin. That the graph is a circle may be seen by referring to Fig. 9-1. According to the Pythagorean theorem, the coordinates of any point of the circle satisfy the equation $x^2 + y^2 = r^2$. Thus the graph of $x^2 + y^2 = 36$ is a circle of radius 6 and center at the origin.

The Ellipse. The graph of an equation of the form

$$ax^2 + by^2 = c$$

with $a, b, c > 0$ and $a \neq b$ is called an *ellipse*.

EXAMPLE 1. Construct the graph of the equation $4x^2 + 9y^2 = 36$.
Solution. Solving for y in terms of x and for x in terms of y, we have

$$y = \pm\tfrac{2}{3}\sqrt{9 - x^2} \quad \text{and} \quad x = \pm\tfrac{3}{2}\sqrt{4 - y^2}$$

From the first equation we see that if $x^2 > 9$, the radicand is negative and hence y is imaginary. This means, for the purpose of graphing, that x may take only values from -3 to 3, inclusive. And the second equation reveals that y is restricted to the values -2 to 2, inclusive. Selecting the first equation, we prepare a table of corresponding values of x and y. The graph drawn through the points thus determined is shown in Fig. 9-2.

x	-3	-2	0	2	3
y	0	± 1.5	± 2	± 1.5	0

The Hyperbola. The graph of an equation of the form

$$ax^2 - by^2 = c \quad \text{or} \quad ay^2 - bx^2 = c$$

where a, b, and c are positive constants is called a *hyperbola*.

EXAMPLE 2. Draw the graph of the equation $16x^2 - 9y^2 = 144$.
Solution. Solving for y, we get

$$y = \pm \tfrac{4}{3}\sqrt{x^2 - 9}$$

We observe that y is imaginary for values of x between -3 and 3. Hence the graph consists of two *branches*, one starting at $(3, 0)$ and extending to the right and the other extending to the left from $(-3, 0)$. The graph, drawn by the use of the pairs of values in the table, is in Fig. 9-3.

x	-5	-4	-3	3	4	5
y	± 5.3	± 3.5	0	0	± 3.5	± 5.3

The graph of $ay^2 - bx^2 = c$, where a, b, and c are positive constants, has one branch opening upward and the other branch opening downward. This is illustrated in Fig. 9-6.

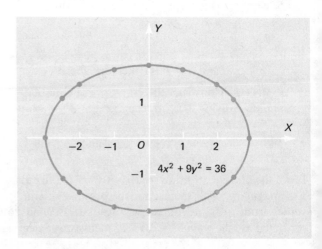

Fig. 9-2

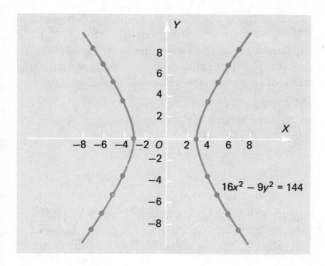

Fig. 9-3

EXAMPLE 3. Draw the graph of the equation $xy = 4$.

Solution. In this equation x may be given any value except zero. When $x = 0$, there is no value for y which makes the product xy equal to 4. The graph is a hyperbola comprising the two branches drawn in Fig. 9-4.

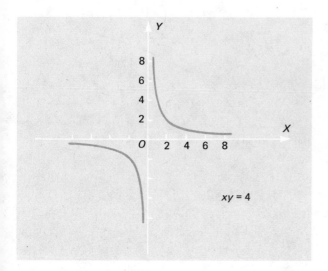

Fig. 9-4

9-2 SOLUTION OF A QUADRATIC SYSTEM BY GRAPHING

In Sec. 6-5 we solved systems of linear equations by graphing each equation of a system and reading the coordinates of the point of intersection. The same procedure may be applied to second-degree equations. However, the graphs are not restricted to straight lines, and there may be more than one point of intersection. Only real solutions are obtained by this process and these, as a rule, are approximations.

EXAMPLE 1. Solve graphically the system of equations

$$3x - 2y + 1 = 0$$
$$y^2 - 4x - 8 = 0$$

Solution. The graph of the first equation is a straight line, and the graph of the second is a parabola. The graphs of the equations are shown in Fig. 9-5. There are two intersection points, whose coordinates we estimate to be $(-1.4, -1.6)$ and $(2.5, 4.2)$. These value pairs are good approximations of the solutions, as may be verified by substitution in the given equations.

EXAMPLE 2. Solve graphically the system of equations

$$4y^2 + 9x^2 = 36$$
$$9y^2 - 4x^2 = 36$$

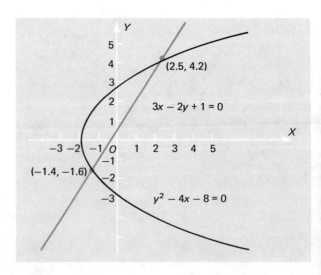

Fig. 9-5

Solution. The graph of the first equation is an ellipse, and the graph of the second equation is a hyperbola. We estimate the coordinates of the intersection points (Fig. 9-6) to be $(-1.4, -2.2)$, $(-1.4, 2.2)$, $(1.4, -2.2)$, and $(1.4, 2.2)$. Each of these number pairs is a close approximation to a solution of the system.

do two

EXERCISE 9-1

Solve the following systems graphically. Estimate the results to one decimal place. (See Table of Powers and Roots, inside the front cover, to find square roots of numbers.)

1. $x^2 - 4y = 0$ $x - y = 0$	**2.** $y^2 - 3x = 0$ $x + y = 1$	**3.** $xy - 12 = 0$ $x - 2y = 2$
4. $x^2 + y^2 = 25$ $y - x = 1$	**5.** $2x^2 + y^2 = 9$ $x - y = 1$	**6.** $x^2 + y^2 = 8$ $x^2 + 4y^2 = 16$
7. $x^2 + y^2 = 10$ $xy - 4 = 0$	**8.** $x^2 - y^2 = 8$ $x^2 + 2y^2 = 14$	**9.** $2x^2 + y^2 = 18$ $x^2 + 2y^2 = 33$
10. $4y^2 - x^2 = 16$ $2xy - 15 = 0$	**11.** $2x^2 + 3y^2 = 20$ $y = x^2 - 2$	**12.** $x^2 + y^2 = 16$ $x^2 = 2y + 8$

9-3 SOLUTION BY ALGEBRAIC METHODS

Any two independent quadratic equations in two variables can be solved by algebraic methods. An algebraic method yields the exact values of all solutions, including both real and imaginary solutions. However, a general

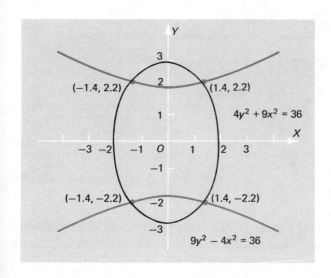

Fig. 9-6

algebraic method leads to a fourth-degree equation in one of the variables. Although an equation of this kind can be solved to obtain its roots exactly, the process is not simple. Consequently we shall consider certain special quadratic systems which can be solved by processes which we have already discussed.

9-4 A SYSTEM WITH NO xy OR FIRST-DEGREE TERMS

If both equations of a system are of the form

$$Ax^2 + Cy^2 + F = 0$$

we may employ one of the methods for solving two linear equations in two variables (Sec. 6-6). If we treat the equations as linear in x^2 and y^2, values for the squares of the variables can be found. Then the variables x and y are obtained by taking square roots.

EXAMPLE. Solve the system of equations

$$4x^2 + 3y^2 = 24$$
$$3x^2 - 2y^2 = 35$$

Solution. We multiply the members of the first equation by 2 and those of the second equation by 3 to get

$$
\begin{aligned}
8x^2 + 6y^2 &= 48 \\
9x^2 - 6y^2 &= 105 \\
\hline
17x^2 &= 153 \\
x &= \pm 3
\end{aligned}
$$

Then substituting for x in $4x^2 + 3y^2 = 24$, we find

$$36 + 3y^2 = 24$$
$$3y^2 = -12$$
$$y = \pm 2i$$

Pairing the values for x and y, we have

$$(3, 2i), (3, -2i), (-3, 2i), (-3, -2i)$$

These solutions are not real, and consequently the graphs of the given equations (Fig. 9-7) do not intersect. These value pairs, however, satisfy each of the given equations.

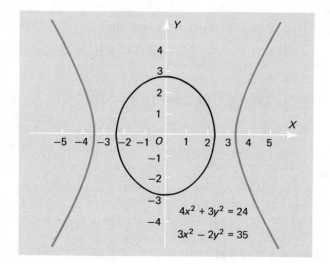

$4x^2 + 3y^2 = 24$

$3x^2 - 2y^2 = 35$

Fig. 9-7

9-5   A SYSTEM WITH A LINEAR AND A QUADRATIC EQUATION

If a system has a linear and a quadratic equation, the solutions can be readily obtained. The process consists in solving the linear equation for one of its unknowns in terms of the other and then substituting in the quadratic equation. This step yields a quadratic equation in one unknown. We illustrate the procedure.

EXAMPLE. Solve the system of equations

$$x^2 - xy + y^2 - 7 = 0$$
$$x - 2y + 1 = 0$$

Solution. Solving the linear equation for x and substituting the result in the quadratic equation, we find

$$(2y - 1)^2 - y(2y - 1) + y^2 - 7 = 0$$
$$3y^2 - 3y - 6 = 0$$

The solutions of this equation are $y = -1$ and $y = 2$. The corresponding values of x, obtained from the given linear equation, are $x = -3$ and $x = 3$. Hence the number pairs $(-3, -1)$ and $(3, 2)$ are the solutions of the given equations.

EXERCISE 9-2

Solve each of the following systems of equations and check your results.

1. $x^2 + y^2 = 25$
    $x^2 - y^2 = 7$

2. $x^2 - y^2 = 11$
    $3x^2 - 2y^2 = 58$

3. $2x^2 + 3y^2 = 23$
    $3x^2 - 2y^2 = 15$

4. $4y^2 - 3x^2 = 29$
    $y^2 + 7x^2 = 46$

5. $5x^2 + 2y^2 = 37$
    $x^2 + y^2 = 5$

6. $6x^2 - 11y^2 = 38$
    $5x^2 - 9y^2 = 31$

7. $8x^2 + 3y^2 = 2$
    $5x^2 - 4y^2 = 13$

8. $5y^2 - 6x^2 = 43$
    $4y^2 - 5x^2 = 35$

9. $4x^2 + 5y^2 = 21$
    $3x^2 + 4y^2 = 16$

10. $x^2 - y = 5$
    $2x + y = 3$

11. $y^2 + 2x = 7$
    $x - 2y = -7$

12. $3x^2 + 2y^2 = 25$
    $x + y = 5$

13. $x^2 + y^2 = 1$
    $2x - y = 4$

14. $x^2 + y^2 = 20$
    $3x - y = 2$

15. $xy + x - y = 13$
    $4x - 3y - 7 = 0$

16. $x^2 - xy + y^2 - 5 = 0$
    $3x - 2y - 5 = 0$

17. $x^2 - 3xy - 2y^2 - 2 = 0$
    $x - y - 2 = 0$

18. $x^2 + 2xy - 2y^2 - 3 = 0$
    $x - 2y - 3 = 0$

19. $y^2 - 4xy - 4y + 8x + 3 = 0$
    $4x - 2y - 3 = 0$

20. $4x^2 - 9y^2 - 8x - 36y = 23$
    $x - y - 2 = 0$

21. $x^2 + xy - 2y^2 - 3y = 1$
    $x - 2y - 1 = 0$

9-6 A SYSTEM WITH NO FIRST-DEGREE TERMS

If both equations of a system are of the form

$$Ax^2 + Bxy + Cy^2 + F = 0$$

where the constant F is not zero, they may be combined to yield two linear equations. The linear equations come from eliminating the constant terms of the given equations. If $F = 0$ in one of the given equations, the left member of that equation may be factored to yield the linear equations.

EXAMPLE 1. Solve the system of equations

$$5x^2 - 2xy - y^2 = 2$$

$$2x^2 + xy - 2y^2 = 1$$

Solution. Multiplying the members of the second equation by 2 and subtracting from the first, we get

$$x^2 - 4xy + 3y^2 = 0 \quad \text{or} \quad (x - y)(x - 3y) = 0$$

From the last equation we obtain the linear equations $x - y = 0$ and $x - 3y = 0$, each of which may be paired with either given equation to form a system. Thus we have the systems

$$\left\{ \begin{array}{l} 2x^2 + xy - 2y^2 = 1 \\ \\ x - y = 0 \end{array} \right. \quad \text{and} \quad \left\{ \begin{array}{l} 2x^2 + xy - 2y^2 = 1 \\ \\ x - 3y = 0 \end{array} \right.$$

These systems may be solved by the method of the preceding section. The solutions are, respectively,

$$(1, 1), (-1, -1) \quad \text{and} \quad \left(\frac{3\sqrt{19}}{19}, \frac{\sqrt{19}}{19}\right), \left(-\frac{3\sqrt{19}}{19}, -\frac{\sqrt{19}}{19}\right)$$

EXAMPLE 2. Solve the system of equations

$$5x^2 + 2xy - y^2 = 3$$
$$2x^2 - xy - y^2 = 0$$

Solution.   Factoring, we express the second equation in the form $(2x + y)(x - y) = 0$. Equating each factor to zero, we have the linear equations $2x + y = 0$ and $x - y = 0$, each of which we pair with the first given equation to yield the systems

$$\begin{cases} 5x^2 + 2xy - y^2 = 3 \\ \\ 2x + y = 0 \end{cases} \quad \text{and} \quad \begin{cases} 5x^2 + 2xy - y^2 = 3 \\ \\ x - y = 0 \end{cases}$$

These systems have the following solutions:

$$(i, -2i), (-i, 2i); \left(\frac{\sqrt{2}}{2}, \frac{\sqrt{2}}{2}\right), \left(-\frac{\sqrt{2}}{2}, -\frac{\sqrt{2}}{2}\right)$$

Each of these number pairs may be checked in the given equations. Substituting $x = i$ and $y = -2i$ in the right members of the equations, we obtain

$$5(i)^2 + 2(i)(-2i) - (-2i)^2 = -5 + 4 + 4 = 3 \qquad \text{check}$$
$$2(i)^2 - (i)(-2i) - (-2i)^2 = -2 - 2 + 4 = 0 \qquad \text{check}$$

As an exercise, the student might check each of the other number pairs.

EXERCISE 9-3

Solve each of the following systems of equations.

1. $2x^2 - xy = 6$
   $x^2 - y^2 = 3$

2. $x^2 - 2xy - 2y^2 = 1$
   $2x^2 + xy + y^2 = 2$

3. $5x^2 + 2xy + y^2 = 16$
   $4x^2 + 4xy = -16$

4. $3x^2 - 5xy - 4y^2 = 8$
   $x^2 - 2xy - y^2 = 2$

5. $2x^2 - 2xy - 3y^2 = -7$
   $x^2 + 3xy + 2y^2 = 4$

6. $2x^2 - 3xy + 2y^2 = 4$
   $4x^2 - 6xy + 3y^2 = 4$

7. $x^2 - xy - y^2 = 5$
   $2x^2 + 2xy + y^2 = 10$

8. $3x^2 - 3xy + y^2 = 7$
   $x^2 - 4xy + 4y^2 = 0$

9. $2xy + y^2 = 0$
   $x^2 - 3xy + 2y^2 = 15$

10. $4x^2 - 5xy - 6y^2 = 0$
   $3x^2 - 3xy - 5y^2 = 9$

11. $2x^2 + xy + y^2 = 16$
    $2x^2 + 7xy + 6y^2 = 0$

12. $x^2 + 2xy - y^2 = 14$
    $5x^2 - 26xy + 5y^2 = 0$

13. $6x^2 - 13xy + 6y^2 = 0$
    $x^2 + 2xy - 5y^2 = -1$

14. $5x^2 + 3xy - 2y^2 = 34$
    $3x^2 + 8xy - 3y^2 = 0$

15. Show that the substitutions $x = u + v$ and $y = u - v$ reduce the system

$$x^2 + B_1xy + y^2 + D_1x + D_1y + F_1 = 0$$

$$x^2 + B_2xy + y^2 + D_2x + D_2y + F_2 = 0$$

to

$$(2 - B_1)v^2 + (2 + B_1)u^2 + 2D_1u + F_1 = 0$$

$$(2 - B_2)v^2 + (2 + B_2)u^2 + 2D_2u + F_2 = 0$$

Observe now that the new system can be solved by first eliminating v^2 to obtain a quadratic equation in u. And then x and y can be easily found, thus giving the solutions of the original system.

Use the information of Problem 15 and solve the following systems.

16. $x^2 + y^2 - 10 = 0$
    $xy + x + y - 7 = 0$

17. $x^2 + y^2 - x - y - 4 = 0$
    $2xy - 3x - 3y + 4 = 0$

18. $x^2 + y^2 - 10x - 10y = -40$
    $xy - 5x - 5y + 22 = 0$

19. $x^2 + y^2 + xy + 3x + 3y = 18$
    $x^2 + y^2 + 2x + 2y = 15$

In the following problems set up a system of two equations in two unknown quantities. Solve the system and check your results.

20. The sum of two numbers is 16; the sum of their squares is 130. Find the numbers.

21. The perimeter of a rectangle is 40, and the area is 96. Find the dimensions of the rectangle.

22. The length of a rectangle is 14 units more than its width, and the diagonal is 34 units. Find the lengths of the sides.

23. The perimeter of an isosceles triangle is 36 and altitude is 12. Find the lengths of the sides.

24. The diagonal of a closed rectangular box is 6 inches and the total surface area is 64 square inches. Find the dimensions of the box if the base is a square.

25. The printed area of a rectangular poster is 704 square inches. The printed area plus the area of the margins is 1200 square inches. Find the dimensions of the poster if each of the four margins is 4 inches wide.

26. A rectangular piece of tin has an area of 144 square inches on one side. A two-inch square is cut from each corner and then a box is made by turning up the sides. Find the dimensions of the piece of tin if the box contains 120 cubic inches.

27. A group of children engaged a skating rink for $24. After engaging the rink, 10 more children joined the party, and thereby reduced the cost per child by 20 cents. Find the cost to each child and the final number of children.

28. Two pipes can fill a pool in 8 hours. The larger of the pipes can fill the pool in 12 hours less time than the smaller. Find the time required for each pipe alone to fill the pool.

29. A car leaves town *A* and travels with uniform speed to town *B*. A second car, traveling 20 miles per hour faster than the first car, goes from town *B* to town *A*. The two cars start at the same time and meet in 1 hour, and one of the cars makes the drive, from town to town, in 50 minutes less time than is required for the other car. Find the speed of each car and the distance between the towns.

30. Three airplanes take off at the same time from an airfield and travel in straight courses to another field. The first airplane flies 40 miles per hour faster than the second airplane and makes the flight in 40 minutes less time. The third airplane flies 40 miles per hour slower than the second airplane and consumes 1 hour more flight time than the second one. Find the speed of the second airplane and the distance between the fields.

31. Mr. Smith sold a field of a certain area and price per acre for $5000. He also sold a second field 25 acres larger than the first field and at a price $200 less per acre. If the second sale brought $10,500, find the acreage and price per acre of the first field.

32. A rectangular feed lot is to be built with a barn 40 feet by 60 feet in one corner. If 300 feet of fencing are available and the area, including the barn area, is 10,000 square feet, find the dimensions of the lot.

33. A page is to contain 36 square inches of printed material. The margins at the top and bottom are each $1\frac{1}{2}$ inches and at each side 1 inch. Find the dimensions of the page if the area is 80 square inches.

34. A rectangular plot contains 5 acres (800 square rods). One side of the plot is along a straight river bank and requires no fence. Find the dimensions of the rectangle if the length of fencing is 350 rods.

35. Mr. Hart sold a batch of wheat at a certain price per bushel. A few days later when the price had increased by $2 per bushel, he sold a 500-bushel larger batch than the first. If the two sales brought $10,500, find the amount of wheat and the price per bushel in the first sale.

36. Mr. Grant sold two square-shaped tracts of land for $24,000, getting $200 per acre for the first tract and $100 per acre for the second tract. The second tract brought $8000 more than the first tract, and the combined areas of the two squares was 200 acres. Find the length of a side of each square. (One acre is equal to 160 square rods.)

37. A ladder 25 feet long leans against the wall of a building at a point 30 feet from the ground and passes over and just touches the top of a fence 9 feet high that is parallel to the wall. If the foot of the ladder is 40 feet from the base of the building, find the distance from the foot of the ladder from the bottom of the fence and the distance of the fence from the building.

38. A rectangular flower plot of 504 square feet area is surrounded by a path 4 feet wide. If the area of the path is 360 square feet, find the dimensions of the plot.

39. The diagonal of a rectangle is 8 inches longer than one of its sides and 4 inches longer than the other side. Find the dimensions of the rectangle.

Chapter 10

INEQUALITIES

10-1 THE ORDER AXIOMS

In Chapter 2 we introduced a set of undefined elements R, called real numbers, which possess the properties embodied in the six field axioms (Sec. 2-2). The real numbers, satisfying the field axioms, are said to form a field. We now ascribe additional properties on the set R by requiring the set to satisfy the so-called *order axioms*. The order symbols are made and described as follows:

$<$ means "is less than"
$>$ means "is greater than"

Axiom 7   *The trichotomy axiom.* If a and $b \in R$, then one and only one of the following relations holds:

$$a < b \qquad a = b \qquad a > b$$

Axiom 8   *The transitive axiom.* If $a, b,$ and $c \in R$ such that $a > b$ and $b > c$, then $a > c$.

Axiom 9   *The addition axiom.* If a, b, and $c \in R$ such that $a > b$, then $a + c > b + c$.

Axiom 10   *The multiplicative axiom.* If a, b, and $c \in R$ such that $a > b$ and $c > 0$, then $ac > bc$.

Definition 10-1   A field which satisfies the order axioms is called an *ordered* field.

Definition 10-2   If a and $b \in R$, then $a < b$ if and only if $b > a$.

Definition 10-3   A real number a is positive if $a > 0$ and negative if $a < 0$.

Definition 10-4   A statement that one quantity is greater than or less than another quantity is called an *inequality*.

If we set $b = 0$ in Axiom 7, it follows that $a < 0$, $a = 0$, or $a > 0$. From this fact and Definition 10-3, we conclude that 0 is neither positive nor negative.

Sometimes it is convenient to combine an inequality with an equality. Thus

$a \leq b$ means "a is less than or equal to b"

$a \geq b$ means "a is greater than or equal to b"

For example, $a \leq 0$ means a is not positive, and $a \geq 0$ means a is not negative. Clearly, if the two statements hold simultaneously, then $a = 0$.

There is a great variety of useful theorems involving inequalities. We shall prove certain theorems to be used later and state other theorems which the student may prove. Where no reason is given for a step in the proof of a theorem, the student should give the reason.

Theorem 10-1   If $a > 0$ and $b > 0$, then $a + b > 0$.

Proof.

$a > 0$	given
$a + b > 0 + b = b$	Axiom 9
$b > 0$	given
$a + b > 0$ ■	Axiom 8

Theorem 10-2   If $a > 0$ and $b > 0$, then $ab > 0$.

These theorems tell us that the sum, or the product, of two positive numbers is positive. In Secs. 3-2 and 3-3, we assumed these closure properties for addition and multiplication of positive numbers. Then we proceeded to prove that the product (or quotient) of two numbers of like signs is positive and the product (or quotient) of two numbers of unlike signs is negative.

Theorem 10-3   If a and $b \in R$, then $a > b$ if and only if $-a < -b$.

Proof. We first prove that if $a > b$, then $-a < -b$.

$a > b$	given
$a + [(-a) + (-b)] > b + [(-a) + (-b)]$	Axiom 9
$-b > -a$	
$-a < -b$ ■	Definition 10-2

Next we need to show that if $-a < -b$, then $a > b$. The steps in the first part of the proof are reversible. Hence the proof would consist in starting with $-a < -b$ and rewriting the inequalities back to $a > b$.

Corollary   If $a > 0$, then $-a < 0$; and if $-a < 0$, then $a > 0$.

Theorem 10-4   Let a, b, and $c \in R$ with $a > b$ and $c < 0$. Then it follows that $ac < bc$.

Proof. Since $c < 0$, we conclude from the preceding corollary that $-c > 0$. Then we have

$$a > b \qquad\qquad \text{given}$$

$$-ac > -bc \qquad\qquad \text{Axiom 10}$$

$$ac < bc \qquad\qquad \text{Theorem 10-3}$$

Axiom 10 tells us that both members of an inequality may be multiplied by a positive number if the *sense* (direction of pointing of the inequality symbol) of the inequality is unchanged. And Theorem 10-4 tells us that the sense must be changed if the multiplier is a negative number.

Theorem 10-5   If a, b, c, and $d \in R$ with $a > b$ and $c > d$, then $a + c > b + d$.

Proof.

$$a > b \quad \text{and} \quad c > d \qquad\qquad \text{given}$$

$$a + c > b + c \quad \text{and} \quad b + c > b + d \qquad\qquad \text{Axiom 9}$$

$$a + c > b + d \quad \blacksquare \qquad\qquad \text{Axiom 8}$$

Theorem 10-6   If $a > b$ and $c > d$, with a, b, c, and $d > 0$, then $ac > bd$.

Theorem 10-7   If $a > 0$, $b > 0$, and $a > b$, then $1/a < 1/b$.

Proof.

$$a > b \qquad\qquad \text{given}$$

$$a\left(\frac{1}{ab}\right) > b\left(\frac{1}{ab}\right) \qquad\qquad \text{Axiom 10}$$

$$\frac{1}{b} > \frac{1}{a}$$

$$\frac{1}{a} < \frac{1}{b} \quad \blacksquare$$

EXERCISE 10-1

1. If $a \in R$, tell why the inequality $a > a$ is impossible.
2. If $b \in R$, prove that $b < 0$ if and only if $-b > 0$. Prove also that $b > 0$ if and only if $-b < 0$. [Hint: Use Theorem 10-4.]
3. If a, b, c, and $d \in R$ with $b > 0$ and $d > 0$, prove that

$$\frac{a}{b} > \frac{c}{d} \quad \text{if and only if} \quad ad > bc$$

4. Verify Theorems 10-1 to 10-7 by using properly chosen integers.
5. Show that $a > b$ if and only if $a - b > 0$.
6. If $a \neq 0$, prove that $a^2 > 0$.
7. If $a > b$ and $c > 0$, show that $a/c > b/c$.
8. If $a > b$ and $c < 0$, show that $a/c < b/c$.

10-2 SOLUTION OF INEQUALITIES

In the remainder of this chapter we shall consider inequalities in which one or both members contain a variable (or variables). In Sec. 6-1 we defined two special types of equations—conditional equations and identities. Analogously, there are in general two special types of inequalities, which we now define.

Definition 10-5 An inequality which is satisfied by some, but not all, of the permissible values of the variables involved is called a *conditional inequality*.*

Definition 10-6 An inequality which is satisfied by all the permissible values of the variables involved is called an *absolute inequality*.

Thus $x + 3 < 4$ is a conditional inequality; it is true only if $x < 1$. But $x^2 + 1 > 0$, valid for all real values of x, is an absolute inequality.

To solve an inequality means to find all the values of the variable or variables which satisfy the inequality. The axioms and theorems of the preceding section may be applied in solving inequalities. In particular, we may use the following properties of inequalities.

1. The sense of an inequality is not changed when the same number is added to, or subtracted from, both members.
2. The sense of an inequality is not changed when both members are multiplied by, or divided by, the same positive number.
3. The sense of an inequality is reversed when both members are multiplied by, or divided by, the same negative number.

Each of these operations changes an inequality into an equivalent inequality (i.e. an inequality with the same solution set). And the solution process consists in obtaining a series of equivalent inequalities which lead to a final inequality whose solution is evident.

EXAMPLE 1. Solve the inequality $2 - 3x \leq 2x + 12$.
Solution. Subtracting $2x + 2$ from both sides, we obtain

$$-5x \leq 10$$

Then dividing by -5 and reversing the sense of the inequality gives

$$x \geq -2$$

* The permissible values include all the values for which both members of an inequality are defined.

This shows that any solution of the given inequality is a solution of the last inequality. We could now start with $x \geq -2$, reverse the operations, and work back to the given inequality, which would establish the equivalence of the given inequality and the final inequality. We conclude, then, that the solution of the given inequality, in set notation, is

$$\{x|x \geq -2\}$$

EXAMPLE 2. Solve the inequality $2x^2 - x - 15 < 0$.

Solution. The first step is to factor the left member. Thus we have

$$(2x + 5)(x - 3) < 0 \quad \text{or} \quad 2(x + \tfrac{5}{2})(x - 3) < 0$$

In the factored form, we see that the zeros of $2x^2 - x - 15$ are $x = -\tfrac{5}{2}$ and $x = 3$. For all other values of x the left side is either positive or negative. Hence we seek the values of x which make one of the factors positive and the other negative. The factor $x + \tfrac{5}{2}$ is negative when $x < -\tfrac{5}{2}$ and positive when $x > -\tfrac{5}{2}$. This is pictured on the number axis (Fig. 10-1) by the minus signs to the left of $-\tfrac{5}{2}$ and the plus signs to the right. Similarly, the factor $x - 3$ is negative to the left of 3 and positive to the right. From the diagram, it is easy to see that the factors have opposite signs for all values of x between $-\tfrac{5}{2}$ and 3. The solution of the given inequality may be expressed by

$$-\tfrac{5}{2} < x < 3 \quad \text{or} \quad \{x| -\tfrac{5}{2} < x < 3\}$$

By referring to Fig. 10-1 we see that the factors of $2x^2 - x - 15$ are both negative when $x < -\tfrac{5}{2}$ and both are positive when $x > 3$. Hence the solution of the inequality

$$2x^2 - x - 15 > 0$$

is $\{x|x < -\tfrac{5}{2} \text{ or } x > 3\}$.

EXERCISE 10-2

Solve the following inequalities.

1. $4x > 12$	**2.** $2x - 4 < 6$	**3.** $5x - 1 > 3x + 7$
4. $6x + 3 < x - 9$	**5.** $x + 6 < 4 - 3x$	**6.** $5x - 7 < 3x + 2$
7. $x^2 - 4 < 0$	**8.** $x^2 - 4 \geq 0$	**9.** $x^2 - 25 \leq 0$
10. $x(x - 2) < 0$	**11.** $x^2 - 4x + 3 < 0$	**12.** $x^2 - 3x - 10 < 0$
13. $2x^2 - x - 10 < 0$	**14.** $3x^2 - 4x - 4 \leq 0$	**15.** $2x^2 - 9x + 7 < 0$
16. $\dfrac{x - 1}{x} < 0$	**17.** $\dfrac{x}{x - 1} > 0$	**18.** $\dfrac{x - 4}{x + 4} < 0$
19. $\dfrac{x - 4}{x + 4} > 0$	**20.** $\dfrac{x - 1}{x} < 1$	**21.** $\dfrac{x - 1}{x} > 1$

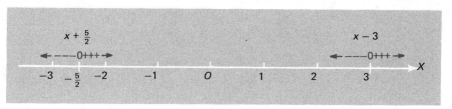

Fig. 10-1

10-3 ABSOLUTE INEQUALITIES

In this section we shall deal with absolute inequalities. To establish an inequality of this type, we need to show that it is true for all permissible values of the variables which appear in the inequality. The examples illustrate the process.

EXAMPLE 1. Show that the inequality $2x^2 - 4x + 3 > 0$ is an absolute inequality.

Solution. We start with the given inequality and reduce it to a form which clearly holds for all real values of x. Thus,

$$2x^2 - 4x + 3 > 0$$

$$2(x^2 - 2x) + 3 > 0$$

$$2(x^2 - 2x + 1) + 3 - 2 > 0$$

$$2(x - 1)^2 + 1 > 0$$

The term $2(x - 1)^2$ of the last inequality cannot be negative when x has any real value. Accordingly, the inequality is true for all values of x. But the left member of the last inequality and the left member of the given inequality are identical, and therefore the given inequality, holding for all values of x, is an absolute inequality.

EXAMPLE 2. Prove the inequality $a + (1/a) \geq 2, a > 0$.

Proof. Since $a > 0$, we multiply both members of the inequality by a and leave the sense unchanged. Hence

$$a^2 + 1 \geq 2a$$

$$a^2 - 2a + 1 \geq 0$$

$$(a - 1)^2 \geq 0$$

The last inequality is true since $(a - 1)^2$ is never negative. Then to complete the proof we start with the last inequality and work back to the first. Thus, by obvious operations, we have

$$(a - 1)^2 \geq 0$$

$$a^2 - 2a + 1 \geq 0$$

$$a^2 + 1 \geq 2a$$

$$a + \frac{1}{a} \geq 2$$

EXERCISE 10-3

Prove that each of the following inequalities is an absolute inequality.

1. $x^2 + 2x + 2 > 0$

2. $x^2 - 10x + 26 > 0$

3. $x^2 + 1 \geq 2x$

4. $4x^2 + 1 \geq 4x$

5. $2x^2 - 4x + 3 > 0$

6. $9x^2 - 6x + 2 > 0$

7. $16x^2 + 1 \geq 8x$

8. $3x^2 - 6x + 4 > 0$

If a, b, and c stand for unequal positive numbers, prove the following inequalities.

9. $(a + b)^2 > a^2 + b^2$

10. $a^2 + b^2 > 2ab$

11. $\dfrac{a}{b} + \dfrac{b}{a} > 2$

12. $\dfrac{a + b}{2} > \sqrt{ab}$

13. $\dfrac{a + b}{4a} > \dfrac{b}{a + b}$

14. $\dfrac{a^3 + b^3}{a + b} > ab$

15. $a^2 + b^2 + c^2 < (a + b + c)^2$

16. $a^2 b + ab^2 < a^3 + b^3$

17. $a^3 + 3ab^2 > b^3 + 3a^2 b, \quad a > b$

18. $\dfrac{a^2}{b} + \dfrac{b^2}{a} > a + b$

19. If x and y are any real numbers, show that

$$\text{(a)} \ |x + y| \leq |x| + |y|, \qquad \text{(b)} \ |x - y| \geq |x| - |y|.$$

[Hint: Consider the cases in which x and y are of the same sign and, next, of different signs.]

10-4 SYSTEMS OF INEQUALITIES

In this section we shall solve certain simple systems of inequalities. The solution set of a system of one variable consists of all numbers which satisfy both inequalities. The solution set of a system of two variables consists of the totality of ordered number pairs which satisfy both inequalities.

EXAMPLE 1. Solve the system of inequalities

$$2x - 7 < 5 - x$$

$$11 - 5x < 1$$

Solution. We seek the set of values for x which make each inequality true. The first inequality is true for all values of $x < 4$ and the second for all values of $x > 2$. Hence the solution of the system is given by the intersection of these two sets, which may be expressed by

$$\{x | 2 < x < 4\} \quad \text{or} \quad 2 < x < 4$$

EXAMPLE 2. Solve the inequality $|2x - 3| < 7$.

Solution. Recalling the definition of the absolute value of a number (Sec. 3-1), we see that the given inequality is equivalent to the system

$$2x - 3 < 7$$
$$2x - 3 > -7$$

For convenience in finding the solution, we first express the system in the equivalent extended form

$$-7 < 2x - 3 < 7$$

Now add 3:

$$-4 < 2x < 10$$

Divide by 2:

$$-2 < x < 5$$

Hence the solution of the given inequality consists of the set of numbers between -2 and 5.

EXAMPLE 3. Solve graphically the inequality $2x - 3y - 6 < 0$.

Solution. This is a linear inequality in the variables x and y. Solving for y, we have the equivalent inequality

$$y > \frac{2x - 6}{3}$$

The graph of $y = (2x - 6)/3$ is the line drawn in Fig. 10-2. The coordinates (x_1, y_1) of any point of the line satisfy this equation. But the coordinates of a point (x_1, y) above the line have $y > y_1$ and consequently satisfy the given inequality. Hence the graph of the solution set of the inequality is the half plane above the line.

In a similar way we could show that the graph of the solution set of $y < (2x - 6)/3$ consists of all points below the line in Fig. 10-2.

EXAMPLE 4. Solve the system of inequalities

$$x^2 - 6x + 2y - 1 < 0$$
$$3x + 7y - 7 > 0$$

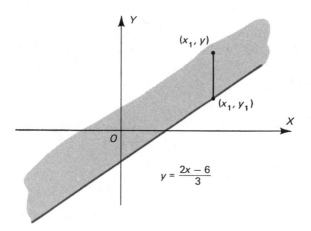

Fig. 10-2

Solution. The graph of the equation $x^2 - 6x + 2y - 1 = 0$ is a parabola and the graph of the equation $3x + 7y - 7 = 0$ is a straight line (Fig. 10-3). By solving each inequality for y, we obtain the equivalent system

$$y < \frac{1 + 6x - x^2}{2}$$

$$y > \frac{7 - 3x}{7}$$

We observe from these inequalities that the coordinates of all points inside, or below, the parabola satisfy the first inequality and the coordinates of all

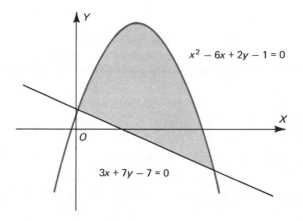

Fig. 10-3

points above the line satisfy the second inequality. The graph of the solution set of the given inequalities is the shaded area in the figure. We write the solution set as

$$\{(x, y)|x^2 - 6x + 2y - 1 < 0\} \cap \{(x, y)|3x + 7y - 7 > 0\}$$

EXERCISE 10-4

Solve each system of inequalities algebraically.

1. $2x + 2 < x + 5$
   $3x + 6 > 2 - x$

2. $8 - 3x > 5x - 10$
   $2x - 15 < 5$

3. $10x + 6 > 2x - 2$
   $4x - 7 < 3 - x$

4. $7x + 50 > 10 - 3x$
   $7x + 4 > 2x - 6$

5. $5x - 6 < x + 2$
   $4x - 7 > 11 - 2x$

6. $6x - 2 > 16$
   $4x - 7 < 1$

7. $|x - 3| < 2$

8. $|3x - 4| \leq 5$

9. $|2x - 9| < 5$

10. $|3 - 2x| \leq 9$

11. $|3 - 5x| \leq 12$

12. $|7 - x| \leq 6$

Solve the following systems of inequalities graphically.

13. $2x - y - 1 > 0$
    $x - 8y + 1 < 0$

14. $x - 2y \leq 4$
    $x + y \geq 1$

15. $2x - y > 2$
    $2x + y > 2$

16. $y + x < 3$
    $y + x > 2$

17. $4x + 3y - 6 > 0$
    $4x + 3y + 1 > 0$

18. $x - 3y \leq 0$
    $x + 3y \leq 0$

19. $3x + 4y + 14 < 0$
    $2x - 5y - 6 > 0$

20. $x + y > 4$
    $x + y < 2$

21. $2x - 4y - 5 > 0$
    $x - 2y + 3 < 0$

22. $x^2 - 4y < 0$
    $y < 4$

23. $x^2 + y^2 < 16$
    $y - 3x + 9 < 0$

24. $y^2 - 4x < 0$
    $x - y < 0$

25. $x^2 + y^2 < 16$
    $x^2 - 4y < 0$

26. $x^2 + y^2 < 16$
    $x^2 - y > 0$

Show that if $a > 0$, then all points inside the parabola determined by the equation $y = ax^2 + bx + c$ satisfy the inequality $y > ax^2 + bx + c$ and all points outside satisfy $y < ax^2 + bx + c$. But if $a < 0$, all points inside the parabola satisfy $y < ax^2 + bx + c$ and all points outside satisfy $y > ax^2 + bx + c$. [Hint: Consider any point (x_1, y_1) on the parabola and the point (y_2, x_1) where $y_2 > y_1$ and also $y_2 < y_1$.]

Solve the following systems of inequality graphically.

27. $y > x^2 - 4$
    $y < -x^2 + 4$

28. $y > x^2 - 4x$
    $y < -x^2 + 4x$

29. $y > x^2$
    $y < -x^2 + 4$

30. $y < -x^2$
    $y > x^2 - 2$

31. $y > x^2$
    $y > 2x^2$

32. $y < -x^2$
    $y < -3x^2$

33. $x^2 + y^2 < 9$
    $x^2 + y^2 > 4$

34. $(3x + 2y - 6)^2 < 0$ **35.** $(2x - 4y + 3)^2 \geq 0$

36. $x > 0$
    $y > 0$
    $y < 4 - 4$

37. $y < x$
    $y > 0$
    $y < -4x + 8$

38. $y > \frac{1}{2}x$
    $y < 3x$
    $y < -2x + 10$

Chapter 11

PROGRESSIONS

11-1 SEQUENCES

A succession of numbers of which one number is designated as the first, another as the second, another as the third, and so on, is called a *sequence*. Each number of the sequence is called a *term*. The set of numbers

$$1, 4, 7, 10, 13, 16$$

is an example of a sequence of six terms.

A sequence which has a fixed number of terms is called a *finite* sequence; otherwise the sequence is an *infinite* sequence. The set of all positive even integers, which we denote by

$$2, 4, 6, 8, \ldots$$

is an infinite sequence.

Since the terms of a sequence have the order first, second, third, and so on, we see that there is a one-to-one correspondence between a set of positive consecutive integers, starting with 1, and the terms of a sequence. Hence the sequence constitutes the range of a function whose domain is the set of integers. For example, pairing the terms of the sequence 1, 3, 5, 7, 9, 11 with the first six positive integers, we form the function

$$\{(1, 1), (2, 3), (3, 5), (4, 7), (5, 9), (6, 11)\}$$

If we let a_1 stand for the first term of a sequence, a_2 for the second term, a_3 for the third term, and so on, we can denote the sequence by

$$a_1, a_2, a_3, \ldots, a_n, \ldots$$

The term a_n is called a *general term* or the *nth term*. Now suppose an expression is given in terms of n by which the members of a sequence can be obtained simply by substituting 1, 2, 3, and so on for n. In such a case we may write as many terms of the sequence as we please. Thus, for example, the equation $a_n = n^2$ defines a function whose values are terms of the sequence

$$1, 4, 9, 16, 25, \ldots, n^2, \ldots$$

157

As another example, let $a_n = n/(n^2 + 1)$. The terms of the sequence thus determined are

$$\tfrac{1}{2}, \tfrac{2}{5}, \tfrac{3}{10}, \tfrac{4}{17}, \tfrac{5}{26}, \cdots, \frac{n}{n^2 + 1}, \cdots$$

The terms of a sequence may be obtained under the following conditions. Let the first term a_1 be given and in addition a formula by which each term of a sequence, after the first, may be computed from the preceding term. We illustrate by choosing $a_n = 3a_{n-1} + 2$ as the formula and letting $a_1 = 3$. The process consists in substituting 2 for n to find the second term, 3 for n to find the third term, and so on. Then for the first four terms, we have

$$a_1 = 3$$
$$a_2 = 3 \cdot 3 + 2 = 11$$
$$a_3 = 3 \cdot 11 + 2 = 35$$
$$a_4 = 3 \cdot 35 + 2 = 107$$

A formula like $a_n = 3a_{n-1} + 2$ of this example is called a *recursion formula*.

EXERCISE 11-1

Write the first four terms of the sequence defined by the given a_n. Find also the 20th term.

1. $a_n = 2n + 1$ **2.** $a_n = \dfrac{1}{n}$ **3.** $a_n = \dfrac{n}{n+1}$

4. $a_n = \dfrac{3n}{n+2}$ **5.** $a_n = \dfrac{1}{n(n+1)}$ **6.** $a_n = 3n - 2$

7. $a_n = \dfrac{2^n}{n^2 + 1}$ **8.** $a_n = \dfrac{(-1)^n}{n}$ **9.** $a_n = (-1)^n 2^n$

Write the first four terms determined by the given conditions in each problem.

10. $a_1 = 1, a_n = a_{n-1} + 2$ **11.** $a_1 = 3, a_n = 2a_{n-1} + 4$

12. $a_1 = 5, a_n = \dfrac{a_{n-1}}{2}$ **13.** $a_1 = 4, a_n = \dfrac{1}{a_{n-1}}$

14. $a_1 = 2, a_n = \dfrac{(-1)^n a_{n-1}}{2}$ **15.** $a_1 = -3, a_n = \dfrac{a_{n-1}}{n}$

11-2 ARITHMETIC PROGRESSIONS

A great variety of sequences is met in mathematics. However, we shall limit our study to three kinds of sequences which, though comparatively simple, are of especial importance.

Definition 11-1 A sequence in which each term after the first is formed by adding a fixed number to the preceding term is called an *arithmetic progression*. The fixed number is called the *common difference*.

EXAMPLE 1. The sequence 2, 5, 8, 11, 14 is an arithmetic progression with common difference 3.

EXAMPLE 2. The sequence 10, 7, 4, 1, -2, -5 is an arithmetic progression with common difference -3.

If a_1 stands for the first term and d for the common difference of an arithmetic progression, we can write the successive terms by repeated addition of d. Thus for n terms, we have

$$a_1, a_1 + d, a_1 + 2d, a_1 + 3d, \ldots, a_1 + (n - 1)d$$

We obtain the last or nth term by observing that the coefficient of d in each term is one less than the corresponding order number of the term. Denoting the nth term by a_n, we have the formula

$$a_n = a_1 + (n - 1)d \tag{1}$$

With a_1 and d given, we may find any desired term of the progression by substituting the proper positive integer for n.

EXAMPLE 1. Find the 30th term of the arithmetic progression 2, 6, 10, ...
Solution. Here $a_1 = 2$, $d = 4$, and $n = 30$. Then using formula (1) for a_n, we find

$$a_{30} = 2 + (30 - 1)4 = 2 + 116 = 118$$

The first and last terms of an arithmetic progression are called *extremes*, and the terms in between are called *arithmetic means*. If we wish to insert a certain number of arithmetic means between two given numbers, we may employ the formula for a_n to find d. Then the terms of the progression can be written by repeated additions of d.

EXAMPLE 2. Insert five arithmetic means between 7 and 25.
Solution. In this case $a_1 = 7$ and $a_n = 25$. The terms of the progression consist of the two given terms plus the five means, giving $n = 7$. Substituting in formula (1), we find

$$25 = 7 + 6d$$

$$d = 3$$

The progression is 7, 10, 13, 16, 19, 22, 25.

We next derive a formula for the sum of the terms of an arithmetic progression. We may write the terms by starting with the first term a_1 and adding d repeatedly or we can start with the last term a_n and subtract d repeatedly.

Then letting S_n stand for the sum of the terms, we have the two equations

$$S_n = a_1 + (a_1 + d) + (a_1 + 2d) + \cdots + (a_n - d) + a_n$$

$$S_n = a_n + (a_n - d) + (a_n - 2d) + \cdots + (a_1 + d) + a_1$$

Adding the members of these equations, we get

$$2S_n = (a_1 + a_n) + (a_1 + a_n) + (a_1 + a_n) + \cdots + (a_1 + a_n) + (a_1 + a_n)$$

Since the progression has n terms, the sum $(a_1 + a_n)$ appears n times, and therefore

$$S_n = \frac{n}{2}(a_1 + a_n)$$

We obtain another formula for S_n by using formula (1) to replace a_n. Thus

$$S_n = \frac{n}{2}[a_1 + a_1 + (n - 1)d]$$

We now have three useful formulas which are applicable to arithmetic progressions. The formulas, rewritten for convenient reference, are

$$a_n = a_1 + (n - 1)d \tag{1}$$

$$S_n = \frac{n}{2}(a_1 + a_n) \tag{2}$$

$$S_n = \frac{n}{2}[2a_1 + (n - 1)d] \tag{3}$$

The five quantities a_1, a_n, d, n, and S_n are called the *elements* of an arithmetic progression. If any three of the elements are given, the remaining two may be found by use of the formulas (1) to (3).

EXAMPLE 3. Given the elements $d = 3$, $a_n = 33$, and $S_n = 195$, find a_1 and a_n.

Solution. Substituting the values of the known elements in formulas (1) and (2) gives the system

$$33 = a_1 + (n - 1)3$$

$$195 = \frac{n}{2}(a_1 + 33),$$

which reduces to

$$a_1 + 3n = 36$$

$$n(a_1 + 33) = 390$$

By solving the first equation for a_1 and substituting the result in the second equation, we get the quadratic equation $n^2 - 23n + 130 = 0$. The roots of this equation are $n = 10$, $n = 13$, and the corresponding values of a_1 are 6 and -3. Hence the two solutions are $n = 10$, $a_1 = 6$ and $n = 13$, $a_1 = -3$.

When a_1 and n are the unknowns, as in this example, and when a_n and n are the unknowns, formulas (1) and (2) give a linear equation and a second-degree equation in the unknowns. This means there may be two solutions. A value of n other than a positive integer, however, must be rejected.

EXERCISE 11-2

Find the last term a_n and the sum S_n of each arithmetic progression.

1. $2, 5, 8, \ldots$ to 8 terms

2. $3, 5, 7, \ldots$ to 21 terms

3. $1.4, 4.5, 7.6, \ldots$ to 51 terms

4. $-11, -7, -3, \ldots$ to 23 terms

5. $10, -2, -14, \ldots$ to 17 terms

6. $x - 2, 4x, 7x + 2, \ldots$ to 12 terms

7. $2x - y, 3x, 4x + y, \ldots$ to 11 terms

8. $y, x, 2x - y, \ldots$ to 10 terms

9. Find the sum of the even integers between 21 and 81.

10. Find the sum of the odd integers between 2 and 100.

11. Find the sum of the integers between 2 and 100 which are divisible by 3.

12. Find the sum of the integers between 50 and 200 which are divisible by 5.

13. Prove that the sum of the first n even positive integers is $n(n + 1)$.

14. Prove that the sum of the first n odd positive integers is n^2.

15. Find the sum of the first n positive integers that are divisible by 4.

16. Insert four arithmetic means between 3 and 8.

17. Insert five arithmetic means between -9 and 9.

18. Insert six arithmetic means between 34 and -43.

Find the two missing elements in each problem.

19. $a_1 = 6, a_n = 33, n = 10$

20. $a_1 = 36, a_n = 8, d = -2$

21. $a_n = 62, d = 6, n = 18$

22. $d = 5, n = 6, S_n = 51$

23. $a_n = 49, n = 14, S_n = 231$

24. $a_n = 6, d = -2, S_n = 50$

25. $d = 4, a_n = 28, S_n = 108$

26. $d = -2, a_n = 8, S_n = 44$

27. $d = -3, n = 10, S_n = -55$

28. $a_1 = -9, d = -2, S_n = -48$

29. A body dropped from a height falls 16 feet during the first second, 48 feet the second second, 80 feet the third second, and so on. Find how far the body falls during the 7th second and how far it falls during the first 7 seconds.

30. An object is dropped from a height of 1600 feet. Find the time required for the object to reach the earth. [See Problem 29.]

31. A man puts $300 into a savings account at the end of each year for fifteen years. If his money draws 3% simple interest, how much will he have in the account at the end of fifteen years?

32. A man buys a car priced at $3300. He agrees to pay each month, beginning one month after the purchase date, $100 on the debt plus an interest charge of 1% of the debt outstanding during the month. Find the total amount of interest which the buyer will pay.

33. A man borrows $16,800. He agrees to settle the debt by paying $280 on the principal at the end of each three months plus an interest charge on the indebtedness during the three-month period. If the interest payments are $210, $206.50, $203, and so on, find the total amount of interest which will be paid in settling the debt.

11-3 GEOMETRIC PROGRESSIONS

As we pointed out in Sec. 11-1, a sequence is a set of functional values whose domain is the set of positive integers or a finite (bounded) subset of positive integers, depending on whether the sequence is infinite or finite. We have introduced the arithmetic progressions as a special type of sequences, and now we consider another important special type of sequences.

Definition 11-2 A sequence in which each term after the first is a fixed number times the preceding term is called a *geometric progression*. The fixed number, denoted by r, is called the *common ratio*.

EXAMPLES.

$$2, 4, 8, 16 \qquad \text{common ratio } 2$$

$$1, -3, 9, -27 \qquad \text{common ratio } -3$$

The first term, the common ratio, the number of terms, the nth term, and the sum of n terms constitute the five elements of a geometric progression. We denote the elements, in the order named, by a_1, r, n, a_n, and S_n. The first and last terms of a geometric progression are called the *extremes* and the terms between are called *means*.

We can write the terms of a geometric progression from the elements a_1, r, and n as

$$a_1, a_1 r, a_1 r^2, a_1 r^3, \ldots, a_1 r^{n-1}$$

We obtain the last, or the nth term, by observing that the exponent of r in each term is 1 less than the number of the term. Denoting the nth term by a_n, we have the formula

$$a_n = a_1 r^{n-1} \tag{1}$$

EXAMPLE 1. Find the 6th term of the geometric progression $9, -6, 4, \ldots$.

Solution. In this case we see that the first term multiplied by $-\frac{2}{3}$ gives the second term, and likewise the product of the second term and $-\frac{2}{3}$ gives the third term. Then using $a_1 = 9$, $r = -\frac{2}{3}$, and $n = 6$ in formula (1), we find

$$a_n = 9(-\tfrac{2}{3})^{6-1} = 9(-\tfrac{32}{243}) = -\tfrac{32}{27}$$

EXAMPLE 2. Insert three geometric means between 81 and 16.

Solution. We have $a_1 = 81$, $a_n = 16$, and, counting the two extremes and the three means, $n = 5$. Hence, from formula (1),

$$16 = 81r^4$$

$$r^4 = \tfrac{16}{81}$$

$$r = \tfrac{2}{3} \quad \text{and} \quad -\tfrac{2}{3}$$

Using these possible values of r, we write the two progressions

$$81, 54, 36, 24, 16 \quad \text{and} \quad 81, -54, 36, -24, 16$$

To derive a formula for the sum S_n of n terms of a geometric progression, we write the two equations

$$S_n = a_1 + a_1 r + a_1 r^2 + a_1 r^3 + \cdots + a_1 r^{n-2} + a_1 r^{n-1}$$

$$rS_n = \quad\;\; a_1 r + a_1 r^2 + a_1 r^3 + \cdots + a_1 r^{n-2} + a_1 r^{n-1} + a_1 r^n$$

The second equation is obtained by multiplying each member of the first equation by r and placing the like terms of the right members one under the other. By subtracting the members of the second equation from the corresponding members of the first equation, we get

$$S_n - rS_n = a_1 - a_1 r^n$$

$$(1 - r)S_n = a_1 - a_1 r^n$$

$$S_n = \frac{a_1(1 - r^n)}{1 - r}$$

This formula gives the sum in terms of a_1, r, and n. The sum in terms of a_1, r, and a_n can be obtained by using formula (1). Thus $a_n = a_1 r^{n-1}$ or $ra_n = a_1 r^n$, and therefore

$$S_n = \frac{a_1 - a_1 r^n}{1 - r} = \frac{a_1 - ra_n}{1 - r}$$

We now have the following formulas involving the elements of a geometric progression.

$$a_n = a_1 r^{n-1} \tag{1}$$

$$S_n = \frac{a_1(1 - r^n)}{1 - r} \tag{2}$$

$$S_n = \frac{a_1 - ra_n}{1 - r} \tag{3}$$

With any three elements known, these formulas suffice to yield the remaining elements. In some cases there will be more than one solution.

We remark that formulas (2) and (3) are not applicable when $r = 1$. The sum for this trivial case, however, is otherwise easily determined.

EXAMPLE 3. Compute the sum of the first six terms of the geometric progression $2, -6, 18, \ldots$, using formulas (2) and (3).

Solution. Here $a_1 = 2$, $r = -3$, $n = 6$, and therefore

$$a_6 = 2(-3)^5 = -486$$

By formula (2),

$$S_6 = \frac{2[1 - (-3)^6]}{1 - (-3)} = \frac{2(1 - 729)}{4} = -364$$

and by formula (3),

$$S_6 = \frac{2 + 3(-486)}{1 - (-3)} = \frac{2 - 1458}{4} = -364$$

EXERCISE 11-3

Find the last term a_n and the sum S_n of each geometric progression.

1. $2, 6, 18, \ldots$ to 6 terms
2. $3, 6, 12, \ldots$ to 7 terms
3. $81, 54, 36, \ldots$ to 5 terms
4. $64, -32, 16, \ldots$ to 6 terms
5. $64, -96, 144, \ldots$ to 5 terms
6. $36, 12, 4, \ldots$ to 6 terms

7. Insert three geometric means between 2 and 162.
8. Insert six geometric means between 128 and 1.
9. Insert four geometric means between $-\frac{1}{4}$ and 8.
10. Insert four geometric means between $\frac{8}{9}$ and $\frac{27}{4}$.

Find the two missing elements in each geometric progression.

11. $a_1 = 6, r = 3, a_n = 486$
12. $a_1 = 3, r = -2, a_n = 768$
13. $a_1 = -3, r = -4, a_n = 3072$
14. $a_n = 81, r = 3, n = 6$
15. $a_1 = \frac{3}{2}, a_n = 48, r = 2$
16. $a_1 = \frac{27}{64}, a_n = \frac{1}{8}, n = 4$
17. $r = -\frac{3}{2}, n = 4, S_n = -\frac{52}{5}$
18. $a_1 = 15, r = \frac{2}{5}, S_n = \frac{117}{5}$

19. If at the end of each year the value of a car is $\frac{2}{3}$ of its value at the beginning of the year, find the value of a $3300 car after 4 years.

20. The population of a city is 30,000. Figuring that each 5 years the population will increase by 50% of what it was at the beginning of the 5 years, find what the population will be in 20 years.

21. The number of bacteria in a certain culture doubles every 3 hours. If there are N bacteria to start with, find the number in 24 hours.

22. If the population of a state is 2,000,000 and if the rate of increase is 10% each year, what will the population be at the end of 6 years?

23. If a stroke of a vacuum pump removes 10% of the air from a container, how much of the original air remains after 10 strokes? How many strokes are required to remove 95% of the air?

24. A ball is thrown vertically upward to a height of 100 feet. Each time the ball strikes the ground the rebound is $\frac{1}{2}$ of the previous height. How far has the ball traveled up and down when it strikes the ground the 5th time?

25. A certain ball when dropped from a height rebounds $\frac{2}{3}$ of the distance from which it fell. Find the total distance traveled by the ball from the time it is dropped from a height of 60 feet until it strikes the ground the 5th time.

26. A 10-quart container is filled with water. One quart of water is drained out and replaced with alcohol. After mixing, a quart of the solution is drained out and replaced with alcohol. This process is continued until 5 quarts of alcohol have been put into the container. The solution in the container is then what percent alcohol?

27. A man can have a position at a beginning salary of $4000 per year with an increase of $450 per year each year, or he can begin at the same annual pay and each following year receive 10% more than during the preceding year. Which proposition will yield the better income during the 5th year? Which proposition will yield the better income for the period of 5 years?

28. According to a story, the owner of a horse made the following agreement. For shoeing the horse, which requires 32 nails, he agreed to pay 1 cent for the first nail, 2 cents for the second nail, 4 cents for the third nail, 8 cents for the fourth nail, and so on. Under this plan, what would be the cost for shoeing the horse?

II-4   INFINITE GEOMETRIC PROGRESSIONS

In dividing 1 by 3, we obtain the unending decimal fraction .3333 .... Each digit of this fraction is a term of a geometric progression. This is evident when we express the values of the digits as

$$.3, .03, .003, .0003, \ldots$$

Denoting the sum of this infinite geometric progression by S, we write

$$S = .3 + .03 + .003 + .0003 + \cdots$$

The right member here is called an *infinite geometric progression* or an *infinite geometric series*. The sum of the first n terms of this series, by formula (2) of the preceding section, is

$$S_n = \frac{.3[1 - (.1)^n]}{1 - .1} = \frac{.3[1 - (.1)^n]}{.9}$$

$$= \tfrac{1}{3}[1 - (.1)^n] = \tfrac{1}{3} - \tfrac{1}{3}(.1)^n$$

Clearly, we can make the last term on the right as small as we please by taking n large enough. When $n = 10$, for example, the last term is equal to $\frac{1}{30,000,000,000}$, and S_{10} therefore differs from $\frac{1}{3}$ by this very small number.

Larger values of n would of course make S_n still closer to $\frac{1}{3}$. Because of this fact, we shall say that the "sum" of the infinite geometric series is $\frac{1}{3}$.

Suppose now we consider the infinite geometric series

$$S = a_1 + a_1 r + a_2 r^2 + \cdots + a_1 r^n + \cdots$$

The formula for the sum of n terms is given by

$$S_n = \frac{a_1[1 - r^n]}{1 - r} = \frac{a_1}{1 - r} - \frac{a \cdot r^n}{1 - r}$$

If $|r| < 1$, or $-1 < r < 1$, the last term of this sum can be made arbitrarily close to zero by taking n sufficiently large. Consequently S_n can be made arbitrarily close to $a_1/(1 - r)$. We describe this situation, using the symbol ∞ (infinity), by writing

$$\lim_{n \to \infty} S_n = \frac{a_1}{1 - r}$$

The left side of this equation is read "the limit of S_n as n approaches infinity." And we define the right side to be the sum of the infinite geometric series. Then, letting S stand for the sum of the series, we have the simple formula

$$S = \frac{a_1}{1 - r}, \quad |r| < 1$$

When $r \le -1$ and when $r \ge 1$, the absolute values of the terms either increase from term to term or remain the same. For these values of r, S_n has no limit as n approaches infinity.

If a_1 and r are rational numbers and $|r| < 1$, the sum of the corresponding infinite geometric series is a rational number.

EXAMPLE 1. Find the sum of the infinite geometric progression $\frac{3}{2}$, -1, $\frac{2}{3}$, $-\frac{4}{9}$, ...

Solution. In this progression $a_1 = \frac{3}{2}$ and $r = -\frac{2}{3}$. Hence

$$S = \frac{\frac{3}{2}}{1 - (-\frac{2}{3})} = \frac{\frac{3}{2}}{1 + \frac{2}{3}} = \frac{9}{10}$$

EXAMPLE 2. Express the repeating of decimal $4.767676\cdots$ as a rational number.

Solution. The fraction $0.767676\cdots$ may be written in the form

$$0.76 + 0.0076 + 0.000076 + 0.00000076 + \cdots$$

To find the sum of this series, we note that $a_1 = 0.76$ and $r = 0.01$, and therefore

$$S = \frac{0.76}{1 - 0.01} = \frac{0.76}{0.99} = \frac{76}{99}$$

Hence

$$4.767676\cdots = 4 + \frac{76}{99} = \frac{472}{99}$$

II-5 HARMONIC PROGRESSIONS

A sequence of numbers whose reciprocals form an arithmetic progression is called a *harmonic progression*. According to this definition, the sequences

$$\frac{1}{3}, \frac{1}{7}, \frac{1}{11}, \frac{1}{15}, \frac{1}{19}, \frac{1}{23}$$

and

$$\frac{1}{a_1}, \frac{1}{a_1 + d}, \frac{1}{a_1 + 2d}, \cdots, \frac{1}{a_1 + (n-1)d}$$

are harmonic progressions. We assume that no denominator in the second sequence is equal to zero.

To insert k harmonic means between two given nonzero numbers p and q, we take the reciprocals of the k arithmetic means between $1/p$ and $1/q$.

EXAMPLE. Insert four harmonic means between $\frac{1}{9}$ and $\frac{1}{5}$.

Solution. We insert four arithmetic means between 9 and 5, and obtain the arithmetic progression

$$9, \frac{41}{5}, \frac{37}{5}, \frac{33}{5}, \frac{29}{5}, 5$$

Hence the required harmonic progression is

$$\frac{1}{9}, \frac{5}{41}, \frac{5}{37}, \frac{5}{33}, \frac{5}{29}, \frac{1}{5}$$

EXERCISE 11-4

Find the sum of each infinite geometric series.

1. $1 + \frac{1}{3} + \frac{1}{9} + \cdots$
2. $-\frac{1}{2} + \frac{1}{4} - \frac{1}{8} + \cdots$
3. $25 + 5 + 1 + \cdots$
4. $7 + 2.1 + 0.63 + \cdots$
5. $8 - 1.6 + 0.32 + \cdots$
6. $5 - 3 + \frac{9}{5} - \cdots$
7. $-7 + 3 - \frac{9}{7} + \cdots$

Express each of the following numbers in rational form. Reduce fractions to lowest terms.

8. $0.777\cdots$
9. $0.888\cdots$
10. $0.3434\cdots$
11. $0.2121\cdots$
12. $0.4747\cdots$
13. $3.1212\cdots$
14. $0.123123\cdots$
15. $0.819819\cdots$
16. $2.987987\cdots$
17. $0.142857142857\cdots$
18. $0.076923076923\cdots$

19. Compare the sum of the infinite geometric series $1 + \frac{1}{2} + \frac{1}{4} + \cdots$ with the sum of the first ten terms.

20. Compare the sum of the infinite geometric series $1 - \frac{1}{9} + \frac{1}{81} - \frac{1}{729} + \cdots$ with the sum of the first four terms.

21. A certain ball rebounds $\frac{4}{5}$ of the distance from which it has fallen. Find how far the ball will travel before coming to rest if it is dropped from a height of 120 feet.

22. Find the total distance which the tip of a pendulum will travel if the distance of the first swing is 5 inches and the distance of each succeeding swing is 0.97 of the previous distance.

23. Find the ninth term of the harmonic progression $\frac{1}{2}, \frac{1}{6}, \frac{1}{10}, \ldots$

24. Find the tenth term of the harmonic progression $\frac{1}{4}, \frac{1}{7}, \frac{1}{10}, \ldots$

25. Insert four harmonic means between $\frac{1}{3}$ and $\frac{1}{14}$.

26. Insert five harmonic means between 2 and 8.

27. Insert three harmonic means between $\frac{1}{17}$ and 1.

28. Insert one harmonic mean between a and b.

29. Insert two harmonic means between $1/a$ and $1/b$.

30. An employer engages a man at a beginning salary of $12,000 per year. He offers to increase the salary $400 per year at the end of each year or, alternatively, he will increase the pay $100 for each six months at the end of each half year. Find the total earnings under each proposition at the end of 4 years.

31. The cost new of a certain tractor is $22,000. Suppose the tractor depreciates in value 20% of its original value the first year, 17% of the original value the second year, and 14% of the original value the third year, and so on. Find the value of the tractor at the end of 7 years.

32. A man accepts a position for a period of 5 years. He can choose either an increase in pay of $30 per month at the end of each 6 months or an increase of $67 per month at the end of each year. How much do the increases alone contribute to the income under each of these propositions?

33. The middle term of an arithmetic progression of three terms is called the *arithmetic mean* of the first and third terms. Show that if a, x, b form an arithmetic progression, then $x = (a + b)/2$.

34. The middle term of a geometric progression of three terms is called the *geometric mean* of the first and third terms. Show that if a, x, b form a geometric progression, then $x = \pm\sqrt{ab}$.

35. If a and b are unequal positive numbers, show that their geometric mean is less than their arithmetic mean.

Find the geometric mean of each pair of numbers.

  36. $81, \frac{1}{9}$ **37.** $\frac{1}{4}, 36$ **38.** $-4, -25$ **39.** $x, 2x$

40. Show that the reciprocals of the terms of a geometric progression also form a geometric progression.

41. Insert five harmonic means between 4 and 2.

Chapter 12

PERMUTATIONS, COMBINATIONS, AND PROBABILITY

12-1 FUNDAMENTAL PRINCIPLE

Permutations, combinations, and probability, which we shall define and study in this chapter, have a strong recreational appeal to many people. But aside from this happy interest, they have important practical and theoretical applications. As a first step, we shall examine a principle which is fundamental in the study of these topics.

An illustration will be helpful in understanding the principle. Suppose a governor is to be chosen from three candidates and a lieutenant governor from four other candidates. Let us determine the number of different ways the two positions can be filled. There are three ways of filling the office of governor and with each of these possible ways there are four choices for a lieutenant governor. Hence, in all, there are $3 \cdot 4 = 12$ different ways of making the two choices. In order to check this result we designate the candidates for governor by A, B, and C and the candidates for lieutenant governor by x, y, z, and w. Then the positions can be filled in any of the following 12 ways:

$$A\,w \quad A\,x \quad A\,y \quad A\,z$$

$$B\,w \quad B\,x \quad B\,y \quad B\,z$$

$$C\,w \quad C\,x \quad C\,y \quad C\,z$$

By generalizing the reasoning in this illustration we may make the following statement:

Fundamental Principle If a thing can be done in any of m different ways and then a second thing can be done in any of n different ways, it follows that the total number of different ways in which the first and then the second thing can be done is m times n.

The truth of this principle is easily seen since for each of the m ways of doing the first thing there are n ways of doing the second. It is also equally evident that the principle may be extended to three or more things.

EXAMPLE 1. How many two-digit numbers can be made from the digits 1, 2, 3, 4, 5 if (a) no digit is repeated in a number; (b) repetitions are allowed?

Solution. (a) There are five choices for the tens' digit, and after a choice is made there remain four choices for the units' digit. Hence the total possibilities are $5 \cdot 4 = 20$. The same result, of course, would be obtained by first selecting the units' digit and then the tens' digit.

(b) If repetitions are allowed there are five choices for the tens' digit and five choices for the units' digit, making $5 \cdot 5$, or 25 possible two-digit numbers.

EXAMPLE 2. In how many ways can five people sit in a car, two in the front seat and three in the back seat, if a particular two of the five are to avoid the driver's position?

Solution. Let us designate the five positions by the letters A, B, C, D, E with A indicating the driver's seat. Taking these places in the order named, let us write above each letter the number which tells how many ways it can be filled.

$$3 \quad 4 \quad 3 \quad 2 \quad 1$$

$$A \quad B \quad C \quad D \quad E$$

We have three choices for a driver, four choices for place B, three choices for place C, two choices for place D, and one choice for place E. Hence the total number of seating arrangements is $3 \cdot 4 \cdot 3 \cdot 2 \cdot 1 = 72$.

EXERCISE 12-1

1. How many two-digit numbers can be made with the digits 2, 4, 6, 8 if (a) repetitions are allowed; (b) no digit is to be repeated in a number?

2. How many three-digit numbers can be formed with the digits 1, 2, 3, 4, 5 if (a) repetitions are allowed; (b) no repetitions are allowed?

3. How many three-digit numbers (zero not the first digit) can be formed with digits 0, 1, 2, 3, 4 if (a) repetitions are allowed; (b) no repetitions are allowed?

4. How many four-digit numbers, each less than 5000, can be formed with the digits 1, 2, 4, 6, 8 if (a) repetitions are allowed; (b) no repetitions are allowed?

5. How many committees consisting of 1 freshman, 1 sophomore, and 1 junior can be selected from 40 freshmen, 30 sophomores, and 25 juniors?

6. Five trails lead to a hilltop. In how many ways can one ascend and descend by (a) a different trail; (b) any trail?

7. Five boys are in a room which has 4 doors. In how many different ways can they leave the room?

8. A building has 6 outside doors. In how many ways can one enter and leave (a) by a different door; (b) by any door?

9. How many signals can be made with 5 different kinds of flags placed on a flagpole one under another from the highest? How many signals can be made if only 4 of the flags are used at a time?

10. License plates are to be distinguished by using 2 letters followed by four digits. How many different plates are possible (a) if letters and digits may be repeated; (b) if no repetitions are allowed?

11. In how many ways can 3 people be seated in a room where there are 7 seats?

12. In how many ways can 5 books be arranged on a shelf? How many ways if a certain 2 are not to be separated?

13. Four couples are to eat at a round table with the men and women alternating. If the hostess reserves a place for herself, in how many ways can she assign seats to the others?

14. How many seating arrangements are possible in problem 13 if the host is not to sit by the hostess?

15. A club has 21 members. In how many ways can a president and a vice-president be elected from the group if no member is eligible for both positions?

16. In how many ways can 5 men and 4 women be seated in a row of 9 seats (a) with the men and women alternating; (b) if the women are not to be separated; (c) if the women are to be together and the men together?

17. Towns A and B are connected by 3 roads and towns B and C by 4 roads. In how many ways can one drive from A to B to C and return to A via B if (a) the same road is not to be traveled twice; (b) the same road may be traveled twice?

18. In how many ways can a club of 21 members choose a president, a secretary, and a treasurer if no member may hold more than one of the positions?

19. In how many ways can 9 books, 6 with green covers and 3 with red covers, be arranged on a shelf if (a) the red books are not to be separated; (b) the green books are not to be separated; (c) books of the same color are kept together?

20. How many odd, three-digit numbers (zero not the first digit) can be formed with the digits 0, 1, 2, 3, 4, 5, 6 if (a) no digit is repeated in a number; (b) repetitions are allowed?

12-2 PERMUTATIONS

The Fundamental Principle of the preceding section can be applied effectively in studying permutations, which we define in the following way.

Definition 12-1 An arrangement of objects in a definite order is called a *permutation* of the objects.

The letters a, b, c, for example, have the following possible permutations or arrangements:

$$a\ b\ c \qquad a\ c\ b \qquad b\ a\ c \qquad b\ c\ a \qquad c\ a\ b \qquad c\ b\ a$$

There are six permutations. It is easy to determine the number of permutations, however, without writing a list. To do so, we notice that there are three positions to be filled. Hence there are 3 choices for the first position, then 2 for the second position, and 1 for the third position. According to the Fundamental Principle, there are $3 \cdot 2 \cdot 1$ different orders in which the letters can be written.

Let us now apply this reasoning to the case of n things where r of the things (r a part or all) are used at a time. We may consider this problem as one in which selections are to be made from n things to fill r positions. Hence there are n choices for the first position, $n - 1$ choices for the second position,

$n - 2$ choices for the third position, and so on. We observe that a continuation of this process leaves $n - (r - 1) = n - r + 1$ choices for the rth position. We discover, then, that the number of permutations of n things taken r at a time is given by the product of r integers from n to $n - r + 1$, inclusive. Denoting this product by $P(n, r)$, we write the formula

$$P(n, r) = n(n - 1)(n - 2) \cdots (n - r + 1) \tag{1}$$

To find the number of permutations of n things taken all at a time, we simply let $r = n$ in the formula. We then have the product of all the integers from 1 to n inclusive. The symbol $n!$ (read "factorial n") is used as a shorthand notation for this product. According to this definition of $n!$, n a positive integer, we have the illustrations

$$1! = 1 \qquad 2! = 1 \cdot 2 \qquad 3! = 1 \cdot 2 \cdot 3 \qquad 4! = 1 \cdot 2 \cdot 3 \cdot 4$$

Using the factorial notation, we express the formula for the number of permutations of n things taken all at a time by

$$P(n, n) = n! \tag{1'}$$

The factorial notation permits us to express formula (1) above in a more compact form. The product in the right member of the formula has as factors all the positive integers from n down to $n - r = 1$ inclusive. If we multiply this product by $(n - r)!$ the result is $n!$. Hence multiplying and dividing the right member of formula (1) by $(n - r)!$, we obtain the formula

$$P(n, r) = \frac{n!}{(n - r)!} \tag{2}$$

If we set $r = n$ in this formula, we have $n!/0!$. Then using the customary definition of factorial zero, namely, $0! = 1$, we have $P(n, n) = n!$. Hence both formulas (1) and (2) are applicable when r is any positive integer from 1 to n, inclusive, and we have the following theorem.

Theorem 12-1 The number of permutations of n things taken r at a time is given by

$$P(n, r) = n(n - 1)(n - 2) \cdots (n - r + 1) \tag{1}$$

or by

$$P(n, r) = \frac{n!}{(n - r)!} \tag{2}$$

EXAMPLE. How many permutations can be made from the letters in the word MONDAY if (a) 4 letters are used at a time; (b) all letters are used; (c) all letters are used but the first is a vowel?

Solution. (a) We want to find the number of permutations of 6 things taken 4 at a time. Using either formula (1) or the fundamental principle, we have

$$P(6, 4) = 6 \cdot 5 \cdot 4 \cdot 3 = 360$$

(b) From formula (1'), or returning to the fundamental principle, we have

$$P(6, 6) = 6 \cdot 5 \cdot 4 \cdot 3 \cdot 2 \cdot 1 = 720$$

(c) The first letter in each permutation is to be either "*o*" or "*a*." Therefore we have two choices for the first letter. But after selecting the first letter, we can select the remaining 5 without restriction, and consequently in 5! ways. The total number of ways of selecting the first letter and then the others is $2 \cdot 5! = 240$.

12-3 PERMUTATIONS OF THINGS NOT ALL DIFFERENT

The formulas of Theorem 12-1 were derived on the assumption that the set of *n* things, or objects, includes objects that are all different. The formulas do not apply if some of the objects are alike. The word ALASKA, for example, has six letters. But we cannot make 6! permutations using the six letters at a time because three letters are alike. To aid in finding the number of different permutations, let us assign subscripts to the three A's. Thus writing

$$A_1 \ L \ A_2 \ S \ K \ A_3$$

we have six distinct objects. Having distinguished the A's, we now have 6! possible permutations. Letting L, S, and K remain in fixed positions, we permute the A's in their positions. This can be done in 3! ways as shown.

$$A_1 \ L \ A_2 \ S \ K \ A_3 \qquad A_1 \ L \ A_3 \ S \ K \ A_2 \qquad A_2 \ L \ A_1 \ S \ K \ A_3$$

$$A_2 \ L \ A_3 \ S \ K \ A_1 \qquad A_3 \ L \ A_1 \ S \ K \ A_2 \qquad A_3 \ L \ A_2 \ S \ K \ A_1$$

These six arrangements would be indistinguishable if the subscripts were removed. Thus we can see that any permutation of the six letters can be replaced by 3! permutations by making the A's distinct. Hence the number of permutations *P* of 6 objects with 3 alike when multiplied by 3! gives the number of permutations of 6 different objects. That is,

$$3!P = 6! \quad \text{and} \quad P = \frac{6!}{3!} = 120$$

This reasoning can be applied to the general case of *n* objects of which n_1 are identical. In this case any arrangement of the *n* objects could be replaced by $n_1!$ permutations by making the n_1 objects distinct. Hence we write the following theorem.

Theorem 12-2 The number of permutations P of n things taken all at a time, where n_1 of the objects are alike and the others distinct, is

$$P = \frac{n!}{n_1!}$$

This formula may be extended to the cases where there are two, three, or more groups of like *objects* among the n objects. Thus we write

$$P = \frac{n!}{n_1!n_2!n_3! \cdots}$$

EXAMPLE 1. How many permutations can be made with the 9 letters in the word TENNESSEE?

Solution. There are 2 s's, 2 N's, and 4 E's, and therefore

$$P = \frac{9!}{2!2!4!} = 3780$$

EXAMPLE 2. A man tosses 1 half dollar, 2 quarters, 3 dimes, and 4 nickels among 10 boys. In how many different ways can the boys profit if each is to get a coin?

Solution. If the coins were all different there would be 10! ways in which each boy could get a coin. But since there are groups of one, two, three, and four like coins, we have

$$P = \frac{10!}{2!3!4!} = 12{,}600$$

In our discussion thus far we have supposed that a permutation is formed by placing one object after another in a line. Thus there is a first and last object. If the objects are arranged in a circle, however, there is no first or last object, and the formulas of Theorem 12-1 are not directly applicable. A circular arrangement presents a new aspect because a mere rotation would not change the position of any object relative to the others. Hence a rotation does not constitute a different permutation.

To find the number of ways of arranging n different objects in a circle, we first select a position for one of the objects. Then the others can be placed in their positions in $(n - 1)!$ different ways. Six people, for example, can form a circle in 5! ways. They could, of course, form a line in 6! ways.

EXAMPLE 3. In how many ways can four couples be seated at a round table with the men and women alternating?

Solution. We first think of assigning one, say a woman, to a chair. This fixes 3 particular places for the other women and 4 particular places for the men. The women, therefore, can be assigned to their places in 3! ways and the men in 4! ways. Applying the fundamental principle, we get the result 3!4! = 144.

EXERCISE 12-2

1. Evaluate (a) $P(7, 1)$; (b) $P(7, 3)$; (c) $P(7, 4)$; (d) $P(7, 7)$.
2. Evaluate $P(4, 1) + P(4, 2) + P(4, 3) + P(4, 4)$.
3. Show that (a) $P(5, 4) = P(5, 5)$; (b) $P(6, 5) = P(6, 6)$; (c) $P(n, n - 1) = P(n, n)$.
4. How many ways can the letters of the word TEXAS be arranged if (a) all letters are used; (b) 2 are used at a time; (c) 3 are used at a time?
5. Find the number of permutations, each with 7 letters, which can be made with the letters in (a) WYOMING; (b) KENTUCKY; (c) WASHINGTON.
6. How many three-digit numbers can be made with the digits 1, 2, 3, 4, 5, 6 if no digit is repeated in a number? How many four-digit numbers?
7. Find how many even numbers of three digits each can be made with the digits 1, 2, 3, 4, 6, 7 if no digit is repeated.
8. How many five-digit numbers, each less than 70,000, can be made with the digits 1, 2, 5, 7, 9 if no digit is repeated?
9. Seven songs are to be given in a program. In how many different orders could they be rendered?
10. In how many ways can 6 people be seated in a row of 9 chairs?
11. In how many ways can 8 books be arranged on a shelf if a certain 3 are (a) not to be separated; (b) not to be separated and left in a particular order?
12. Find the number of possible batting orders for a baseball team of 9 players if the pitcher is to bat last.
13. Five different mathematics books, 4 different physics books, and 2 different history books are to be placed on a shelf with the books of each subject together. Find the number of ways in which the books can be placed.
14. How many permutations can be made by using all the letters in (a) OHIO; (b) INDIANA; (c) ASAMAYAMA?
15. Find the number of permutations which can be formed by using all the letters in (a) ALABAMA; (b) ARKANSAS; (c) MISSISSIPPI.
16. How many signals can be made by arranging 9 flags in a line if 4 are red, 3 are white, and 2 are blue?
17. Five copies of a mathematics book, 4 copies of a physics book, and 4 copies of a history book are to be placed on a shelf. If the covers are of different colors for each kind of book, how many noticeably different ways can the books be placed on the shelf?
18. In how many ways is it possible for 10 customers to buy 5 cans of peaches, 3 cans of pineapple, and 2 cans of tomatoes if each customer gets 1 can?
19. Find the number of ways 2 half dollars, 4 quarters, and 6 dimes can be distributed among 12 boys if each gets a coin?
20. Seven children join hands and form a circle. In how many ways could the circle be formed? In how many ways could they make a line?
21. Five boys and five girls form a line with the boys and girls alternating. Find the number of ways of making the line. How many different ways could they form a circle in which the boys and girls alternate?
22. Ten people are to sit at a round table. Find the number of seating arrangements if the host and hostess are to sit opposite each other.

23. In how many ways can 8 persons be seated at a round table if a certain 2 are (a) to sit next to each other; (b) not to sit next to each other?

24. Four couples are to be seated at a round table. Find the number of seating arrangements if each couple is not to be separated.

12-4 COMBINATIONS

A part or all of a set of objects is called a *combination*. A combination, in distinction from a permutation, does not involve the order of selection of the members of the combination or their arrangement. The interest in a committee selected from a group of people is in the individuals rather than any arrangement of the committeemen. When all of a set of objects are taken at a time, there is of course only one combination. But a part of a set of objects obviously can be chosen in more than one way. We shall be interested in learning how to determine the number of combinations in situations of this kind.

The letters a, b, c, d taken 3 at a time have the combinations

$$a\ b\ c \qquad a\ b\ d \qquad a\ c\ d \qquad b\ c\ d$$

Changing the order of one of these combinations does not make a new combination. Thus $a\ b\ c$ and $b\ c\ a$ are two permutations of the same things, or of a single combination.

We next consider the general case of the number of combinations of n things taken r at a time.

Theorem 12-3 If $C(n, r)$ stands for the number of combinations of n things taken r at a time $(r \leq n)$, then

$$C(n, r) = \frac{n(n - 1)(n - 2) \cdots (n - r + 1)}{r!} \tag{1}$$

or, alternatively,

$$C(n, r) = \frac{n!}{r!(n - r)!} \tag{2}$$

Proof. Now each combination of r things can be permuted in $r!$ ways. But if each of the combinations is thus permuted, then all possible permutations of n things taken r at a time are obtained. Hence by Theorem 12-1

$$r!C(n, r) = n(n - 1)(n - 2) \cdots (n - r + 1)$$

and

$$C(n, r) = \frac{P(n, r)}{r!} = \frac{n(n - 1)(n - 2) \cdots (n - r + 1)}{r!}$$

or

$$C(n, r) = \frac{n!}{r!(n - r)!} \quad \blacksquare$$

When r things are taken from a group of n things, $n - r$ things are left. Thus we see that for each combination of r things there is a corresponding combination of $n - r$ things. We conclude, therefore, that the number of combinations of n things taken r at a time is equal to the number of combinations of n things taken $n - r$ at a time. This conclusion can be reached also by replacing r by $n - r$ in formula (2) above. This gives

$$C(n, n - r) = \frac{n!}{[n - (n - r)]!(n - r)!} = \frac{n!}{r!(n - r)!}$$

Hence,

$$C(n, r) = C(n, n - r)$$

EXAMPLE 1. In how many ways can a committee be selected from 18 persons if the committee is to have (a) 3 members; (b) 14 members?

Solution. (a) We wish to find the number of combinations of 18 things taken 3 at a time. Hence, by formula (1),

$$C(18, 3) = \frac{18 \cdot 17 \cdot 16}{3!} = 816$$

(b) We may substitute in formula (1) and evaluate $C(18, 14)$, but it is easier to recall that $C(18, 14) = C(18, 4)$ and write

$$C(18, 4) = \frac{18 \cdot 17 \cdot 16 \cdot 15}{1 \cdot 2 \cdot 3 \cdot 4} = 3060$$

EXAMPLE 2. In how many ways can 7 mathematics teachers be employed from 10 men applicants and 7 lady applicants if (a) 3 are to be men; (b) 3 or 4 are to be men?

Solution. (a) The 3 men can be had in $C(10, 3)$ different ways. The remaining 4 can be had from the lady applicants in $C(7, 4)$ ways. Hence the total number of ways is

$$C(10, 3) \cdot C(7, 4) = \frac{10 \cdot 9 \cdot 8}{1 \cdot 2 \cdot 3} \cdot \frac{7 \cdot 6 \cdot 5 \cdot 4}{1 \cdot 2 \cdot 3 \cdot 4} = 4200$$

(b) To find the number of ways of filling the vacancies with either 3 or 4 men among those employed, we add the number of ways of filling the vacancies with 3 men and 4 ladies to the number of ways of employing 4 men and 3 ladies. That is, we have

$$C(10, 3) \cdot C(7, 4) + C(10, 4) \cdot C(7, 3) = 4200 + 7350 = 11,550$$

1. Evaluate (a) $C(31, 2)$; (b) $C(26, 3)$; (c) $C(24, 4)$; (d) $C(13, 5)$.
2. Taking full advantage of possible cancellations, evaluate $C(16, 4)$ and $C(16, 12)$ by formula (1). Evaluate both by use of formula (2).
3. How many different ways can a tennis team of 4 be chosen from 17 players?
4. Twelve people meet in a room and each shakes hands with all the others. Find the number of handshakes.
5. Nine points, no three of which are on a straight line, are marked on a blackboard. How many lines, each through 2 of the points, can be drawn? How many triangles are determined by the points?
6. Find the number of choices which a student has for doing 6 problems from a set of 11 problems.
7. Seven different coins are tossed simultaneously. In how many ways can 3 heads and 4 tails come up?
8. In how many ways can a committee of 5 be chosen from a club of 12 members if (a) the president must be on the committee; (b) the president is not eligible; (c) a certain two members refuse to serve together on the committee?
9. In how many ways can a court of 9 judges make (a) a 5 to 4 decision; (b) a majority decision?
10. A presiding officer is authorized to appoint a committee of 5 from 4 elders and 8 deacons. How many different committees are possible if 3 elders are to be members?
11. A committee of 7 is to be composed of 4 Republicans and 3 Democrats. Find the number of choices if 7 Republicans and 8 Democrats are available for appointment.
12. A grocery store manager wishes to employ 6 checkers from 4 men applicants and 5 lady applicants. Find the number of ways of employing the 6 if (a) 3 are to be men; (b) 2 or 3 are to be men.
13. How many different groups of 9 baseball players can be chosen from 3 who play only as pitchers, 2 who play only as catchers, and 14 who play in any position other than pitcher or catcher?
14. Find the number of ways of selecting 8 books from 10 different mathematics books and 7 different physics books if (a) 5 are to be mathematics books; (b) 3 or 4 are to be physics books.
15. Find the number of ways of selecting 9 balls from 6 red balls, 5 white balls, and 4 blue balls if each selection consists of 3 balls of each color.
16. A committee of 9 is to be chosen from 10 seniors, 8 juniors, and 7 sophomores. Find the number of choices if the committee is to have 4 seniors, 3 juniors, and 2 sophomores.
17. Find how many sums of money, each consisting of 3 or more coins, can be formed from 6 different kinds of coins.
18. In how many ways can a group of 3 or more be chosen from 8 people?
19. Find the number of ways in which a librarian can check out 8 different kinds of books to two students so that each gets 4 books. Suggestion: Find the number of ways of giving 4 books to the first student.
20. In how many ways can 6 different kinds of books be distributed among three children if each child is to have 2 books?

21. In how many ways can 12 people be separated into two groups of 5 and 7?

22. Find the number of ways in which 11 people can be separated into two groups of 5 and 6 if a certain one of the 11 is to be in the smaller group.

12-5 PROBABILITY IN SINGLE EVENTS

Probability is a fascinating subject and has held the interest of mathematicians during the past two or three centuries. Receiving its first impetus from games of chance, the theory of probability has been highly developed, and now has wide and important applications in the fields of insurance, annuities, and the social sciences. Many people who have made no formal or philosophical study of probability have some understanding of its meaning and applications. As is commonly understood, probability has to do with the likelihood of the occurrence of events. We often hear remarks like "the probability of rain is 30%," "he has a fifty–fifty chance of being elected," and "the probability of a knockout in the first round is small." But statements of this kind, though giving some general evaluation, are far from explicit. The conclusions are drawn without full knowledge or significance of the variables involved.

In our study of probability, we shall definitely spell out the conditions which favor the happening of an event and those which oppose the happening. Although the treatment will be quite elementary, the principles introduced are fundamental in the theory of probability. We begin our study by examining outcomes of some simple experiments.

First, consider the experiment of tossing a coin. There are obviously two possible outcomes—the head H will show or the tail T will show. So the set of possible outcomes is $\{H, T\}$. Next, in the fair toss of a die, with faces numbered 1 to 6, the set of possible outcomes is $\{1, 2, 3, 4, 5, 6\}$. The sets of outcomes in these experiments are called the sample space for the toss of a coin and the sample space for the toss of a die. The idea of a sample space, which we now define generally, is fundamental in the theory of probability.

Definition 12-2 The *sample space* U for an experiment is the set of all possible outcomes of the experiment.

EXAMPLE 1. Suppose that an urn contains a white (W) ball, a red (R) ball, and a green (G) ball, and let the experiment be that of drawing one ball from the urn and then another ball. To determine the sample space, we note that the first ball removed will be W, R, or G. If the first ball is W, the next ball will be either R or G. Hence there are two possible outcomes, if the first ball is W. Similarly, there are two possible outcomes if either of the other colors shows up first. Hence the sample space for this experiment is $\{WR, WG, RW, RG, GW, GR\}$. Each element of this set represents the order in which the two balls are drawn. Thus, for example, WR and RW are distinct elements.

We shall often be interested in certain combinations of the elements of a sample space. For example, we may wish to focus our attention on the elements

1, 2, 3 of the sample space for the toss of a die. Then this combination of elements, according to the following definition, is an event.

Definition 12-3 Any subset of a sample space is called an *event*.

Thus, the subsets for the experiment of tossing a coin, $U = \{H, T\}$, are

$$\emptyset, \{H\}, \{T\}, \{H, T\}$$

As we see here, four events are associated with a sample space of two elements. The subset $\emptyset$, however, having no element, represents an impossible event. Each of the two subsets with a single element is a possible event in the toss of the coin, and the subset with two elements is an event certain to occur.

If A is any subset of a sample space U, we shall denote the probability of the occurrence of event A in a performance of the experiment by $P(A)$. Then we require that $P(A)$ shall conform to the axioms:

Axiom 12-1 $0 \leq P(A) \leq 1$

Axiom 12-2 $P(U) = 1$

Axiom 12-3 $P(A \cup B) = P(A) + P(B)$

provided A and B are disjoint subsets of U.

Axiom 12-1 sets the bounds for the measure of the probability of any event so that the least probability is 0 and the greatest is 1. Hence we assign 0 to the probability of any result which cannot occur and 1 to the probability of an event which is certain to occur. Accordingly, the probability of drawing a red ball from a bag which contains only white balls is 0 and the probability of a white ball is 1. Axiom 12-2 tells us that one of the possible outcomes in a trial of an experiment is certain to occur. Axiom 12-3 states that the probability of the occurrence of either event A or event B for an experiment is equal to the sum of the probabilities of the separate events, provided A and B have no common element. Axiom 12-3 can be extended to include three or more disjoint subsets. Thus for three disjoint subsets the probability of event A or event B or event C is given by the equation

$$P(A \cup B \cup C) = P[(A \cup B) \cup C]$$

$$= P(A \cup B) + P(C)$$

$$= P(A) + P(B) + P(C)$$

We are led, intuitively, to think that in some experiments the elements of a sample space all have the same probability. This situation is described by saying the outcomes are all "equally likely." Thus for the experiment of tossing a die, we have

$$P(U) = P(1) + P(2) + P(3) + P(4) + P(5) + P(6) = 1$$

If the six probabilities are all equally likely, then the probability that a particular face will be up is $\frac{1}{6}$. Similarly, in the toss of a coin the probabilities are $P(H) = \frac{1}{2}$ and $P(T) = \frac{1}{2}$. And, in general, if a sample space has n elements each with the same probability, then the probability of the occurrence of any one of the elements is $1/n$.

Theorem 12-4 Let a sample space U have n elements with each element possessing the same probability, and let A, consisting of k elements, be a subset of U. Then the probability of the occurrence of some one of the elements of A is equal to k/n.

Proof. Denoting the elements of A by $a_1, a_2, a_3, \ldots a_k$, we have

$$P(a_1 \cup a_2 \cup a_3 \cup \cdots \cup a_k) = P(a_1) + P(a_2) + P(a_3) + \cdots P(a_k)$$

$$= \frac{1}{n} + \frac{1}{n} + \frac{1}{n} + \cdots + \frac{1}{n}$$

$$= \frac{k}{n} \quad \blacksquare$$

This theorem states that the probability of an event is equal to the number of possible ways in which the event can occur divided by the number of possible outcomes of the experiment. We remark that k in the theorem may be any integer from 1 to n inclusive.

EXAMPLE 2. Six pennies are tossed simultaneously. What is the probability that (a) all are heads, (b) 3 are heads and 3 are tails?

Solution. (a) There are two ways for a penny to fall, and by the Fundamental Principle (Sec. 12-1) six pennies can fall in $2^6 = 64$ ways. All heads can occur in only 1 way. Therefore the probability of obtaining all heads is $\frac{1}{64}$.

(b) A particular three of the pennies can fall heads and the remaining pennies fall tails in only 1 way. But three pennies can be chosen in $C(6, 3) = 20$ ways. Hence the probability of 3 heads and 3 tails is $\frac{5}{16}$.

EXAMPLE 3. The names of 5 boys and 4 girls are written on cards. If 6 of the cards are drawn at random, what is the probability of getting the names of 3 boys and 3 girls?

Solution. There are $C(5, 3)$ ways of drawing the names of 3 boys and $C(4, 3)$ ways of drawing the names of 3 girls. Hence $C(5, 3) \cdot C(4, 3)$ is the number of ways the event can occur. But the total number of ways of drawing 6 names is $C(9, 6)$. Hence

$$P(3 \text{ boys and } 3 \text{ girls}) = \frac{C(5, 3) \cdot C(4, 3)}{C(9, 6)} = \frac{40}{84} = \frac{10}{21}$$

Theorem 12-5 If A and A' are complementary subsets of a sample space U, then $P(A) = 1 - P(A')$.

Proof. Since $U = A \cup A'$ and $A \cap A' = \varnothing$, we have

$$P(A) + P(A') = 1$$
$$P(A) = 1 - P(A') \quad \blacksquare$$

We note that $P(A)$ is the probability that event A will occur and $P(A')$ is the probability the event will not occur.

EXAMPLE 4. A pair of dice is tossed. Find the probability that the sum of the numbers coming up is less than 11.

Solution. The sum of the numbers can be from 2 to 12 inclusive. To find the probabilities of getting the sums 2 to 10 inclusive requires several calculations. So we attack this problem by first finding the probabilities of getting 11 or 12. Then for this approach

$$A' = \{(5, 6), (6, 5), (6, 6)\}$$

which means there are two ways for 11 and one way for 12. In this experiment $U = 6 \cdot 6 = 36$, and consequently

$$P(A') = \tfrac{3}{36} = \tfrac{1}{12} \quad \text{and} \quad P(A) = 1 - \tfrac{1}{12} = \tfrac{11}{12}$$

EXERCISE 12-4

1. Eight white balls and 2 black balls are in a box. What is the probability that in drawing 1 ball it is (a) black, (b) white?
2. If a single die is cast, find the probability that it falls (a) a one or two, (b) a four, five, or six.
3. Two pennies are tossed simultaneously. Find the probability that (a) both fall heads, (b) one falls heads and the other tails.
4. In a throw of 2 dice, find the probability that (a) both are aces, (b) the sum of the spots is 7, (c) just one is an ace.
5. Five pennies are tossed simultaneously. Find the probability that (a) all are heads, (b) 4 are heads and 1 is tails, (c) 2 are heads and 3 are tails.
6. In an examination, the questions and answers are given but are not paired. A student succeeds in pairing all but 5 questions and 5 answers. These he does wholly by guessing. What is the probability that he pairs (a) all of them correctly, (b) only 3 correctly?
7. A baby in playing with the wooden digits 1, 2, 3, 4, 5 places them in a row. What is the probability that the number thus formed is (a) less than 20,000, (b) more than 40,000?
8. At a party, 8 men draw for partners from 8 cards which have the names of their wives. What is the probability that each man will draw the name of his wife?
9. Three balls are drawn from a bag containing 6 red balls and 5 black balls. Find the probability that (a) all are red, (b) all are black, (c) 2 are red and 1 is black.
10. Five coins are tossed simultaneously. Find the probability that (a) at least 1 head is up, (b) at least 2 heads are up.

11. A committee of 5 is chosen by lot from a group of 5 seniors and 4 juniors. Find the probability that the committee will have (a) 3 seniors and 2 juniors, (b) at least 2 seniors, (c) at least 1 junior.

12. Three balls are drawn from a bag containing 6 green balls, 7 black balls, and 8 yellow balls. Find the probability that they are (a) of different colors, (b) all of the same color.

13. If two cards are drawn from a deck of 52 cards, what is the probability that both are (a) queens, (b) hearts, (c) of the same suit?

14. Three cards are drawn from a deck of 52 cards. Find the probability that they are (a) hearts, (b) of the same suit, (c) of different suits.

15. Eight children, 4 boys and 4 girls, draw for places to form a line. Find the probability that (a) the boys will be together in the line, (b) a particular couple will be together.

16. Ten people, 5 boys and 5 girls, are grouped into pairs as partners. If the pairs are formed by drawing, find the probability that each boy has a girl for his partner.

17. A party of 5 men and 5 women are to eat at a round table. If the hostess reserves a place for herself and the others take places at random, find the probability that the men and women alternate.

18. Eight books, 5 red and 3 green, are taken from a table one after the other at random. Find the probability that (a) all the red books are taken first, (b) all the books of one color are taken first.

12-6 PROBABILITY IN MULTIPLE EVENTS

In the preceding section we employed Theorem 12-4 to find the probability of an event A. In this section we shall consider the probability of the occurrence of two or more events.

Theorem 12-6 If A and B are subsets of a sample space U, then

$$P(A \cup B) = P(A) + P(B) - P(A \cap B)$$

Proof. If A and B are disjoint sets, $A \cap B = \varnothing$ and the equation is true by Axiom 12-3, Sec. 12-5. We assume then that A and B contain one or more common elements of U. Under this condition the sum $P(A) + P(B)$ includes the probability of the common elements twice. This sum therefore needs to be diminished by $P(A \cap B)$. Hence the probability that at least one of two events occurs is equal to the sum of the probabilities of the separate events minus the probability of the occurrence of both events. ∎

EXAMPLE 1. A card is drawn from a deck of 52 cards. Find the probability that the card is a queen or a heart.

Solution. Let A be the set of queens and B the set of hearts. Then $A \cap B = \{$queen of hearts$\}$, and therefore

$$P(A \cup B) = P(A) + P(B) - P(A \cap B)$$

$$= \tfrac{4}{52} + \tfrac{13}{52} - \tfrac{1}{52} = \tfrac{4}{13}$$

Definition 12-4 If the occurrence of an event A does not affect the occurrence of another event B, then event B is said to be *independent* of event A. If, however, the probability of event B is affected by the prior occurrence of event A, then B is said to *depend* on A.

If an event B depends on event A, the probability of B after the occurrence of A is denoted by $P(B|A)$. If B is independent of A, then

$$P(B|A) = P(B)$$

EXAMPLE 2. Suppose a bag contains 4 green balls and 3 red balls, and that A represents the event of drawing a red ball from the bag and B the event of drawing a green ball after the occurrence of event A. After a red ball is drawn, the bag has 4 green balls and only 2 red balls. Hence the probability that the second ball drawn is green is

$$P(B|A) = \tfrac{4}{6} = \tfrac{2}{3}$$

If, however, the red ball is replaced before the second drawing, then event B is independent of event A. So in this case B represents the event of drawing a green ball from a bag containing 4 green balls and 3 red balls, and consequently

$$P(B|A) = P(B) = \tfrac{4}{7}$$

Theorem 12-7 If A and B are events in a sample space, then

$$P(A \text{ and } B) = P(A \cap B) = P(A) \cdot P(B|A)$$

Proof. If n_1 is the number of ways A can occur and N_1 is the number of ways A can occur or fail to occur, then $P(A) = n_1/N_1$. Following the occurrence of A, let n_2 be the number of ways B can occur and N_2 the number of ways B can occur or fail to occur. Then $P(B|A) = n_2/N_2$. By the Fundamental Principle (Sec. 12-1), both events can occur $n_1 n_2$ ways out of a total of $N_1 N_2$ ways of occurring or failing to occur. Therefore

$$P(A \cap B) = \frac{n_1 n_2}{N_1 N_2} = \frac{n_1}{N_1} \cdot \frac{n_2}{N_2} = P(A) \cdot P(B|A) \quad \blacksquare$$

We note that if event B is independent of event A, then $P(B|A) = P(B)$, and therefore the formula becomes

$$P(A \cap B) = P(A) \cdot P(B)$$

EXAMPLE 3. A card is drawn from a deck of 52 cards and then a second card is drawn. Find the probability that the first card is a spade and the second a club if (a) the first card is replaced before the second card is drawn, (b) the first card is not replaced.

Solution. (a) The probability of drawing a spade the first time is $\tfrac{1}{4}$, and the probability of a club on the second drawing is also $\tfrac{1}{4}$. Hence the probability of the occurrence of both events is $\tfrac{1}{16}$.

(b) After the event of drawing a spade, the deck has 51 cards, 13 of which are clubs. Hence the probability of a spade S and then a club C is

$$P(S \cap C) = P(S) \cdot P(C|S) = \tfrac{1}{4} \cdot \tfrac{13}{51} = \tfrac{13}{204}$$

Theorem 12-7 can be extended to include three or more events. Thus letting $P(B|A)$ stand for the probability of event B after the occurrence of A and $P(C|A \cap B)$ stand for the probability of the occurrence of event C after the occurrence of A and then B, we have

$$P(A \cap B \cap C) = P(A) \cdot P(B|A) \cdot P(C|A \cap B)$$

If the events are independent, the formula for the probability of A, B, and C becomes

$$P(A \cap B \cap C) = P(A) \cdot P(B) \cdot P(C)$$

EXERCISE 12-5

1. Events A and B of a sample space U are

$$A = \{3, 4, 5\}, \qquad B = \{1, 3, 4, 6\}, \qquad U = \{1, 2, 3, 4, 5, 6\}$$

Find $P(A \cup B)$ by (a) using Theorem 12-6, (b) using the fact that $A \cup B = \{1, 3, 4, 5, 6\}$.
2. A ball is drawn from a box containing 4 red balls and 5 green balls. It is then returned to the box and a second ball is drawn. Find the probability that (a) both balls are red, (b) both are green, (c) the first is red and the second green.
3. Repeat Problem 2 if the first ball is not returned.
4. A coin is tossed three times in succession, find the probability that it falls (a) heads each time, (b) heads first and tails on the other throws.
5. One card is drawn from each of three decks of 52 cards. What is the probability that (a) all are aces, (b) all are face cards, (c) all are spades?
6. Three cards are drawn from a deck one at a time and are not replaced. What is the probability that (a) all are aces, (b) all are face cards, (c) all are spades?
7. If the probability that a brush salesman will make a sale at one house is $\tfrac{2}{3}$ and the probability that he will make a sale at a second house is $\tfrac{1}{2}$, find the probability that he will make (a) both sales, (b) neither sale, (c) a sale at the first house but not at the second house.
8. A store entrance has 4 doors. What is the probability that the next three customers, coming singly, will enter by (a) the same door, (b) different doors?
9. The names of 9 freshmen, 8 sophomores, and 7 juniors are written on cards and placed in a box. If three cards are drawn from the box, one after the other, find the probability of getting the names of (a) 3 freshmen, (b) one from each class, (c) all from one class.
10. A student is to take 3 tests. If the separate probabilities of his passing the tests are $\tfrac{4}{5}$, $\tfrac{3}{4}$, and $\tfrac{2}{3}$, find the probability that he will pass (a) all tests, (b) none of the tests, (c) exactly two of the tests.
11. To be initiated into an organization a candidate must pass at least two of three tests. If the separate probabilities of his passing the tests are $\tfrac{2}{3}$, $\tfrac{1}{4}$, and $\tfrac{4}{5}$, find the probability that he will win the right to be initiated.

12. Three balls are drawn from a box one at a time and are not replaced. If the box had 5 white balls and 7 green balls, find the probability that the 3 balls are drawn in the order (a) white, green, green, (b) white, white, green, (c) green, green, green.

13. The names of 6 girls and 5 boys are drawn, one at a time, from a box. Find the probability that (a) the girls' names are drawn first, (b) the names of the girls and boys alternate in the drawing.

14. Three balls are drawn from a box one at a time and are not replaced. If the box had 5 red balls, 6 white balls, and 7 blue balls, find the probability that the three balls drawn are (a) of the same color, (b) of different colors.

15. To determine which will do a task, three men match coins on the basis that the odd man is to do the task. Find the probability that the decision will be made on (a) the first trial, (b) the second trial.

16. Repeat Problem 15 if the odd man is relieved of the task and the others then match to determine the issue.

17. An urn contains 5 black balls and 2 green balls and another urn contains 3 black balls and 3 green balls. If one ball is to be drawn, what is the probability of drawing (a) a black ball, (b) a green ball?

18. A, B, and C are to decide on one of them to do·a task. A and B first draw lots, and the loser then draws with C. Find each man's probability of having to do the task.

If the probability that a person will receive a sum of A dollars is p, then pA is called the *value of his expectation.*

19. A man has three entries in a horse race. He figures that the separate probabilities for his horses to be first are $\frac{1}{3}$, $\frac{1}{4}$, and $\frac{1}{6}$. On this basis, find the probability that one of his horses wins. If there is a prize of $500 for the winner, what is the value of the man's expectation?

20. The separate probabilities that marksmen A, B, and C will hit a target are $\frac{2}{3}$, $\frac{3}{4}$, and $\frac{4}{5}$. Shooting in the order A, B, and C, the first to hit the target gets a prize of $100, but not more than a total of 3 shots are allowed. Find the value of each man's mathematical expectation.

21. A person draws 3 balls in succession from a box containing 5 red balls, 6 yellow balls, and 7 green balls. He is to receive a prize of $100 if the balls are drawn in the order red, yellow, and green. Find the value of his expectation if (a) each ball is replaced before the next is drawn, (b) if the balls are not replaced.

22. Find the value of the man's expectation in Problem 21 if a second chance is granted in case of failure in the first attempt.

12-7 PROBABILITIES IN REPEATED TRIALS

Let us consider a common type experiment in which the outcome of any trial is independent of the outcome of any other trial. Then let us find the probability that an event A will occur 0, 1, 2, 3, and so on, times in n trials.

We let p be the probability that A will occur in a single trial and q be the probability that A fails to occur. Clearly the probability that A fails to occur on all of the n trials is q^n. The probability of the occurrence of A on a specified trial and failures on all the other trials is pq^{n-1}. But the occurrence could happen on any of the n trials, and therefore in $C(n, 1)$ ways. So the probability for exactly one occurrence of A is $C(n, 1)pq^{n-1}$. Similarly, the probability that A occurs on r specified trials and fails to occur on the remaining $n - r$ trials is $p^r q^{n-r}$. Since the r occurrences of A could come in $C(n, r)$ ways, we conclude that the probability that A occurs exactly r times in n trials is $C(n, r)p^r q^{n-r}$. We state this result as a theorem.

Theorem 12-8 If p is the probability that an event will occur in a single trial of an experiment and q is the probability the event will fail to occur, then the probability that the event will occur exactly r times in n trials is

$$C(n, r)p^r q^{n-r}$$

EXAMPLE. In 4 throws of a die, what is the probability of (a) exactly 2 aces, (b) 2 or more aces?

Solution. (a) the probability of one ace on a single throw is $\frac{1}{6}$, and of failure $\frac{5}{6}$. Hence the probability of two aces is

$$C(n, r)p^r q^{n-r} = C(4, 2)(\tfrac{1}{6})^2(\tfrac{5}{6})^2 = \tfrac{25}{216}$$

(b) the probability of 2, 3, or 4 aces is equal to the sum of the separate probabilities of those events. Hence we have

$$C(4, 2)(\tfrac{1}{6})^2(\tfrac{5}{6})^2 + C(4, 3)(\tfrac{1}{6})^3(\tfrac{5}{6}) + (\tfrac{1}{6})^4 = 150(\tfrac{1}{6})^4 + 20(\tfrac{1}{6})^4 + (\tfrac{1}{6})^4$$

$$= \frac{171}{6^4} = \frac{171}{1296}$$

EXERCISE 12-6

1. A coin is tossed 4 times. Find the probability of (a) exactly 2 heads, (b) exactly 3 heads, (c) fewer than 2 heads.

2. Find the probability that in 3 throws of a die there will be (a) no six, (b) 1 six, (c) 2 or 3 sixes.

3. A box contains 4 red balls and 3 white balls. If 4 balls are drawn in succession and each is replaced before the next is taken, find the probability that (a) all are white, (b) exactly 2 are white, (c) 3 or more are red.

4. In 6 tosses of a coin, find the probabilities of exactly 0, 1, 2, 3, 4, 5, and 6 heads. What number of heads is most likely?

5. The probability that a certain ball player will get a hit in one time at bat is 0.3. Find the probability that in 4 times at bat he will (a) go hitless, (b) make exactly 1 hit, (c) make more than 1 hit.

6. If the probability that a rifleman will hit a target in a single trial is 0.9, find the probability that in 4 trials he will make (a) no hit, (b) exactly 3 hits, (c) 4 hits.

7. The probability that team A will win a basketball game from team B is 0.4. In a series of 5 games, what number of games is A most likely to win? Is A more likely to win exactly 1 or exactly 3 games?

Chapter 13

THE BINOMIAL THEOREM AND MATHEMATICAL INDUCTION

13-1 THE BINOMIAL EXPANSION

In Chapter 3 we saw how to write the square and cube of a binomial. We wish now to discover how to write any positive integral power of a binomial as a polynomial.

We may show by multiplying that

$$(a + x)^1 = a + x$$

$$(a + x)^2 = a^2 + 2ax + x^2$$

$$(a + x)^3 = a^3 + 3a^2x + 3ax^2 + x^3$$

$$(a + x)^4 = a^4 + 4a^3x + 6a^2x^2 + 4ax^3 + x^4$$

$$(a + x)^5 = a^5 + 5a^4x + 10a^3x^2 + 10a^2x^3 + 5ax^4 + x^5$$

The right members of these identities are the expanded forms of the positive integral powers of $a + x$ from the first through the fifth. In each of the expansions we may observe the following facts:

1. The number of terms in each expansion is one more than the exponent of the binomial.
2. The first term is a with an exponent the same as the exponent of the binomial, and the exponent decreases by 1 from term to term.
3. The exponent of x in the second term is 1, and it increases by 1 from term to term.
4. The sum of the exponents of a and x in any term is equal to the exponent of the binomial.
5. The coefficients of the terms equidistant from the first term and the last term are equal.

6. The coefficient of the second term is the same as the exponent of the binomial. The coefficient of any term farther on may be computed from the previous term by multiplying that term's coefficient by the exponent of a and dividing by 1 more than the exponent of x.

The student should verify item 6 in the foregoing expansions. To check the expansion of $(a + x)^5$, we observe that the second term is $5a^4x$. Then the coefficient of the next term is $(5 \cdot 4) \div 2 = 10$, which gives $10a^3x^2$ for the third term. A continuation of this simple scheme will yield the expansion.

It is natural to ask if items 1 to 6 are true for $(a + x)^n$, where n is any positive integer. We answer the question by proving the so-called *binomial theorem*.

Theorem 13-1   If n is any positive integer, then

$$(a + x)^n = a^n + na^{n-1}x + \frac{n(n-1)}{2!}a^{n-2}x^2 + \frac{n(n-1)(n-2)}{3!}a^{n-3}x^3$$

$$+ \frac{n(n-1)(n-2)(n-3)}{4!}a^{n-4}x^4 + \cdots + x^n \qquad (1)$$

Proof. By definition $(a + x)^n$ means the product of n factors each of which is $a + x$. Thus

$$(a + x)^n = (a + x)(a + x)(a + x) \cdots (a + x) \qquad n \text{ factors}$$

From the distributive axiom we know that the expansion consists of the sum of all possible products formed by taking one term from each of the n factors. Thus choosing a from every factor, we have the product a^n. Clearly, there is only one way of forming this product. If we choose a from all but one of the factors and x from the remaining factor, we obtain the product $a^{n-1}x$. But x can come from any of the n factors. Hence $a^{n-1}x$ occurs $n = C(n, 1)$ times, giving $C(n, 1)a^{n-1}x$ as a term of the expansion. Now, more generally, suppose we choose a from $n - r$ of the factors and x from the remaining r factors. This can be done $C(n, r)$ times, which gives $C(n, r)a^{n-r}x^r$ as a term of the expansion. Using the definition of $C(n, r)$, we may write

$$C(n, r)a^{n-r}x^r = \frac{n(n-1)(n-2) \cdots (n-r+1)}{r!}a^{n-r}x^r \qquad (2)$$

If we give r, successively, the values $1, 2, 3, \ldots, n$, we obtain in order the terms of the expansion following the first term a^n. We note that the expansion thus obtained agrees with the right member of equation (1), and therefore the theorem is proved. ■

The coefficients in (1) are called the *binomial coefficients*. The binomial coefficients, as we have seen, may be expressed in the notation of combinations.

This permits us to write (1) in the more compact form

$$(a + x)^n = a^n + C(n, 1)a^{n-1}x + C(n, 2)a^{n-2}x^2 + \cdots$$

$$+ C(n, r)a^{n-r}x^r + \cdots + C(n, n - 1)ax^{n-1} + C(n, n)x^n \qquad (3)$$

The binomial theorem provides a formula for writing the expansion of any positive integral power of a binomial. Examining formula (1), we see that the binomial coefficients after the second term may be computed in accordance with item 6. And, as we surmised, the coefficients equidistant from the first term and the last term are equal. This conclusion follows from the fact that $C(n, r) = C(n, n - r)$. Hence the coefficients in the first half of an expansion may be used, in reverse order, as the coefficients in the second half, provided the expansion has an even number of terms. If the expansion has an odd number of terms, the coefficients equidistant from the middle term are equal. This property of the binomial coefficients may be used to advantage in expanding binomials.

Examining the right member of equation (3), we see that x, x^2, x^3, and so on, are the respective factors of the 2nd, 3rd, 4th, and so on, terms of the expansion. Accordingly, the rth term of the expansion is given by $C(n, r - 1)a^{n-r+1}x^{r-1}$. Hence we may write

$$r\text{th term} = \frac{n(n - 1)(n - 2)\cdots(n - r + 2)}{(r - 1)!}a^{n-r+1}x^{r-1} \qquad (4)$$

We now have sufficient information for finding any particular term of a binomial expansion without finding the previous terms. The first and second terms, of course, may be written at once, and formula (4) will yield readily any other chosen term of the expansion by assigning to r the appropriate value.

EXAMPLE 1. Expand $(2 + x)^6$ and simplify each term.
Solution. Applying formula (1) with $a = 2$ and $n = 6$, we get

$$(2 + x)^6 = 2^6 + 6(2)^5x + \frac{6 \cdot 5}{2!}(2)^4x^2 + \frac{6 \cdot 5 \cdot 4}{3!}(2)^3x^3 + \frac{6 \cdot 5 \cdot 4 \cdot 3}{4!}(2)^2x^4$$

$$+ \frac{6 \cdot 5 \cdot 4 \cdot 3 \cdot 2}{5!}(2)x^5 + x^6$$

$$= 64 + 192x + 240x^2 + 160x^3 + 60x^4 + 12x^5 + x^6$$

In this example we expressed the binomial coefficient of each term independently of any other term. The expansion may be more conveniently written by computing the coefficients after the second term in accordance with item 6. We illustrate this procedure in the following examples.

EXAMPLE 2. Expand $(a^2 + 3y)^7$ and simplify each term.
Solution.

$$(a^2 + 3y)^7 = (a^2)^7 + 7(a^2)^6(3y) + 21(a^2)^5(3y)^2 + 35(a^2)^4(3y)^3$$
$$+ 35(a^2)^3(3y)^4 + 21(a^2)^2(3y)^5 + 7(a^2)(3y)^6 + (3y)^7$$
$$= a^{14} + 21a^{12}y + 189a^{10}y^2 + 945a^8y^3 + 2835a^6y^4$$
$$+ 5103a^4y^5 + 5103a^2y^6 + 2187y^7$$

EXAMPLE 3. Expand $(a - x)^6$ and simplify each term.
Solution. Noting that $(a - x) = a + (-x)$, we have

$$[a + (-x)]^6 = a^6 + 6a^5(-x) + 15a^4(-x)^2 + 20a^3(-x)^3$$
$$+ 15a^2(-x)^4 + 6a(-x)^5 + (-x)^6$$

Hence

$$(a - x)^6 = a^6 - 6a^5x + 15a^4x^2 - 20a^3x^3 + 15a^2x^4 - 6ax^5 + x^6$$

The signs alternate here, as will be the case where the binomial is the difference of two quantities.

EXAMPLE 4. Write and simplify the 7th term of the expansion of $(2x - y)^{11}$
Solution. Employing formula (4) with $n = 11$ and $r = 7$, we obtain

$$7\text{th term} = \frac{11 \cdot 10 \cdot 9 \cdot 8 \cdot 7 \cdot 6}{6!}(2x)^5(-y)^6$$

$$= 462(2x)^5(-y)^6 = 14{,}784x^5y^6$$

EXAMPLE 5. Find the term involving x^9 in the expansion of $[x^2 + (1/x)]^{12}$.
Solution. Using $n = 12$, we find that the rth term of the expansion, except for the numerical coefficient, is given by

$$(x^2)^{12-r+1}\left(\frac{1}{x}\right)^{r-1} = x^{26-2r} \cdot x^{1-r} = x^{27-3r}$$

Then $27 - 3r$ must be equal to 9, and therefore $r = 6$. Using this value of r and formula (4), we have

$$\frac{12 \cdot 11 \cdot 10 \cdot 9 \cdot 8}{5 \cdot 4 \cdot 3 \cdot 2 \cdot 1}(x^2)^7\left(\frac{1}{x}\right)^5 = 792x^9$$

EXERCISE 13-1

Write the expansion of each binomial and simplify the terms.

1. $(2a + y)^3$ 2. $(3a - 2y)^4$ 3. $(a - x)^5$
4. $(2m - 3n)^5$ 5. $(3m + n^2)^5$ 6. $(1 + 2x)^6$
7. $(a + x)^8$ 8. $(n^2 + 3m)^6$ 9. $(x - y)^8$

10. $\left(\dfrac{x}{2} - \dfrac{2}{x}\right)^6$ **11.** $\left(\dfrac{a}{b} + \dfrac{b}{a}\right)^6$ **12.** $\left(x - \dfrac{1}{x}\right)^7$

13. $\left(\dfrac{x}{3} - \dfrac{y}{2}\right)^5$ **14.** $\left(\dfrac{y}{x} - \dfrac{1}{y}\right)^6$ **15.** $\left(\dfrac{a^2}{4} - \dfrac{2}{a}\right)^6$

Find the specified term in each of the following powers. Do not find the preceding terms.

16. $(a + x)^8$; 5th term   **17.** $(a - x)^7$; 4th term

18. $(m - 3)^7$; 4th term   **19.** $(x - 1)^{11}$; 8th term

20. $(2x - y)^{10}$; 7th term   **21.** $(2a^2 + 1)^{12}$; 9th term

22. $(2x^2 - y^2)^{13}$; 10th term   **23.** $(\tfrac{1}{2}a - \tfrac{1}{3}b)^{10}$; middle term

24. $(a + \tfrac{1}{2}b)^{10}$; middle term   **25.** $\left(a + \dfrac{b}{2}\right)^{12}$; middle term

26. $(x^2 - y^2)^{14}$; middle term   **27.** $(3x^2 + 2y)^8$; middle term

28. $\left(x - \dfrac{1}{x}\right)^8$; term free of x **29.** $\left(x^2 + \dfrac{1}{x}\right)^9$; term free of x

30. $\left(3x - \dfrac{1}{x^2}\right)^9$; term free of x **31.** $\left(x^3 + \dfrac{1}{x}\right)^{13}$; term involving x^3

Use enough terms of the binomial expansion to evaluate each of the following expressions to three decimal places.

32. $(1.01)^7 = (1 + .01)^7$ **33.** $(1.02)^6$ **34.** $(0.99)^5$

35. $(0.98)^6$ **36.** $(1.01)^{10}$ **37.** $(0.98)^8$

38. Prove that the total number of combinations of n things taken successively 1 at a time, 2 at a time, 3 at a time, and so on, to n at a time, is equal to $2^n - 1$. Hint: Expand $(1 + 1)^n$, expressing the binomial coefficients as in formula (3).

The binomial coefficients of the expansions of $(a + x)^n$ from $n = 0$ to $n = 6$ are arranged below in a triangular form known as *Pascal's triangle*. The first and last numbers of each row is 1; and each of the other numbers, starting with the third row, is equal to the sum of the two nearest numbers in the row just above.

$(a + x)^0$						1						
$(a + x)^1$					1		1					
$(a + x)^2$				1		2		1				
$(a + x)^3$			1		3		3		1			
$(a + x)^4$		1		4		6		4		1		
$(a + x)^5$	1		5		10		10		5		1	
$(a + x)^6$	1	6		15		20		15		6		1

39. Form a Pascal triangle of ten rows. Use the numbers of the tenth row to write the expansion of $(a + x)^9$.

13-2 THE BINOMIAL SERIES

Thus far we have considered binomials whose exponents are positive integers. We wish now to let n be a number other than a positive integer. Expanding in the usual way, we have

$$(1 + x)^n = 1 + nx + \frac{n(n - 1)}{2!}x^2 + \cdots$$

$$+ \frac{n(n - 1)(n - 2)\cdots(n - r + 1)}{r!}x^r + \cdots$$

Observing the coefficient of the $(r + 1)$st term here, we see that the factor $(n - r + 1)$ is equal to 1 when $r = n$, provided n is a positive integer, and is equal to 0 when $r = n + 1$. Hence the expansion ends with the $(n + 1)$st term. But no binomial coefficient could have 0 as a factor if n is different from a positive integer. In this case the expansion does not terminate, and is called a *binomial series*. We can, however, write as many terms as we like of such a series. There are mathematical situations in which the first few terms of a binomial series yield useful results, provided values of x are chosen between -1 and 1. We shall not give a proof of this statement. The question is considered in calculus.

EXAMPLE 1. Find the first four terms of the expansion of $(1 + 2x)^{-\frac{1}{2}}$.
Solution. Applying the binomial formula (1), Sec. 13-1, we have

$$(1 + 2x)^{-\frac{1}{2}} = 1 - \tfrac{1}{2}(2x) + \frac{(-\frac{1}{2})(-\frac{3}{2})}{2!}(2x)^2 + \frac{(-\frac{1}{2})(-\frac{3}{2})(-\frac{5}{2})}{3!}(2x)^3 + \cdots$$

$$= 1 - x + \tfrac{3}{2}x^2 - \tfrac{5}{2}x^3 + \cdots$$

EXAMPLE 2. Find approximately $\sqrt[4]{15}$.
Solution. In order to apply the binomial theorem, we write

$$\sqrt[4]{15} = (16 - 1)^{1/4} = [16(1 - \tfrac{1}{16})]^{1/4} = 2(1 - \tfrac{1}{16})^{1/4}$$

Then

$$\left(1 - \frac{1}{16}\right)^{1/4} = 1 + \frac{1}{4}\left(-\frac{1}{16}\right) + \frac{\frac{1}{4}(-\frac{3}{4})}{2!}\left(-\frac{1}{16}\right)^2 - \cdots$$

$$= 1 - \frac{1}{64} + \frac{-3}{8192} - \cdots$$

$$= 1 - 0.0156 - 0.0004 - \cdots$$

$$= 0.9840 \text{ approximately}$$

and $\sqrt[4]{15} = 2(0.9840) = 1.97$ to two decimal places.

Expand to four terms and simplify.

1. $(1 - x)^{-1}$
2. $(1 + y)^{1/2}$
3. $(1 + y)^{-1}$
4. $(1 + x)^{-2}$
5. $(1 - x)^{-3}$
6. $(1 + y)^{3/2}$
7. $(1 - 3x)^{-3/2}$
8. $(1 + 2y)^{2/3}$
9. $(1 + 4x^2)^{-4/3}$
10. $(x + y)^{-1}$
11. $(a + 2x)^{-3}$
12. $(3 + y)^{-4}$
13. $(2 - x)^{1/4}$
14. $(4 - x)^{-5}$
15. $(3 + y)^{-4/3}$

Approximate to two decimal places, using the necessary number of terms of each expansion.

16. $\sqrt{24} = (25 - 1)^{1/2}$
17. $\sqrt[3]{28}$
18. $\sqrt[4]{80}$
19. $\sqrt{0.98}$
20. $26^{-1/2}$
21. $33^{-1/5}$
22. $(1.01)^{-4}$
23. $(0.99)^{-3}$
24. $(1.03)^{-4}$

13-3 MATHEMATICAL INDUCTION

Certain mathematical theorems and formulas can be proved by a process of reasoning called mathematical induction. This method of proof is based on the following axiom.

Axiom 11 *The mathematical induction axiom.* Let A stand for a set of positive integers such that:
1. The integer 1 belongs to A.
2. If k belongs to A, it follows that $k + 1$ belongs to A.
Then the set A consists of all positive integers.

As this axiom indicates, there are two essential parts in the proof of a theorem by mathematical induction. The method may best be explained with illustrative examples.

EXAMPLE 1. Prove that the sum of the first n positive odd integers is n^2. That is, show that

$$1 + 3 + 5 + \cdots + (2n - 1) = n^2$$

Proof. The proof has two separate parts.

Part I. For the formula to be true it must hold in particular for the smallest integer which may be assigned to n. The smallest integer for which the formula would have meaning is $n = 1$. Then checking the formula for $n = 1$ constitutes the first part of the proof. To illustrate the meaning of the formula we substitute also $n = 2$ and $n = 3$.

For $n = 1$, $1 = 1^2$ or $1 = 1$

For $n = 2$, $1 + 3 = 2^2$ or $4 = 4$

For $n = 3$, $1 + 3 + 5 = 3^2$ or $9 = 9$

Thus it is seen that the formula is true for $n = 1$, $n = 2$, and $n = 3$. But this does not prove the formula for other values of n, and further checking would make the proof only for the particular values used. To complete the proof we proceed to the next part.

Part II. This part of the proof consists in showing that if the formula holds for any value of n, say $n = k$, then consequently it also holds for $n = k + 1$. That is, if k^2 is the sum of the first k positive odd integers, then it must be shown that $(k + 1)^2$ is the sum of the first $k + 1$ positive odd integers. Hence, substituting k for n, we *assume* that

$$1 + 3 + 5 + \cdots + (2k - 1) = k^2$$

Now if we add $2k + 1$, the next consecutive odd integer, to both sides of the equation we have

$$1 + 3 + 5 + \cdots + (2k - 1) + (2k + 1) = k^2 + (2k + 1)$$

The left member of this equation is the sum of the first $k + 1$ positive odd integers. The right member, when factored, is $(k + 1)^2$. Hence the given formula is true for $n = k + 1$, provided it is true for $n = k$. ∎

Having established the two parts of the proof, let us notice what conclusions can be drawn. By Part I the formula holds for $n = 1$. Then by applying Part II repeatedly, we see that the formula holds for $n = 2, 3, 4$, and so on. Hence we conclude (Axiom 11) that the formula holds for n any positive integer.

It should be noticed that both parts of a proof by mathematical induction are essential. The failure of either part in a supposed formula means that the formula holds for no value of n or, at most, for only some values of n. For example, the relation

$$1 + 2 + 3 + \cdots + n = 2n - 1$$

holds for $n = 1$ and for $n = 2$. It does not hold for any other value of n; and hence Part II could not be established.

Let us next test the formula

$$5 + 10 + 15 + \cdots + 5n = \tfrac{5}{2}n(n + 1) - 1$$

Assuming that the formula holds for $n = k$, we have

$$5 + 10 + 15 + \cdots + 5k = \tfrac{5}{2}k(k + 1) - 1$$

Add $5k + 5$ to both sides of the equation:

$$5 + 10 + 15 + \cdots + 5k + (5k + 5) = \tfrac{5}{2}k(k + 1) - 1 + (5k + 5)$$

$$= \tfrac{5}{2}(k + 1)(k + 2) - 1$$

The right member of this equation is just what we would get by substituting $n = k + 1$ in the given formula. Thus Part II of the proof is established. But if we try Part I by substituting $n = 1$ in the supposed formula, we have

$$5 \neq \tfrac{5}{2}(1)(2) - 1 = 5 - 1 = 4$$

Part I fails; the relation as given holds for no value of n.

EXAMPLE 2. Prove by mathematical induction that

$$1^2 + 2^2 + 3^2 + \cdots + n^2 = \tfrac{1}{6}n(n + 1)(2n + 1)$$

Proof. We must establish the two parts of the proof.
Part I. If we try the formula for $n = 1$, we have

$$1^2 = \tfrac{1}{6}(1)(1 + 1)(2 + 1) = \tfrac{1}{6}(2)(3) = 1$$

Thus we see that the formula holds for $n = 1$. While this constitutes the demonstration of Part I, a few other small values of n may be tried if one so desires.
Part II. We next assume that the formula is true for $n = k$. That is,

$$1^2 + 2^2 + 3^2 + \cdots + k^2 = \tfrac{1}{6}k(k + 1)(2k + 1)$$

Now add the square of $k + 1$ to both sides of the equation:

$$1^2 + 2^2 + 3^2 + \cdots + k^2 + (k + 1)^2 = \tfrac{1}{6}k(k + 1)(2k + 1) + (k + 1)^2$$

$$= \tfrac{1}{6}(k + 1)(k + 2)(2k + 3)$$

The left member of this equation is the sum of the squares of the first $k + 1$ positive integers. And the expression in the right member is exactly that which the given formula yields for $n = k + 1$. That is,

$$\tfrac{1}{6}n(n + 1)(2n + 1) = \tfrac{1}{6}(k + 1)(k + 1 + 1)[2(k + 1) + 1]$$

$$= \tfrac{1}{6}(k + 1)(k + 2)(2k + 3)$$

Hence, by Axiom 11, we conclude that the given formula is true. ∎

EXERCISE 13-3

Prove that the following formulas are true for all positive integral values of n. Use mathematical induction.

1. $1 + 2 + 3 + \cdots + n = \dfrac{n(n + 1)}{2}$

2. $2 + 4 + 6 + \cdots + 2n = n(n + 1)$

3. $1^2 + 3^2 + 5^2 + \cdots + (2n - 1)^2 = \dfrac{n(2n - 1)(2n + 1)}{3}$

4. $1^3 + 2^3 + 3^3 + \cdots + n^3 = \dfrac{n^2(n + 1)^2}{4}$

5. $2^2 + 4^2 + 6^2 + \cdots + (2n)^2 = \dfrac{2n(n + 1)(2n + 1)}{3}$

6. $1^3 + 3^3 + 5^3 + \cdots + (2n - 1)^3 = n^2(2n^2 - 1)$

7. $\dfrac{1}{1 \cdot 2} + \dfrac{1}{2 \cdot 3} + \dfrac{1}{3 \cdot 4} + \cdots + \dfrac{1}{n(n + 1)} = \dfrac{n}{n + 1}$

8. $\dfrac{1}{1 \cdot 3} + \dfrac{1}{3 \cdot 5} + \dfrac{1}{5 \cdot 7} + \cdots + \dfrac{1}{(2n - 1)(2n + 1)} = \dfrac{n}{2n + 1}$

9. $a + (a + d) + (a + 2d) + \cdots + [a + (n - 1)d] = \dfrac{n[2a + (n - 1)d]}{2}$

10. $a + ar + ar^2 + \cdots + ar^{n-1} = \dfrac{a - ar^n}{1 - r}$

11. Prove that $2^{3n} - 1$ is divisible by 7 if n is any positive integer.
   Proof. (a) When $n = 1$, the expression $2^{3n} - 1 = 7$. (b) Assume that $2^{3k} - 1 = 7x$, when x is some positive integer. Then, multiplying both members of this equation by 8, we obtain

$$2^3 \cdot 2^{3k} - 8 = 56x$$
$$2^{3(k+1)} - 1 = 56x + 7$$
$$= 7(8x + 1)$$

In the following problems, assume that n stands for any positive integer.

12. Prove that $2^{2n} - 1$ is divisible by 3.
13. Prove that $3^n - 1$ is divisible by 2.
14. Prove that $7^n - 1$ is divisible by 6.
15. Prove that $3^{2n} - 1$ is divisible by 8.
16. Prove that $3^{3n} - 1$ is divisible by 26.
17. Prove that $x^n - y^n$ is divisible by $x - y$, where $x - y \neq 0$.
Hint: $x^{k+1} - y^{k+1} = x^{k+1} - xy^k + xy^k - y^{k+1}$
$$= x(x^k - y^k) + y^k(x - y)$$
18. Prove that $x^{2n+1} + y^{2n+1}$ is divisible by $x + y$, where $x + y \neq 0$.

13-4 PROOF OF THE BINOMIAL THEOREM BY MATHEMATICAL INDUCTION

We now apply mathematical induction to obtain a second proof of the binomial theorem. Since the theorem has been tested for $n = 1$ we proceed to the second part of the proof.

Part II. We assume that the formula is true for $n = k$. Then we have

$$(a + x)^k = a^k + ka^{k-1}x + \cdots + \frac{k(k - 1)\cdots(k - r + 2)}{(r - 1)!}a^{k-r+1}x^{r-1}$$

$$+ \frac{k(k - 1)\cdots(k - r + 2)(k - r + 1)}{r!}a^{k-r}x^r + \cdots + kax^{k-1} + x^k \qquad (1)$$

To obtain an expression for $(a + x)^{k+1}$, we multiply both members of equation (1) by $a + x$. This gives

$$(a + x)^{k+1} = a^{k+1} + (k + 1)a^k x + \cdots$$

$$+ \frac{(k + 1)(k)(k - 1)\cdots(k - r + 2)}{r!}a^{k-r+1}x^r + \cdots + x^{k+1}$$

The student should verify this multiplication. It is easy to check the first, second, and last terms. The term containing $a^{k-r+1}x^r$ is obtained by multiplying the rth term of the expansion in (1) by x and the $(r + 1)$st term by a and then adding these two products. Thus the coefficient of $a^{k-r+1}x^r$ is the sum of the coefficients of the rth term and the $(r + 1)$st term of the right member of (1). That is,

$$\frac{k(k - 1)\cdots(k - r + 2)}{(r - 1)!} + \frac{k(k - 1)\cdots(k - r + 2)(k - r + 1)}{r!}$$

$$= \frac{rk(k - 1)\cdots(k - r + 2) + k(k - 1)\cdots(k - r + 2)(k - r + 1)}{r!}$$

$$= \frac{k(k - 1)\cdots(k - r + 2)[r + (k - r + 1)]}{r!}$$

$$= \frac{(k + 1)(k)(k - 1)\cdots(k - r + 2)}{r!}$$

The expression which we have obtained for $(a + x)^{k+1}$ is exactly what the binomial formula yields by substituting $n = k + 1$. Hence for n any positive integer we have

$$(a + x)^n = a^n + na^{n-1}x + \frac{n(n - 1)}{2!}a^{n-2}x^2 + \cdots$$

$$+ \frac{n(n - 1)\cdots(n - r + 2)}{(r - 1)!}a^{n-r+1}x^{r-1} + \cdots + x^n \quad ■$$

REVIEW EXERCISE

Write the expansion of each binomial and simplify the terms (Sec. 13-1).

1. $(3a + 2y)^4$ 2. $(2m + 3n)^5$ 3. $(1 - 2x)^6$

4. $(n^2 - 3m)^6$ 5. $\left(\dfrac{x}{2} + \dfrac{2}{x}\right)^6$ 6. $\left(x + \dfrac{1}{x}\right)^7$

7. $\left(\dfrac{y}{x} + \dfrac{1}{y}\right)^6$

Find the specified term in each of the following powers. Do not find the preceding terms (Sec. 13-1).

8. $(a - x)^8$; 5th term

9. $(m + 3)^7$; 4th term

10. $(2x + y)^{10}$; 7th term

11. $(2x^2 + y^2)^{13}$; 10th term

12. $(a - \frac{1}{2}b)^{10}$; middle term

13. $(x^2 + y^2)^{14}$; middle term

14. $\left(x + \dfrac{1}{x}\right)^8$; term free of x

15. $\left(3x + \dfrac{1}{x^2}\right)^9$; term free of x

16. $\left(x^2 - \dfrac{1}{x^3}\right)^{10}$; term involving x^5

Use enough terms of the binomial expansion to evaluate each of the following expressions to three decimal places (Sec. 13-1).

17. $(1.02)^8$ **18.** $(0.98)^9$ **19.** $(1.01)^8$

Expand each binomial to four terms and simplify the coefficients (Sec. 13-2).

20. $(1 - y)^{1/2}$ **21.** $(1 - x)^{-2}$ **22.** $(1 + y)^{-1/2}$

23. $(1 - 2y)^{2/3}$ **24.** $(x + y)^{-2}$ **25.** $(3 - y)^{-4}$

26. $(4 + x)^{-5}$

Approximate to two decimal places, using the necessary number of terms of each expansion (Sec. 13-2).

27. $\sqrt{26}$ **28.** $\sqrt[4]{82}$ **29.** $24^{-1/2}$

30. $(0.99)^{-4}$ **31.** $(0.97)^{-5}$

Chapter 14

THEORY OF EQUATIONS

14-1 POLYNOMIALS AND POLYNOMIAL EQUATIONS

An expression of the form

$$a_0x^n + a_1x^{n-1} + a_2x^{n-2} + \cdots + a_{n-1}x + a_n$$

where n is a positive integer and $a_0, a_1, a_2, \ldots, a_n$ are any constants with $a_0 \neq 0$, is called a *polynomial*, or *rational integral expression*, of the nth degree in x. Denoting the polynomial by $f(x)$ and writing $f(x) = 0$, we form the corresponding polynomial equation. A value of x which satisfies the equation is called a *root* or *solution* of the equation and also a *zero* of the polynomial.

In this chapter we shall study certain properties of polynomials with particular reference to their zeros or the roots of the corresponding polynomial equations. We have developed methods for solving polynomial equations of the first and second degrees. Finding roots of equations of higher degrees is, in general, less simple. In fact, the problem is much too involved for a complete investigation at this stage. We shall, however, establish certain theorems relating to polynomials and the roots of the corresponding equations.

14-2 THE REMAINDER THEOREM AND THE FACTOR THEOREM

If we divide $f(x) = 2x^3 - 3x^2 + x - 1$ by $x - 2$, we get the quotient $2x^2 + x + 3$ with 5 as the remainder. Now substituting 2 for x in the dividend, we have

$$f(2) = 2(2)^3 - 3(2)^2 + 2 - 1 = 5$$

Thus we obtain the remainder 5 in two ways. This is an illustration of a theorem which, as we shall see, has important consequences.

Theorem 14-1 *The Remainder Theorem.* If a polynomial $f(x)$ is divided by $x - r$ until a remainder free of x is obtained, the resulting remainder is equal to $f(r)$.

Proof. Let $Q(x)$ stand for the quotient when $f(x)$ is divided by $x - r$, and let R stand for the remainder. Then

$$f(x) = (x - r)Q(x) + R$$

$$(\text{dividend} = \text{divisor} \times \text{quotient} + \text{remainder})$$

Since this is an identity, true for all values of x, we may substitute r for x, and obtain

$$f(r) = (r - r)Q(r) + R$$
$$f(r) = 0 \cdot Q(r) + R$$
$$f(r) = R$$

This establishes the remainder theorem. ■

Theorem 14-2 *The Factor Theorem and its Converse.* If r is a root of the polynomial equation $f(x) = 0$, then $x - r$ is a factor of $f(x)$; and conversely, if $x - r$ is a factor of $f(x)$, then r is a root of the equation $f(x) = 0$.

Proof. From the remainder theorem, we have

$$f(x) = (x - r)Q(x) + R = (x - r)Q(x) + f(r)$$

If r is a root of the equation $f(x) = 0$, then $f(r) = 0$. Hence we may present $f(x)$ in the factored form

$$f(x) = (x - r)Q(x)$$

To prove the converse of the theorem we need but to notice that with $x - r$ given as a factor, we may write the identity

$$f(x) = (x - r)Q(x)$$

which is true for all the values of x. Whence, replacing x by r, we have

$$f(r) = (r - r)Q(r) = 0 \cdot Q(r) = 0 \quad ■$$

EXAMPLE 1. Show by the factor theorem that $x^3 + x^2 - 7x - 3$ is divisible by $x + 3$.

Solution. To determine if $x + 3$ is a factor we need only to see if -3 is a root of the equation

$$x^3 + x^2 - 7x - 3 = 0$$

Substituting -3 for x, we obtain

$$(-3)^3 + (-3)^2 - 7(-3) - 3 = -27 + 9 + 21 - 3 = 0$$

The equation has -3 as a root; hence $x + 3$ is a factor of $x^3 + x^2 - 7x - 3$.

EXAMPLE 2. Form a cubic equation $f(x) = 0$ whose roots are $2, -3, \frac{1}{5}$.

Solution. From the factor theorem we know that $x - 2$, $x + 3$, and $x - \frac{1}{5}$ are factors of $f(x)$. Hence we write the equation

$$(x - 2)(x + 3)(x - \tfrac{1}{5}) = 0$$

This equation has the required roots. To avoid fractional coefficients, however, we replace $x - \frac{1}{5}$ by $5x - 1$ and obtain an equivalent equation. That is,

$$(x - 2)(x + 3)(5x - 1) = 0 \quad \text{or} \quad 5x^3 + 4x^2 - 31x + 6 = 0$$

EXAMPLE 3. Determine if $x - y$ and $x + y$ are factors of $x^{2n} - y^{2n}$, where n is a positive integer.

Solution. If y and $-y$ are roots of the equation $f(x) = x^{2n} - y^{2n} = 0$, then, by the factor theorem, $x - y$ and $x + y$ are factors of $x^{2n} - y^{2n}$. We see at once that

$$f(y) = y^{2n} - y^{2n} = 0 \quad \text{and} \quad f(-y) = (-y)^{2n} - y^{2n} = 0,$$

and consequently $x - y$ and $x + y$ are factors of the given expression.

EXERCISE 14-1

Find the remainder when the indicated division is performed. Use both long division and the remainder theorem.

1. $(x^2 - 4x + 7) \div (x - 2)$
2. $(3x^2 - 5x - 3) \div (x + 3)$
3. $(x^3 - 4x^2 - 2x + 5) \div (x - 1)$
4. $(3x^3 - 4x^2 - 5x + 3) \div (x + 1)$
5. $(2x^4 + 5x^3 + x - 7) \div (x - 2)$
6. $(x^4 - 5x^2 + 7x - 2) \div (x + 2)$

By means of the factor theorem, determine if the second expression is a factor of the first.

7. $x^3 - 4x^2 + 3x + 8, x + 1$
8. $2x^3 - 3x^2 + 2x - 8, x - 3$
9. $3x^3 + 5x^2 - 6x + 17, x + 2$
10. $4x^3 - 6x^2 + 2x + 11, x - 1$
11. $x^3 + 2x^2 - 6x - 72, x - 4$
12. $x^3 + 8x^2 + 8x - 32, x + 3$
13. $2x^4 + x^3 + x - \frac{3}{4}, x - \frac{1}{2}$
14. $3x^4 - x^3 - 3x^2 + \frac{1}{3}, x - \frac{1}{3}$
15. $x^4 - 16, x + 2$
16. $x^3 + 27, x - 3$
17. $x^5 - 1, x + 1$
18. $x^5 + 1, x + 1$
19. $x^8 - 1, x + 1$
20. $x^7 - y^7, x + y$
21. $x^9 + y^9, x + y$
22. $x^6 - y^6, x + y$
23. $x^6 - y^6, x + y$
24. $x^n - y^n, x - y, \quad n$ a positive integer
25. $x^n - y^n, x + y, \quad n$ a positive even integer
26. $x^n + y^n, x + y, \quad n$ a positive odd integer

Write equations with integral coefficients which have the following numbers, and no others, as roots.

27. $1, -2, 3$
28. $-1, 2, 4$
29. $-1, -2, -3$
30. $\frac{1}{2}, 1, 2$
31. $\frac{1}{3}, -1, 1$
32. $\frac{1}{3}, -\frac{2}{3}, 1$

14-3 SYNTHETIC DIVISION

A much-used operation in finding the roots of polynomial equations is that of dividing a polynomial $f(x)$ by a linear expression of the form $x - r$. This kind of division can be done rapidly by a process called *synthetic division*.

The process is simply an abbreviated form of the usual long division. An explanation can best be made by an example.

EXAMPLE 1. Divide $2x^3 + x^2 - 18x + 7$ by $x - 2$.

Solution. We first use the long division method.

$$
\begin{array}{r|l}
2x^3 + x^2 - 18x + 7 & x - 2 \quad \text{(DIVISOR)} \\
\underline{2x^3 - 4x^2} & \\
5x^2 - 18x & \underline{2x^2 + 5x - 8} \quad \text{(QUOTIENT)} \\
\underline{5x^2 - 10x} & \\
- 8x + 7 & \\
\underline{- 8x + 16} & \\
- 9 \quad \text{(REMAINDER)} &
\end{array}
$$

In the steps of the division the terms $-18x$ and 7 were rewritten. Now let us leave these terms in the first line and write beneath them the terms to be subtracted. The division then appears as

$$
\begin{array}{r|l}
2x^3 + x^2 - 18x + 7 & x - 2 \\
\underline{2x^3 - 4x^2 - 10x + 16} & \underline{2x^2 + 5x - 8} \\
5x^2 - 8x - 9 &
\end{array}
$$

The third line has in order the coefficients 5, -8, and -9. Placing the first coefficient of the first line before these, we have

$$2, 5, -8, -9$$

These values, in proper order, are the coefficients of the quotient $2x^2 + 5x - 8$ and the remainder -9.

The powers of x are not essential in forming the coefficients of the second and third lines. So we next delete all the x's and bring the coefficient of x^3 in the second line down to the third line. The division then takes the form

$$
\begin{array}{r|l}
2 + 1 - 18 + 7 & -2 \\
\underline{- 4 - 10 + 16} & \\
2 + 5 - 8 - 9 &
\end{array}
$$

The numbers in the third line, after the first, are obtained by subtraction. If we change -2 of the divisor to 2, the signs of the numbers in the second line will be reversed. Then addition, rather than subtraction, completes the numbers of the third line. Making these changes, we finally present the division in the abbreviated form

$$
\begin{array}{r|l}
2 + 1 - 18 + 7 & 2 \\
\underline{+ 4 + 10 - 16} & \\
2 + 5 - 8 - 9 &
\end{array}
$$

The quotient may be written, by supplying the proper powers of x, as

$$2x^3 + 5x - 8$$

with -9 as the remainder.

Following the process above, we may formulate directions for dividing a polynomial $f(x)$ by $x - r$.

1. Arrange the coefficients of $f(x)$ in a line in order of descending powers of x, supplying zero as the coefficient of each missing power of x. At the end of this line write r.
2. Rewrite the first coefficient of this line underneath as the first number of the third line. Multiply this number by r and place the product immediately beneath the second number of the first line, and write the sum of these two second position numbers as the second number of the third line. Continue this process until the second and third lines are completed to the position of the last number of the first line, which is the constant term of $f(x)$.
3. The numbers of the third line give the coefficients of the quotient in order of descending powers of x, the last number being the remainder. The degree of the quotient is, of course, one less than the degree of the dividend.

EXAMPLE 2. Divide $3x^4 - 8x^3 - 7x - 8$ by $x - 3$.

Solution. We place zero in the first line as the coefficient of the missing power of x.

$$
\begin{array}{r}
3 - 8 + 0 - 7 - 8 \ \lfloor 3 \\
+ 9 + 3 + 9 + 6 \\
\hline
3 + 1 + 3 + 2 - 2
\end{array}
$$

The quotient is $3x^3 + x^2 + 3x + 2$ and the remainder is -2.

EXAMPLE 3. Divide $2x^4 + 5x^3 - 8x^2 - 7x - 9$ by $x + 2$.

Solution. The divisor is $x - r = x + 2$ and therefore $r = -2$.

$$
\begin{array}{r}
2 + 5 - \ 8 - \ 7 - \ 9 \ \lfloor -2 \\
- 4 - \ 2 + 20 - 26 \\
\hline
2 + 1 - 10 + 13 - 35
\end{array}
$$

The quotient is $2x^3 + x^2 - 10x + 13$ and the remainder is -35.

EXAMPLE 4. Find the value of $f(x) = 4x^3 - 7x^2 + 13x$ when x is replaced by -3.

Solution. Recalling that when $f(x)$ is divided by $x - r$ the remainder is $f(r)$, we may use synthetic division to find the value of a polynomial corresponding to a given value of x. Hence

$$
\begin{array}{r}
4 - 7 + 13 + 0 \enspace \underline{-3} \\
-12 + 57 - 210 \\
\hline
4 - 19 + 70 - 210
\end{array}
$$

The remainder is -210; and therefore $f(-3) = -210$.

EXERCISE 14-2

Find the quotient and remainder in each problem by the use of synthetic division.

1. $(2x^3 - 7x^2 + 3x + 1) \div (x + 1)$ 2. $(x^3 + 5x^2 - 5x - 6) \div (x - 1)$
3. $(x^3 - x^2 - 9x + 11) \div (x - 2)$ 4. $(3x^3 + 3x^2 + 4x + 21) \div (x + 2)$
5. $(4x^3 + 4x^2 + x + 75) \div (x + 3)$ 6. $(2x^3 - 8x^2 + 6x - 3) \div (x - 3)$
7. $(5x^3 + 8x^2 + 7x - 4) \div (x + 2)$ 8. $(4x^3 - 4x^2 + 3x + 6) \div (x + 2)$
9. $(x^4 - x^3 + 3x + 1) \div (x - 1)$ 10. $(x^4 + x^2 - 5x + 5) \div (x - 4)$
11. $(2x^4 + 7x^3 + x + 11) \div (x + 3)$ 12. $(x^4 - x^3 - 7x^2 + 3) \div (x - 4)$
13. $(x^5 - x^4 - x^2 + 3) \div (x - 2)$ 14. $(x^5 + x^3 + x - 1) \div (x + 2)$
15. $(x^5 + 2x^4 - 5x^3 + 1) \div (x + 4)$ 16. $(x^5 - 1) \div (x - 1)$
17. $(x^5 + 32) \div (x + 2)$ 18. $(x^6 - y^6) \div (x - y)$
19. $(x^7 - y^7) \div (x + y)$ 20. $(x^8 + y^8) \div (x + y)$

Evaluate by synthetic division.

21. If $f(x) = x^3 - 7x^2 + 12x + 2$, find $f(-2), f(3), f(4)$.
22. If $f(x) = x^4 - 7x^3 + 7x - 18$, find $f(2), f(-2), f(-3)$.
23. If $f(x) = 2x^4 + 6x^3 + 7x^2 - x + 5$, find $f(-1), f(-2), f(3)$.
24. If $f(x) = x^3 - 4x^2 - 20x + 50$, find $f(5), f(6), f(7)$.

14-4 THEOREMS CONCERNING ROOTS

Having found roots of certain polynomial equations, particularly equations of the first and second degrees, we might expect that every equation of this kind has a root. In this connection, we state the following theorem.

Theorem 14-3 *The Fundamental Theorem of Algebra.* Every polynomial equation of degree $n \geq 1$ has at least one root, real or imaginary.

This important theorem was first proved in 1797 by a German mathematician named Gauss. The theorem cannot be established by elementary algebraic methods, and consequently we do not include a proof. We shall, however, assume the theorem to be true and from it establish other theorems.

Theorem 14-4 Every polynomial $f(x)$ of degree $n \geq 1$ can be expressed as the product of n linear factors.

Proof. Let $f(x) = a_0 x^n + a_1 x^{n-1} + \cdots a_{n-1} x + a_n$, with $a_0 \neq 0$. From the fundamental theorem, $f(x) = 0$ has at least one root. Denoting this root by

r_1 and recalling the factor theorem, we conclude that $x - r_1$ is a factor of $f(x)$. Hence,

$$f(x) = (x - r_1)Q_1(x)$$

where the factor $Q_1(x)$ is a polynomial of degree $n - 1$, its highest degree term being $a_0 x^{n-1}$. Using the fundamental theorem and the factor theorem again, we have

$$Q_1(x) = (x - r_2)Q_2(x)$$

Whence

$$f(x) = (x - r_1)(x - r_2)Q_2(x)$$

where r_2 is a root of $Q_1(x) = 0$, and $Q_2(x)$ is a polynomial with $a_0 x^{n-2}$ its highest degree term. This process may be continued until n linear factors are obtained, the last factor having the term $a_0 x$. Then factoring the constant a_0 from the last factor we may write it as $a_0(x - r_n)$ and have

$$f(x) = a_0(x - r_1)(x - r_2)\cdots(x - r_n) \quad \blacksquare$$

The numbers $r_1, r_2, \ldots, r_n$ need not all be distinct, or even real. That is, certain factors may be repeated and some, or all, may contain imaginary constants.

Theorem 14-5 Every polynomial equation $f(x) = 0$ of degree n has exactly n roots.

Proof. By the preceding theorem we may factor $f(x)$ and write the equation as

$$a_0(x - r_1)(x - r_2)\cdots(x - r_n) = 0$$

The roots of the equation are $r_1, r_2, \ldots, r_n$ because any one of these values when substituted for x makes a factor equal to zero, and hence the product of the factors is equal to zero. A value of x different from these quantities will obviously make no factor equal to zero. Hence the equation has only the n roots $r_1, r_2, \ldots, r_n$. $\blacksquare$

If no two factors of $f(x)$ are alike, the n roots are all different. Each root is then called a *single*, or *simple*, *root*. If the same factor occurs twice, a root occurs twice. A root which occurs twice is called a *double root*. If a factor occurs three times, there is a *triple root*. In general, if a factor occurs m times, the corresponding root is said to be of *multiplicity m*. It is customary to count a double root as two roots, a triple root as three roots, and a root of multiplicity m as m roots. Counting the roots in this way, we say that a polynomial equation of degree n has exactly n roots.

EXAMPLE 1. The equation $(x - 1)(x + 2)(x - 5) = 0$ is of degree three, and it has the three distinct roots, 1, -2, and 5.

EXAMPLE 2. The equation $(x - 4)^2(x - 3)^4(x + 2) = 0$ is of degree seven. It has 4 as a double root, 3 as a root of multiplicity four, and -2 as a simple root. The total number of roots, giving each a count according to its multiplicity, is seven.

As we have seen, a quadratic equation with real coefficients which has the complex number $a + bi$, $b \neq 0$, as a root, also has $a - bi$ as a root. That is, conjugate complex roots occur in pairs. We now prove a general theorem concerning roots of this kind.

Theorem 14-6 If the complex number $a + bi$, $b \neq 0$, is a root of a polynomial equation with real coefficients, then the complex number $a - bi$ is also a root.

Proof. Designating the equation by $f(x) = 0$, we have $f(a + bi) = 0$ since $(a + bi)$ is a root of the equation. Then $(x - a - bi)$ is a factor of $f(x)$. We shall show that $(x - a + bi)$ is also a factor by showing that the product

$$(x - a - bi)(x - a + bi) = x^2 - 2ax + (a^2 + b^2)$$

is a factor of $f(x)$. Let $D(x)$ stand for this product. Now $f(x)$ can be divided by the quadratic expression $D(x)$ until a remainder of degree lower than two is obtained. Then indicating the quotient by $Q(x)$ and the remainder by $Rx + S$, we may write

$$f(x) = D(x)Q(x) + Rx + S$$

If we substitute $a + bi$ for x in this identity, we get

$$f(a + bi) = D(a + bi)Q(a + bi) + R(a + bi) + S$$

But, by hypothesis, $f(a + bi) = 0$. And $D(a + bi) = 0$ since $(x - a - bi)$ is a factor of $D(x)$. Thus we obtain

$$0 = 0 \cdot Q(a + bi) + R(a + bi) + S$$

or

$$(Ra + S) + Rbi = 0$$

A complex number is equal to zero if, and only if, the real and imaginary components are each equal to zero. Hence we must have

$$Ra + S = 0 \quad \text{and} \quad Rb = 0$$

Since $a + bi$ is an imaginary number, $b \neq 0$; and therefore $R = 0$. Substituting $R = 0$ in $Ra + S = 0$ gives $S = 0$. Hence the remainder $Rx + S$ is equal to zero, and

$$f(x) = D(x)Q(x) = (x - a - bi)(x - a + bi)Q(x)$$

Then $(x - a + bi)$ is a factor of $f(x)$, and consequently $a - bi$ is a root of $f(x) = 0$. ∎

It is quite easy to see that the interchange of $a + bi$ and $a - bi$ in the steps of the preceding proof reveals that $a + bi$ is a root of $f(x) = 0$ if $a - bi$ is a root. We conclude, then, that a polynomial equation with real coefficients which has a complex root has also the conjugate complex number as a root.

If the coefficients of a polynomial equation $f(x) = 0$ are rational numbers, then we have a similar theorem concerning quadratic surd roots. The numbers $a + \sqrt{b}$ and $a - \sqrt{b}$, where a and b are rational and $\sqrt{b}$ (b positive) is irrational, are called conjugate quadratic surds.

Theorem 14-7   If the quadratic surd $a + \sqrt{b}$ (or $a - \sqrt{b}$) is a root of a polynomial equation $f(x) = 0$ with rational coefficients, then the conjugate surd $a - \sqrt{b}$ (or $a + \sqrt{b}$) is also a root.

The proof of this theorem is much like that for conjugate complex roots and is left to the student.

EXAMPLE 1.   Solve the equation $x^4 - 4x^3 + 10x^2 + 12x - 39 = 0$, if $(2 - 3i)$ is one of the roots.

Solution.   Since $(2 - 3i)$ is a root, $(2 + 3i)$ is also a root.   Then the product

$$(x - 2 + 3i)(x - 2 - 3i) = x^2 - 4x + 13$$

is a factor of the left member of the given equation. The other factor, obtained by division, is $x^2 - 3$. The roots of $x^2 - 3 = 0$ are $\pm\sqrt{3}$. Hence the solution of the given equation is

$$x = \pm\sqrt{3}, \quad 2 \pm 3i$$

EXAMPLE 2.   Form an equation with rational coefficients and of the lowest possible degree if two of the roots are $2 - 3\sqrt{5}$ and $4 + i$.

Solution.   In addition to the given roots, $2 + 3\sqrt{5}$ and $4 - i$ must also be roots. Hence the required equation is

$$(x - 2 + 3\sqrt{5})(x - 2 - 3\sqrt{5})(x - 4 - i)(x - 4 + i) = 0$$

or, multiplying,

$$x^4 - 12x^3 + 8x^2 + 260x - 697 = 0$$

14-5   COEFFICIENTS IN TERMS OF ROOTS

If r_1, r_2, and r_3 are the roots of the cubic equation

$$x^3 + a_1 x^2 + a_2 x + a_3 = 0 \tag{1}$$

then the equation may be written in the factored form

$$(x - r_1)(x - r_2)(x - r_3) = 0 \tag{2}$$

By multiplication we get

$$x^3 - (r_1 + r_2 + r_3)x^2 + (r_1r_2 + r_1r_3 + r_2r_3)x - r_1r_2r_3 = 0 \qquad (3)$$

The left members of equations (1) and (3) are identically equal; hence we may equate the coefficients of the like powers of x and obtain

$$a_1 = -(r_1 + r_2 + r_3)$$

$$a_2 = (r_1r_2 + r_1r_3 + r_2r_3)$$

$$a_3 = -(r_1r_2r_3)$$

In the same manner it could be shown that the coefficients $a_1, a_2, \ldots, a_n$ of the equation

$$x^n + a_1x^{n-1} + \cdots + a_{n-1}x + a_n = 0 \qquad (4)$$

may be expressed as follows:

$a_1 = -$(*the sum of the roots*),
$a_2 = $ (*the sum of the products of the roots taken two at a time*),
$a_3 = -$(*the sum of the products of the roots taken three at a time*),
$\vdots$
$a_n = (-1)^n$ (*product of the roots*).

If the coefficient of the highest degree term is a_0, different from 1, then the equation can be put in the form (4) by dividing by a_0.

EXAMPLE. Form a cubic equation which has 2, -3, and 5 as roots.
Solution. Write the equation as $x^3 + a_1x^2 + a_2x + a_3 = 0$. Then

$$a_1 = -(2 - 3 + 5) = -(4) = -4$$

$$a_2 = (2)(-3) + (2)(5) + (-3)(5) = -6 + 10 - 15 = -11$$

$$a_3 = -(2)(-3)(5) = 30$$

Hence we obtain the equation $x^3 - 4x^2 - 11x + 30 = 0$, which has the given numbers as roots.

EXERCISE 14-3

Find the roots of each equation which are not given.

1. $x^3 + 4x^2 + x + 4 = 0$, i is a root
2. $x^3 - 3x^2 + 9x + 13 = 0$, $2 + 3i$ is a root
3. $x^3 - 8x^2 + 6x + 52 = 0$, $5 - i$ is a root
4. $3x^3 - 7x^2 - x + 1 = 0$, $1 - \sqrt{2}$ is a root
5. $x^4 - 4x^3 + x^2 + 28x - 56 = 0$, $\sqrt{7}$ is a root

6. $x^4 - 4x^3 + 28x - 49 = 0$, $\sqrt{7}$ is a root

7. $x^4 - 2x^3 + 2x^2 + 4x - 8 = 0$, $1 + i\sqrt{3}$ is a root

8. $x^4 + 4x^3 + 2x^2 + 4x + 1 = 0$, $-2 + \sqrt{3}$ is a root

9. $x^4 + 6x^3 + 17x^2 + 36x + 66 = 0$, $-3 + i\sqrt{2}$ is a root

Find the equation of lowest possible degree with integral coefficients which has the given numbers among its roots.

10. $1 + i, 2$ **11.** $6 - 3i, 4$ **12.** $i, 2 - i$

13. $5 - 2i, \sqrt{2}$ **14.** $3i, 4 + \sqrt{2}$ **15.** $2 + 4i, 4 - 3i$

16. $\sqrt{7}, 3 - \sqrt{5}$ **17.** $1 - 2i, 1 + 3i$ **18.** $1 + \sqrt{2}, 1 + i\sqrt{2}$

Find the sum and product of the roots of each equation.

19. $x^3 - 3x^2 + x - 5 = 0$ **20.** $x^4 + x^2 - 5x + 1 = 0$

21. $x^4 - 7x^3 + 2x^2 + 3x = 0$ **22.** $2x^3 - 5x^2 + 8x + 10 = 0$

23. $x^6 - x^4 + x^2 + 3 = 0$ **24.** $x^5 + 5x^4 - x - 12 = 0$

Use the method of Sec. 14-5 to find an equation having the given roots.

25. $2, -1, -4$ **26.** $0, -3, 5$ **27.** $-2, -1, 3$

28. $1, i, -i$ **29.** $-1, 0, 1, 2$ **30.** $1, 1 + i, 1 - i$

31. Prove Theorem 14-7.

32. Prove that a cubic equation with real coefficients has either three real roots or one real root and a pair of conjugate imaginary roots.

33. Prove that a fourth-degree equation with real coefficients has four real roots, or four imaginary roots, or two real roots and two imaginary roots.

34. Prove that a polynomial equation of fifth degree has at least one real root, and state the possibilities for the nature of the remaining roots.

35. State the possibilities for the distribution of the real and imaginary roots of a sixth-degree polynomial equation with real coefficients.

14-6 RATIONAL ROOTS

A rational number which is the root of an equation is called a *rational root*. We shall consider a quite simple method of finding the rational roots, if any, of a polynomial equation with rational coefficients.

Theorem 14-8 If a rational number b/c in its lowest terms is a root of the equation

$$a_0 x^n + a_1 x^{n-1} + \cdots + a_{n-1} x + a_n = 0$$

where the coefficients $a_0, a_1, a_2, \ldots, a_n$ are all integers with $a_0 \neq 0$, then b is a factor of a_n and c is a factor of a_0.

Proof. Assuming that b/c is a root, we have

$$a_0\left(\frac{b}{c}\right)^n + a_1\left(\frac{b}{c}\right)^{n-1} + \cdots + a_{n-1}\left(\frac{b}{c}\right) + a_n = 0 \tag{1}$$

Multiplying this equation by c^n/b and transposing the last term gives

$$a_0 b^{n-1} + a_1 b^{n-2}c + a_2 b^{n-3}c^2 + \cdots + a_{n-1}c^{n-1} = -\frac{a_n c^n}{b} \tag{2}$$

The left member of equation (2), being made up of terms whose factors are integers, has an integral value. Hence, the right member

$$-\frac{a_n c^n}{b}$$

likewise yields an integer. By hypothesis b and c have no common integral factor other than ± 1; then it must follow that a_n is divisible by b.

Now returning to equation (1), we multiply by c^{n-1}, transpose the first term, and obtain

$$a_1 b^{n-1} + a_2 b^{n-2}c + a_3 b^{n-3}c^2 + \cdots + a_{n-1}bc^{n-2} + a_n c^{n-1} = -\frac{a_0 b^n}{c}$$

The left member of this equation, having an integral value, means that the fraction in the right member is expressible as an integer. Since b and c have no common integral factor other than ± 1, a_0 is divisible by c. ∎

For the case in which $a_0 = 1$, the rational root b/c is an integer. This is true because c, being a factor of a_0, would be equal to 1 or -1. Hence we have the following corollary.

Corollary *Any rational root of the equation*

$$x^n + a_1 x^{n-1} + a_2 x^{n-2} + \cdots + a_{n-1}x + a_n = 0$$

where the coefficients are integers, is an integer and a factor of a_n.

EXAMPLE 1. Find all the rational roots of

$$f(x) = x^4 + x^3 + 4x^2 + 6x - 12 = 0$$

Solution. The rational roots, if any, must be factors of -12; and the possibilities are

$$\pm 1, \quad \pm 2, \quad \pm 3, \quad \pm 4, \quad \pm 6, \quad \pm 12$$

Testing 1 by synthetic division, we find $f(1) = 0$.

$$\begin{array}{r}
1 + 1 + 4 + 6 - 12 \ \underline{1} \\
1 + 2 + 6 + 12 \\
\hline
1 + 2 + 6 + 12 + 0
\end{array}$$

Then $f(x) = (x - 1)(x^3 + 2x^2 + 6x + 12)$. Additional roots of $f(x) = 0$ may be more readily found from the equation

$$Q_1(x) = x^3 + 2x^2 + 6x + 12 = 0$$

$Q_1(x) = 0$ is called the *first depressed equation*. This equation has no positive roots; for a positive value for x makes all the terms on the left positive and their sum, of course, not equal to zero. Trying negative values, and starting with -1, we find $Q_1(-1) = 7$. Passing to -2, we get

$$
\begin{array}{r|r}
1 + 2 + 6 + 12 & \underline{-2} \\
-2 + 0 - 12 & \\
\hline
1 + 0 + 6 + \ \ 0 &
\end{array}
$$

Hence $Q_1(-2) = 0$, and consequently

$$Q_1(x) = (x + 2)(x^2 + 6)$$

The equation $Q_2(x) = x^2 + 6 = 0$ is called the *second depressed equation*. Its roots are $\pm\sqrt{-6}$, which are not real. We conclude, then, that -2 and 1 are the rational roots of the given equation.

EXAMPLE 2. Find the rational roots, if any, of the equation

$$f(x) = 2x^4 - 3x^3 + 10x^2 - 13x + 9 = 0$$

Solution. The signs of the coefficients alternate here, and consequently all terms become positive for any negative value of x. Hence any real root of the equation must be positive. The possibilities for rational roots, then, are

$$\tfrac{1}{2}, \quad 1, \quad \tfrac{3}{2}, \quad 3, \quad \tfrac{9}{2}, \quad 9$$

We find $f(\tfrac{1}{2}) = -\tfrac{19}{4}$ and $f(1) = 5$. Testing $\tfrac{3}{2}$, we have

$$
\begin{array}{r|r}
2 - 3 + 10 - 13 + \ \ 9 & \underline{\tfrac{3}{2}} \\
+ 3 + \ \ 0 + 15 + \ \ 3 & \\
\hline
2 + 0 + 10 + \ \ 2 + 12 &
\end{array}
$$

We see that $\tfrac{3}{2}$ is not a root. Further, we see that testing a number greater than $\tfrac{3}{2}$ will yield a number greater than 12 in the last number of the third line. Therefore we need not test the values $3, \tfrac{9}{2}$, and 9. The given equation has no rational root.

As is illustrated in this example, the following theorem may in some instances be applied to advantage in testing for rational roots.

Theorem 14-9 If in the synthetic division of a polynomial $f(x)$ by $x - r$, where r is positive, each term of the third line is positive (some may be zero), then r is an upper limit for the real roots of the equation $f(x) = 0$. If r is negative and the terms of the third line are alternately positive and negative, then r is a lower limit for the roots.

The truth of the theorem may be observed immediately. If a positive r produces positive terms in the third line in the division process, a number larger than r would increase all terms of the line after the first. The last term of the line could not then be zero. Consequently a value larger than r is not a root of the equation.

By quite similar reasoning the second part of the theorem can be established.

EXAMPLE 3. Find upper and lower limits for the roots of the equation

$$2x^3 - 5x^2 - 7x + 4 = 0$$

Solution. We show the tests for $r = 3$ and $r = 4$

$$
\begin{array}{r}
2 - 5 - 7 + 4 \;\lfloor\underline{3} \\
+ 6 + 3 - 12 \\
\hline
2 + 1 - 4 - 8
\end{array}
\qquad
\begin{array}{r}
2 - 5 - 7 + 4 \;\lfloor\underline{4} \\
+ 8 + 12 + 20 \\
\hline
2 + 3 + 5 + 24
\end{array}
$$

These tests show that the smallest integral upper limit, as determined by the theorem, is 4.

We next try negative integers, starting with -1

$$
\begin{array}{r}
2 - 5 - 7 + 4 \;\lfloor\underline{-1} \\
- 2 + 7 + 0 \\
\hline
2 - 7 + 0 + 4
\end{array}
\qquad
\begin{array}{r}
2 - 5 - 7 + 4 \;\lfloor\underline{-2} \\
- 4 + 18 - 22 \\
\hline
2 - 9 + 11 - 18
\end{array}
$$

Since -2 makes the terms of the third line alternate in signs, this number is a lower limit. Hence all the real roots of the given equation lie between -2 and 4.

EXERCISE 14-4

Find all the rational roots, if any, of each equation. Solve completely if the last depressed equation is quadratic.

1. $x^3 + x^2 - 8x - 12 = 0$ 2. $x^3 + 2x^2 - 7x + 4 = 0$
3. $x^3 - 6x^2 + 21x - 26 = 0$ 4. $3x^3 - x^2 + 27x - 9 = 0$
5. $4x^3 + x^2 + 16x + 4 = 0$ 6. $8x^3 + 4x^2 - 2x - 1 = 0$
7. $8x^3 - 4x^2 - 2x + 1 = 0$ 8. $2x^3 - 7x^2 + 16x - 15 = 0$
9. $3x^3 + 8x^2 + 19x + 10 = 0$ 10. $x^4 - 2x^3 + x^2 - x - 2 = 0$
11. $x^4 + 2x^3 + x^2 + 3x + 2 = 0$ 12. $x^4 - 8x^2 - 5x + 6 = 0$
13. $6x^4 + x^3 + 5x^2 + x - 1 = 0$ 14. $6x^4 + 5x^3 + 5x - 6 = 0$
15. $2x^4 - 13x^3 + 30x^2 - 28x + 8 = 0$ 16. $2x^4 + 13x^3 + 36x^2 + 70x + 75 = 0$
17. $3x^4 - 4x^3 - 8x^2 + 9x - 2 = 0$ 18. $12x^4 - 7x^3 + 13x^2 - 7x + 1 = 0$
19. $x^5 - 9x^4 + 31x^3 - 51x^2 + 40x - 12 = 0$
20. $2x^5 + 9x^4 + 21x^3 + 28x^2 + 21x + 9 = 0$
21. $2x^5 - 3x^4 + 8x^3 - 12x^2 + 6x - 9 = 0$
22. $x^5 - 20x^4 + 70x^3 - 100x^2 + 65x - 16 = 0$

Factor the following polynomials.

23. $6x^3 - 5x^2 - 7x + 4$ **24.** $4x^3 - 4x^2 - 3x - 10$
25. $2x^3 - 7x^2 + 18x + 11$ **26.** $5x^4 - 2x^3 - 2x$
27. $x^4 + 4x^3 + 3x^2 - 4x - 4$ **28.** $x^4 - 4x^3 + x^2 + 16x - 20$
29. $x^5 + x^4 - 5x^3 - x^2 + 8x - 4$ **30.** $x^5 + x^4 - 5x^3 - 5x^2 + 4x + 4$

14-7 THE GRAPH OF A POLYNOMIAL FUNCTION

In Sec. 5-3 we introduced graphs of functions and relations and in Sec. 8-7 we constructed graphs of quadratic functions, or polynomial functions of the second degree. From the graphs of quadratic functions we estimated the real zeros of the functions, or the roots of the corresponding quadratic equations. The graphical method may be employed to obtain approximately the real roots, rational and irrational, of a polynomial equation of higher degree than the second. Computing functional values in the case of higher degree equations may be somewhat tedious, but the work can be facilitated by the use of synthetic division. Here, as previously, we shall assume the following theorem to be true.

Theorem 14-10   If $f(x)$ denotes a polynomial with real coefficients and if a and b are real numbers such that $f(a)$ and $f(b)$ are of opposite signs, then the equation $f(x) = 0$ has at least one real root between a and b.

Interpreted geometrically, this theorem means that the graph of $y = f(x)$ crosses the x axis at least once between any two of its points which are on opposite sides of the x axis.

EXAMPLE. Estimate, to one decimal place, the roots of the equation

$$x^3 - 4x^2 + x + 3 = 0$$

Solution. To draw the graph of the function defined by $y = x^3 - 4x^2 + x + 3$, we prepare the following table. The functional values should be checked by synthetic division. When $x = 3.5$, for example, we have

$$\begin{array}{r} 1 - 4.0 + 1.00 + 3.000 \big|\,3.5 \\ + 3.5 - 1.75 - 2.625 \\ \hline 1 - 0.5 - 0.75 + 0.375 \end{array}$$

x	-1	-0.5	0	1.5	2	3	3.5
y	-3	1.4	3	-1.1	-3	-3	0.4

From the graph (Fig. 14-1), we estimate the zeros of the function, or the roots of the given equation to be -0.7, 1.2, and 3.5.

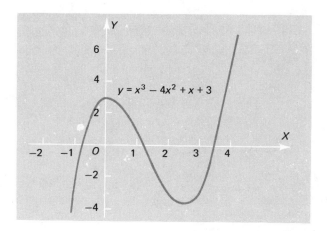

Fig. 14-1

EXERCISE 14-5

Solve for the real roots by graphing. Estimate the roots to one decimal place.

1. $x^3 + 3x^2 + 1 = 0$

2. $x^3 - x^2 - 5x - 6 = 0$

3. $x^3 - 3x + 4 = 0$

4. $x^3 + 4x^2 + x - 4 = 0$

5. $x^3 - 7x - 5 = 0$

6. $x^3 - 3x^2 + 3 = 0$

7. $x^4 - 2x^2 - 3 = 0$

8. $x^4 - 4x^2 + 3 = 0$

9. $x^4 - 3x + 5 = 0$

10. $x^4 + 4x^3 - 16x + 5 = 0$

I4-8   A METHOD OF APPROXIMATING IRRATIONAL ROOTS

There are several processes by which the irrational roots of a polynomial equation may be found to any desired degree of accuracy. We shall consider a method which is called the solution by *successive enlargements*. This method, as in the previous section, is based on the principle that if $f(a)$ and $f(b)$ have opposite signs, the equation $f(x) = 0$ has a root between a and b. The process consists in finding pairs of values of x closer and closer together for which $f(x)$ has opposite signs. We shall illustrate with some examples.

EXAMPLE 1. Find an approximation of the real root of the equation

$$f(x) = x^3 - 2x^2 + x - 1 = 0$$

Solution. The graph (Fig. 14-2) of $y = f(x)$ shows that the given equation has a root between $x = 1$ and $x = 2$. The next step is to find consecutive tenths of units between which the curve crosses the x axis. Computing, we find

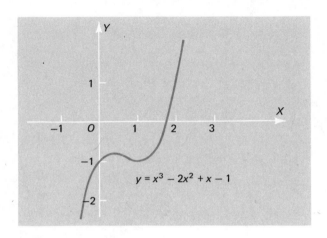

Fig. 14-2

$f(1.6) = -0.424$. Next we try $x = 1.7$ and $x = 1.8$:

$$
\begin{array}{r|l}
1 - 2.0 + 1.00 - 1.000 & 1.7 \\
 1.7 - 0.51 + 0.833 & \\
\hline
1 - 0.3 + 0.49 - 0.167 &
\end{array}
$$

$$
\begin{array}{r|l}
1 - 2.0 + 1.00 - 1.000 & 1.8 \\
 1.8 - 0.36 + 1.152 & \\
\hline
1 - 0.2 + 0.64 + 0.152 &
\end{array}
$$

We see that $f(1.7) = -0.167$ and $f(1.8) = 0.152$. Hence the points $(1.7, -0.167)$ and $(1.8, 0.152)$ are on the graph of $y = f(x)$, one below the x axis and the other above the x axis. Consequently the desired root is between 1.7 and 1.8.

We next locate the root between consecutive hundredths of units. For this purpose we plot the points of the graph for which $x = 1.7$ and $x = 1.8$ on an enlarged scale (Fig. 14-3). The line segment joining these points crosses the x axis at about $x = 1.75$, and the graph likely crosses the x axis at a point quite near. By synthetic division, as here shown, we find $f(1.75) = -0.01526$ and $f(1.76) = 0.016576$.

$$
\begin{array}{r|l}
1 - 2.00 + 1.0000 - 1.000000 & 1.75 \\
 1.75 - 0.4375 + 0.984375 & \\
\hline
1 - 0.25 + 0.5625 - 0.015625 &
\end{array}
$$

$$
\begin{array}{r|l}
1 - 2.000 + 1.0000 - 1.000000 & 1.76 \\
 1.76 - 0.4224 + 1.016576 & \\
\hline
1 - 0.24 + 0.5776 + 0.016576 &
\end{array}
$$

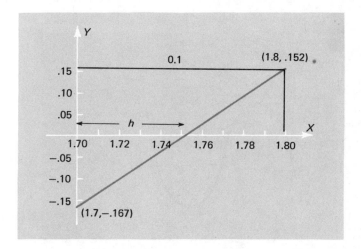

Fig. 14-3

Thus we see that the root is between 1.75 and 1.76. But we do not know at this stage which of these values is the better approximation. We could continue the process by next locating the root between consecutive thousandths of units. The computations, of course, would become increasingly tedious. It is easy, however, to estimate the root to one more digit. Figure 14-4 shows the

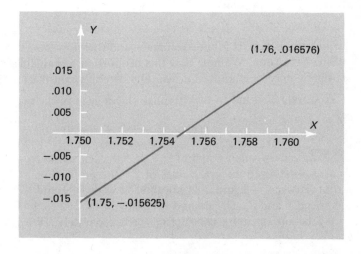

Fig. 14-4

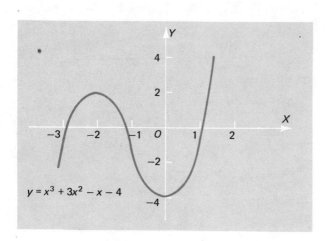

$$y = x^3 + 3x^2 - x - 4$$

Fig. 14-5

points on the graph for which $x = 1.75$ and $x = 1.76$. The line connecting these points crosses the x axis at about $x = 1.755$. Hence this value is approximately equal to a root of the given equation.

In this example we used Fig. 14-3 to estimate the third digit of the desired root, and Fig. 14-4 to estimate the fourth digit. We could, however, obtain these estimates algebraically. Thus, from the similar triangles in Fig. 14-3, we have the proportion

$$\frac{h}{0.1} = \frac{0.167}{0.167 + 0.152}$$

which yields $h = 0.05$, approximately, and therefore $x = 1.75$ is an approximation of the root. We note that this proportion, obtainable without the use of a figure, may be employed in lieu of a carefully constructed figure.

EXAMPLE 2. Find to two decimal places each of the real roots of

$$f(x) = x^3 + 3x^2 - x - 4 = 0$$

Solution. A rough graph of $y = f(x)$ may be made by plotting a few points corresponding to integral values of x. The graph (Fig. 14-5) shows there is a root between -3 and -2, another between -2 and -1, and a third root between 1 and 2. The desired accuracy in the approximations to these roots may be obtained by considering each separately. We show the necessary computations for the smallest of the three roots.

$$
\begin{array}{r}
1 + 3.0 - 1.00 - 4.000 \quad \underline{|-2.9} \\
- 2.9 - 0.29 + 3.741 \\
\hline
1 + 0.1 - 1.29 - 0.259
\end{array}
$$

$$
\begin{array}{r}
1 + 3.0 - 1.00 - 4.000 \quad \underline{|-2.8} \\
- 2.8 - 0.56 + 4.368 \\
\hline
1 + 0.2 - 1.56 + 0.368
\end{array}
$$

The root lies between $x = -2.9$ and $x = -2.8$. Figure 14-6 shows the points of the graph corresponding to these values of x. The line segment connecting the points crosses the x axis at about $x = -2.86$. We are not certain, however, that 6 is the best digit for the second decimal place. But, computing, we find $f(-2.86) = 0.005144$ and $f(-2.85) = -0.068375$. These results reveal that the root is between $x = -2.85$ and $x = -2.86$, with the root surely closer to -2.86.

The other two roots to two decimal places, as the student should verify, are $x = -1.25$ and $x = 1.12$.

Illustrating with the equation of the preceding example, we describe an alternate procedure for approximating negative roots. If we replace x by $-x$ in the $x^3 + 3x^2 - x - 4 = 0$, we get

$$-x^3 + 3x^2 + x - 4 = 0 \quad \text{or} \quad x^3 - 3x^2 - x + 4 = 0$$

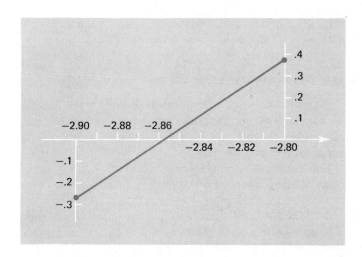

Fig. 14-6

The roots of this new equation are the negatives of the roots of the original equation. Hence the approximations to the negative roots of the original equation may be found by reversing the signs of the corresponding positive roots of the new equation.

In the two examples we have found approximations to the real roots of cubic equations. The process, of course, is applicable to equations of higher degrees. Further, a root can be obtained to as many digits as desired. It is evident, however, that the computations increase in tediousness with the degree of the equation and with the accuracy required in the result.

EXERCISE 14-6

Find to two decimal places the real root of each of the following equations.

1. $x^3 + 3x - 8 = 0$
2. $x^3 + x - 14 = 0$
3. $x^3 - x^2 - 2 = 0$
4. $x^3 - 3x^2 + 4x - 3 = 0$
5. $x^3 + 3x^2 + 3x - 27 = 0$
6. $x^3 + 2x - 5 = 0$
7. $x^3 + 3x^2 + 5x - 2 = 0$
8. $x^3 - 2x^2 - 8 = 0$
9. $x^3 + 3x^2 + 3x - 8 = 0$
10. $2x^3 + 6x^2 + 6x - 3 = 0$
11. $3x^3 + 4x^2 - x - 14 = 0$
12. $2x^3 - 3x^2 - 3x - 6 = 0$

Find three roots to two decimal places.

13. $x^3 - 3x^2 - x + 4 = 0$
14. $x^3 - 3x^2 - 2x + 1 = 0$
15. $x^3 - 6x^2 + 8x + 1 = 0$
16. $x^3 - 5x^2 + 3x + 8 = 0$
17. $x^3 - 4x^2 + 3x + 1 = 0$
18. $x^3 - 9x^2 + 24x - 17 = 0$
19. $4x^3 - 12x^2 + 4x + 5 = 0$
20. $3x^3 - 5x^2 - 4x + 3 = 0$

Find the indicated root to two decimal places.

21. $x^4 - x - 3 = 0$; the root between 1 and 2
22. $x^4 + 2x^3 - 2 = 0$; the root between 0 and 1
23. $x^4 + x^3 - 6x - 15 = 0$; the root between 2 and 3
24. $x^4 - x^3 + x^2 - 3x - 4 = 0$; the root between 1 and 2

14-9 DESCARTES' RULE OF SIGNS

A polynomial $f(x)$ with the terms written in the order of descending powers of x is said to have a variation of sign if two consecutive terms have opposite signs. Any missing power of x is to be disregarded (a zero coefficient is not supplied) in this definition.

Illustration: The polynomial $2x^5 - x^4 - 3x^3 + x^2 - 2$ has three variations of sign. There is one variation from $2x^5$ to $-x^4$, another from $-3x^3$ to $+x^2$, and a third from $+x^2$ to -2.

The variations of sign furnish some information concerning the roots of a polynomial equation. In this connection we state a theorem which is sometimes useful. Because of the difficulty, the proof is not given.

Theorem 14-11 *Descartes' rule of signs.* The number of positive roots of a polynomial equation $f(x) = 0$ is equal to the number of variations of sign in $f(x)$ or is less than that number by an even integer. The number of negative roots of $f(x) = 0$ is equal to the number of variations of sign in $f(-x)$ or is less than that number by an even integer.

Although the proof of this rule is omitted, we shall apply it to some equations. According to the rule, one variation of sign means there is exactly one positive root. For variations in excess of one the rule does not furnish definite information. Four variations, for example, reveal that there are four, two, or no positive roots.

EXAMPLE 1. Apply Descartes' rule of signs to the equation

$$x^3 + 7x^2 + 8x - 17 = 0.$$

Solution. The left member $f(x)$ has one variation of sign. Hence the equation has exactly one positive root. To test for negative roots, we replace x by $-x$ and have

$$f(-x) = -x^3 + 7x^2 - 8x - 17.$$

Here there are two variations of sign. According to the rule, the given equation has either two negative roots or no negative root. We conclude, therefore, that the original equation has one positive root and that the two remaining roots are either both negative or both imaginary. That is, the possibilities are: (a) one positive root and two negative roots; or (b) one positive root and two imaginary roots.

EXAMPLE 2. Determine the information which Descartes' rule provides for the equation

$$x^4 + 3x^3 + 4x^2 + 3x + 5 = 0.$$

Solution. There is no variation of sign in the left member, and consequently the equation has no positive root. We have $f(-x) = x^4 - 3x^3 + 4x^2 - 3x + 5$, which contains four variations of sign. Hence, by the rule, the following possibilities exist: (a) all roots negative; (b) two roots negative and two imaginary; or (c) all roots imaginary.

EXERCISE 14-7

Apply Descartes' rule of signs to each equation. Where the rule does not yield definite information, state the various possibilities as to the types of roots.

1. $2x^3 + 5x - 1 = 0$ **2.** $x^3 + 3x + 3 = 0$

3. $3x^3 + 3x^2 + 4 = 0$

5. $3x^3 - x^2 + x - 4 = 0$

7. $x^4 + x^3 - 1 = 0$

9. $x^4 + 2x^3 + x^2 + x + 4 = 0$

11. $x^6 - 2x^5 + 4x^3 - 5 = 0$

4. $x^3 + 2x^2 - 3x + 5 = 0$

6. $5x^3 - x^2 + x - 1 = 0$

8. $x^4 + x^3 + x^2 + x - 7 = 0$

10. $x^4 + x^3 + 4x^2 - x + 8 = 0$

Chapter 15

COMPLEX NUMBERS

15-1 THE FIELD OF COMPLEX NUMBERS

In Sec. 7-7 we gave a brief introduction to complex numbers, and in this chapter we continue the treatment of this system of numbers. Although either of the two customary notations for a complex number could be used in this further study, we shall employ the form $a + bi$.* Using this notation, we begin by listing the following definitions.

Definition 15-1 A number of the form $a + bi$, with a and b real constants and $i = \sqrt{-1}$, is called a *complex number*. The number a is called the *real component* and b is called the *imaginary component*.

Definition 15-2 The sum, difference, product, and quotient of two complex numbers are defined by the following equations.

Sum: $(a + bi) + (c + d)i = (a + c) + (b + d)i$

Difference: $(a + bi) - (c + d)i = (a - c) + (b - d)i$

Product: $(a + bi)(c + d)i = (ac - bd) + (bc + ad)i$

Quotient: $\dfrac{a + bi}{c + di} = \dfrac{(ac + bd) + (bc - ad)i}{c^2 + d^2}$

Definition 15-3 Two complex numbers $a + bi$ and $c + di$ are equal if and only if $a = c$ and $b = d$.

Definition 15-4 The numbers $a + bi$ and $a - bi$ are called *conjugate complex numbers*. Each is the conjugate of the other.

The system of complex numbers constitutes a field. This property of complex numbers follows from the Definitions 15-1 and 15-2, and the assumption that real numbers obey the field axioms (Sec. 2-2). In this connection we have the following theorems. The student may supply the proofs which are missing.

Theorem 15-1 The number $1 + 0 \cdot i$ is a unique multiplicative identity element of any complex number $a + bi$.

* The student should review Sec. 7-7 at this point.

Proof. Let $c + di$ be a number such that

$$(a + bi)(c + di) = a + bi$$

Multiplying and applying Definition 15-3, we have

$$(ac - bd) + (ad + bc)i = a + bi$$

and

$$ac - bd = a$$

$$ad + bc = b$$

The solution of this system of linear equations in c and d is $c = 1, d = 0$. Since there is no other solution of the equations, we conclude that $1 + 0 \cdot i$ or 1 is a unique identity element for multiplication. ■

Theorem 15-2　A complex number $a + bi$, with a and b not both zero, has a unique multiplicative inverse.

Proof. Let $c + di$ be a number such that

$$(a + bi)(c + di) = 1$$

Then

$$(ac - bd) + (ad + bc)i = 1$$

and

$$ac - bd = 1$$

$$ad + bc = 0$$

The unique solution of this system of linear equations in c and d is

$$c = \frac{a}{a^2 + b^2} \quad \text{and} \quad d = \frac{-b}{a^2 + b^2}$$

Hence the multiplicative inverse of $a + bi$ is the complex number

$$\frac{a}{a^2 + b^2} - \frac{b}{a^2 + b^2}i \quad \text{or} \quad \frac{a - bi}{a^2 + b^2} \quad ■$$

Theorem 15-3　The associative law for multiplication holds for complex numbers. That is,

$$[(a + bi)(c + di)](e + fi) = (a + bi)[(c + di)(e + fi)]$$

Proof. By employing the closure, commutative, and associative laws for addition and multiplication of real numbers (Sec. 2-2), we reduce the left member of this equation to the form of the right member. All steps should be followed.

$[(a + bi)(c + di)](e + fi)$

$$= [(ac - bd) + (ad + bc)i](e + fi)$$

$$= (ac - bd)e - (ad + bc)f + [(ac - bd)f + (ad + bc)e]i$$

$$= a(ce - df) - b(cf + de) + [a(cf + de)i + b(ce - df)i]$$

$$= (a + bi)[(c + di)(e + fi)] \quad \blacksquare$$

Theorem 15-4 Complex numbers obey the closure and commutative laws for addition and multiplication.

Theorem 15-5 Complex numbers obey the distributive law of multiplication with respect to addition. That is,

$$(a + bi)[(c + di) + (e + fi)] = (a + bi)(c + di) + (a + bi)(e + fi)$$

Referring to Definition 15-2, we see that the sum, difference, and product of two complex numbers are obtained by carrying out the operations as though i were a real number and, for the product, replacing i^2 by -1. We note also that the quotient of two complex numbers may be obtained by multiplying the dividend and divisor by the conjugate of the divisor. Thus

$$\frac{a + bi}{c + di} = \frac{(a + bi)(c - di)}{(c + di)(c - di)} = \frac{(ac + bd) + (bc - ad)i}{c^2 + d^2}$$

We have seen that the imaginary unit $i = \sqrt{-1}$ is vital in the building of the system of complex numbers. By joining this unit to the system of real numbers, we have the elements for passing to an enlarged set of numbers which contain the set of real numbers as a proper subset. The integral powers of i furnish additional interesting properties of this unit. By definition $i^2 = -1$, and consequently we see that

$$i^3 = i^2 \cdot i = (-1)i = -i$$
$$i^4 = i^2 \cdot i^2 = (-1)(-1) = 1$$

Since $i^4 = -1$, it follows that i with an exponent which is an integral multiple of 4 is also equal to 1. Hence we discover that integral powers of i yield only the numbers $i, -1, -i, 1$. Thus, as examples,

$$i^{13} = i^{12} \cdot i = i$$
$$i^{-6} = i^{-6} \cdot i^8 = i^2 = -1$$

EXAMPLE 1. Find the product of $\sqrt{-3}$ and $\sqrt{-5}$.

Solution. We first express each number in the form i times a real number and then multiply. Thus

$$\sqrt{-3} \cdot \sqrt{-5} = i\sqrt{3} \cdot i\sqrt{5} = i^2\sqrt{15} = -\sqrt{15}$$

This example illustrates the fact that the law for multiplying two radicals of like order involving real numbers must not be applied to imaginary numbers. Thus $\sqrt{3} \cdot \sqrt{5} = \sqrt{15}$, but $\sqrt{-3} \cdot \sqrt{-5}$ is not equal to $\sqrt{15}$. An error in this kind of multiplying is less likely to be made if each factor is first expressed as the product of i and a real number.

EXAMPLE 2. If $3x - 5 + (4x + 3)i = y - 4 + (3x + y)i$, find the values of x and y.

Solution. By Definition 15-3, the real components of the members of the equation must be equal and the imaginary components must be equal. Hence we have the system of equations

$$3x - 5 = y - 4$$

$$4x + 3 = 3x + y$$

The solution of these equations is $x = 2$ and $y = 5$.

EXAMPLE 3. Find the quotient of $2 + 3i$ divided by $4 - 5i$.

Solution. We multiply the numerator and denominator by the conjugate of the denominator. Thus, remembering $i^2 = -1$, we have

$$\frac{2 + 3i}{4 - 5i} = \frac{(2 + 3i)(4 + 5i)}{(4 - 5i)(4 + 5i)}$$

$$= \frac{8 + 10i + 12i + 15i^2}{16 - 25i^2}$$

$$= \frac{-7 + 22i}{41}$$

15-2 GRAPHICAL REPRESENTATION OF COMPLEX NUMBERS

Complex numbers are usually represented graphically by points in the plane of a rectangular coordinate system. A number of the form $a + bi$ is represented by the point whose coordinates are (a, b). Hence any complex number determines a point in the plane. Conversely, any point of the plane determines a complex number. The points on the axes correspond to special complex numbers. Points on the x axis represent real numbers because the ordinates of the points are zero. Similarly, the points on the y axis have abscissas equal

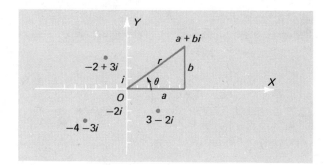

Fig. 15-1

to zero and therefore correspond to pure imaginary numbers. Thus the point $(a, 0)$ corresponds to the real number $a = a + 0 \cdot i$, and the point $(0, b)$ corresponds to the pure imaginary number $bi = 0 + bi$.

When a rectangular coordinate plane is used to represent complex numbers, the x axis is called the *real axis*, the y axis the *pure imaginary axis*, and the plane the *complex plane*.

In Fig. 15-1 the correspondence between some complex numbers and points is indicated.

The points corresponding to the numbers $a + bi$, $c + di$, and their sum $(a + c) + (b + d)i$ are plotted in Fig. 15-2. These points and the origin are vertices of a parallelogram, as can be easily verified by examining the figure.

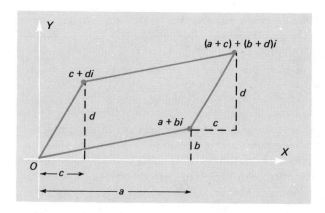

Fig. 15-2

Perform the indicated operations, leaving each result in the form $a + bi$.

1. $(1 + \sqrt{-4}) + (12 - \sqrt{-9})$
2. $(3 - \sqrt{-25}) + (5 - \sqrt{-16})$
3. $(4 + \sqrt{-1}) - (6 + \sqrt{-2})$
4. $(7 - \sqrt{-8}) - (6 + \sqrt{-18})$
5. $(5 + \sqrt{-12}) - (2 + \sqrt{-27})$
6. $(-3 - \sqrt{-45}) + (8 - \sqrt{-80})$
7. $(2 - 8i)(1 - 2i)$
8. $(3 + i\sqrt{2})(2 - i\sqrt{2})$
9. $(4 + i)(-4 - i)$
10. $(5 + 2i)(2 + 5i)$
11. $(3 + \sqrt{-48})(6 - \sqrt{-75})$
12. $(7 - \sqrt{-50})(3 - \sqrt{-72})$

13. $\dfrac{5 - i}{i}$

14. $\dfrac{6 + 12i}{3i}$

15. $\dfrac{4}{4 - 3i}$

16. $\dfrac{7i}{4 + i}$

17. $\dfrac{2 + i}{6 - i}$

18. $\dfrac{3 - i}{3 + 2i}$

19. $\dfrac{17 + 6i}{2 + 3i}$

20. $\dfrac{4 - 3i}{4 + 3i}$

21. $\dfrac{9 + 7i}{7 - 9i}$

Simplify the following expressions.

22. $3i^8 - 2i^3 - 4i^6$
23. $5i^{11} - 6i^7 + 7i^9$
24. $i^{19} + 2i^{14} + 5i^{23}$
25. $10i^{10} - 9i^{11} + 8i^{12}$
26. $4i^7 - 9i^{11} + 5i^{15}$
27. $6i^3 + 2i^{16} - 6i^{23}$

Solve the following equations for real values of x and y.

28. $4 - ix = y - 3i$
29. $x - iy = 2 + 5i$
30. $2y - 3 + (4x + 8)i = 0$
31. $3x + (y - x)i = 6$
32. $x + 2y + 3i = 3 + (2x - y)i$
33. $7x + (x - 3y)i = 3y + 9i$

Plot each complex number, its negative, and its conjugate.

34. $3 + 2i$
35. $6 - 5i$
36. $0 + 2i$
37. $-4 + 0 \cdot i$
38. $-2 + 6i$
39. $7 - 4i$
40. $-4 - 3i$
41. $-3i + 4$

Plot the points corresponding to the two given complex numbers and their sum. Then draw the parallelogram determined by the three points and the origin.

42. $1 + 3i, 4 + 3i$
43. $2 - 2i, 2 + 2i$
44. $-2 + 4i, 3 + i$
45. $-3 - 4i, 5 - i$

46. Prove that the sum and the product of two conjugate complex numbers are both real.
47. If the product of two complex numbers is zero, prove that at least one of the factors is zero.
48. If the quotient of two complex numbers is a real number, prove that one of the given numbers is a constant times the other.
49. Prove that the conjugate of the sum of two complex numbers is equal to the sum of their conjugates.

15-3 TRIGONOMETRIC FORM OF A COMPLEX NUMBER†

A complex number expressed as $a + bi$ is said to be in *algebraic*, or *rectangular, form.* We next show how to express a complex number in another useful form.

The point P (Fig. 15-3) corresponding to the number $a + bi$ has the coordinates (a, b). The distance from the origin to P is denoted by r, and the angle from the positive real axis to the distance segment is denoted by θ. The distance r, chosen positive when P is any point other than the origin, is called the *absolute value*, or *modulus* of $a + bi$. The angle θ is called the *amplitude* or *argument.*

Referring to the diagram, we have the relations

$$r = \sqrt{a^2 + b^2}, \qquad \tan \theta = \frac{b}{a}$$

and

$$a = r \cos \theta, \qquad b = r \sin \theta$$

Hence it follows that

$$a + bi = r(\cos \theta + i \sin \theta)$$

The expression $r(\cos \theta + i \sin \theta)$ is called the *trigonometric*, or *polar, form* of a complex number. The number r (modulus) is never negative. The factor $(\cos \theta + i \sin \theta)$ reveals the direction of P from the origin. This is true because θ must satisfy the equations $r \cos \theta = a$ and $r \sin \theta = b$. The preceding formulas, then, are sufficient for changing a number from algebraic form to trigonometric form and vice versa.

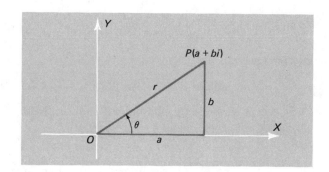

Fig. 15-3

---

† An understanding of trigonometry is needed in the remainder of this chapter.

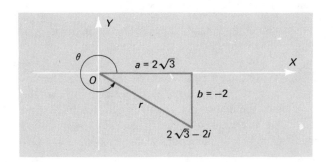

Fig. 15-4

EXAMPLE 1. Plot the number $2\sqrt{3} - 2i$ and change to the trigonometric form.

Solution. The point $(2\sqrt{3}, -2)$ represents the number (Fig. 15-4). Letting $a = 2\sqrt{3}$ and $b = -2$ in the preceding formulas, we obtain

$$r = \sqrt{12 + 4} = 4, \qquad 4\cos\theta = 2\sqrt{3}, \qquad 4\sin\theta = -2$$

We choose the smallest positive value of θ which satisfies the last two equations and have

$$2\sqrt{3} - 2i = 4(\cos 330° + i\sin 330°)$$

EXAMPLE 2. Plot the number $5(\cos 120° + i\sin 120°)$ and change to algebraic form.

Solution. The point corresponding to this number is located by drawing the angle 120° in standard position and measuring 5 units off along the terminal side. From Fig. 15-5, we have

$$a = 5\cos 120° = 5(-\tfrac{1}{2}) = -\tfrac{5}{2}$$

$$b = 5\sin 120° = 5\left(\frac{\sqrt{3}}{2}\right) = \frac{5\sqrt{3}}{2}$$

and therefore

$$5(\cos 120° + i\sin 120°) = -\frac{5}{2} + \frac{5\sqrt{3}}{2}i$$

EXAMPLE 3. Express the number $4 + 3i$ in trigonometric form.

Solution. Since $a = 4$ and $b = 3$, we have

$$r = \sqrt{16 + 9} = 5, \qquad \cos\theta = \tfrac{4}{5}, \qquad \sin\theta = \tfrac{3}{5}$$

Referring to the table of trigonometric functions, inside the front cover, we find that θ, to the nearest degree, is 37°. Hence

$$4 + 3i = 5(\cos 37° + i\sin 37°)$$

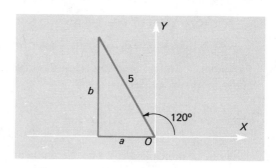

Fig. 15-5

Plot each complex number and write the corresponding trigonometric form.

1. $1 + i$ **2.** $4 - 4i$ **3.** $4 + 0 \cdot i$
4. $1 + i\sqrt{3}$ **5.** $2 - 2i\sqrt{3}$ **6.** $-\sqrt{3} + i$
7. $0 + 2i$ **8.** $-3 + 0 \cdot i$ **9.** $0 - i$
10. $\sqrt{2} + i\sqrt{2}$ **11.** $8\sqrt{3} - 8i$ **12.** $-\sqrt{15} + i\sqrt{5}$
13. $\sqrt{7} - i\sqrt{21}$ **14.** $3 + 4i$ **15.** $-2 + i$
16. $5 - 2i$ **17.** $-7 - 24i$ **18.** $8 + 6i$

Change each number to algebraic form.

19. $2(\cos 30° + i \sin 30°)$ **20.** $8(\cos 60° + i \sin 60°)$ **21.** $\sqrt{2}(\cos 45° + i \sin 45°)$
22. $\cos 135° + i \sin 135°$ **23.** $5(\cos 150° + i \sin 150°)$ **24.** $7(\cos 180° + i \sin 180°)$
25. $6(\cos 270° + i \sin 270°)$ **26.** $20(\cos 90° + i \sin 90°)$ **27.** $\cos 128° + i \sin 128°$
28. $2(\cos 51° + i \sin 51°)$

15-4   MULTIPLICATION AND DIVISION OF NUMBERS IN POLAR FORM

The product and quotient of two complex numbers may be written immediately if the numbers are in polar form.

Theorem 15-6   The absolute value of the product of two complex numbers is equal to the product of their absolute values. The amplitude of the product of two complex numbers is equal to the sum of their amplitudes.

Proof. Let the two complex numbers be denoted by

$$z_1 = r_1(\cos \theta_1 + i \sin \theta_1)$$

and

$$z_2 = r_2(\cos \theta_2 + i \sin \theta_2)$$

Then, multiplying in the usual way, we obtain

$$z_1 z_2 = r_1(\cos \theta_1 + i \sin \theta_1) \cdot r_2(\cos \theta_2 + i \sin \theta_2)$$

$$= r_1 r_2[(\cos \theta_1 \cos \theta_2 + i^2 \sin \theta_1 \sin \theta_2) + i(\sin \theta_1 \cos \theta_2 + \sin \theta_2 \cos \theta_1)]$$

Replacing i^2 by -1 and applying the identities for the cosine and the sine of the sum of two angles, we get

$$r_1(\cos \theta_1 + i \sin \theta_1) \cdot r_2(\cos \theta_2 + i \sin \theta_2)$$

$$= r_1 r_2[\cos (\theta_1 + \theta_2) + i \sin (\theta_1 + \theta_2)] \quad \blacksquare$$

This law of multiplication may be applied repeatedly to give the product of three or more complex numbers. For the three numbers

$$z_1 = r_1(\cos \theta_1 + i \sin \theta_1)$$

$$z_2 = r_2(\cos \theta_2 + i \sin \theta_2)$$

$$z_3 = r_3(\cos \theta_3 + i \sin \theta_3)$$

we have

$$z_1 z_2 = r_1 r_2[\cos (\theta_1 + \theta_2) + i \sin (\theta_1 + \theta_2)]$$

and then

$$z_1 z_2 z_3 = r_1 r_2 r_3[\cos (\theta_1 + \theta_2 + \theta_3) + i \sin (\theta_1 + \theta_2 + \theta_3)]$$

The product of n complex numbers is given by

$$z_1 z_2 \cdots z_n = r_1 r_2 \cdots r_n[\cos (\theta_1 + \theta_2 + \cdots + \theta_n) + i \sin (\theta_1 + \theta_2 + \cdots + \theta_n)]$$

We suggest that the student establish this general formula by mathematical induction. We suggest also that the student prove the associative property of complex numbers for multiplication; that is, prove $(z_1 z_2)z_3 = z_1(z_2 z_3)$.

Theorem 15-7   The absolute value of the quotient of two complex numbers is the quotient of their absolute values. The amplitude is the amplitude of the dividend minus the amplitude of the divisor.

Proof. To find the quotient of z_1 divided by z_2, we multiply the dividend and divisor by the conjugate of the divisor. Thus we have

$$\frac{z_1}{z_2} = \frac{r_1(\cos \theta_1 + i \sin \theta_1)(\cos \theta_2 - i \sin \theta_2)}{r_2(\cos \theta_2 + i \sin \theta_2)(\cos \theta_2 - i \sin \theta_2)}$$

The product of the factors in the numerator is

$$r_1[(\cos \theta_1 \cos \theta_2 + \sin \theta_1 \sin \theta_2) + i(\sin \theta_1 \cos \theta_2 - \cos \theta_1 \sin \theta_2)]$$

and the product of the factors of the denominator is

$$r_2(\cos^2 \theta_2 - i^2 \sin^2 \theta_2) = r_2$$

We now apply the trigonometric identities for the cosine and sine of the difference of two angles and express the numerator in the form $r_1[\cos(\theta_1 - \theta_2) + i\sin(\theta_1 - \theta_2)]$. Then we obtain the formula

$$\frac{r_1(\cos\theta_1 + i\sin\theta_1)}{r_2(\cos\theta_2 + i\sin\theta_2)} = \frac{r_1}{r_2}[\cos(\theta_1 - \theta_2) + i\sin(\theta_1 - \theta_2)]$$

The two preceding formulas may be applied graphically to give the product and quotient of two complex numbers.   To find the product of $r_1(\cos\theta_1 + i\sin\theta_1)$ and $r_2(\cos\theta_2 + i\sin\theta_2)$, draw the angle $(\theta_1 + \theta_2)$ in standard position and measure off a distance on the terminal side equal to r_1r_2. To find the quotient of $r_1(\cos\theta_1 + i\sin\theta_1)$ divided by $r_2(\cos\theta_2 + i\sin\theta_2)$, draw the angle $(\theta_1 - \theta_2)$ in standard position and measure off a distance along the terminal side equal to r_1/r_2. ■

EXAMPLE 1. Express the numbers $1 + i\sqrt{3}$ and $3 - 3i$ in polar forms and find their product.

Solution. We have for the polar forms

$$1 + i\sqrt{3} = 2(\cos 60° + i\sin 60°)$$

and

$$3 - 3i = 3\sqrt{2}(\cos 315° + i\sin 315°)$$

Hence,

$$(1 + i\sqrt{3})(3 - 3i) = 6\sqrt{2}(\cos 375° + i\sin 375°)$$

$$= 6\sqrt{2}(\cos 15° + i\sin 15°)$$

EXAMPLE 2. Find the quotient of $12(\cos 30° + i\sin 30°)$ divided by $3(\cos 70° + i\sin 70°)$.

Solution. Applying the law for division, we obtain

$$\frac{12(\cos 30° + i\sin 30°)}{3(\cos 70° + i\sin 70°)} = 4[\cos(-40°) + i\sin(-40°)]$$

$$= 4(\cos 320° + i\sin 320°)$$

EXERCISE 15-3

Perform the indicated operations, leaving each result in the form $a + bi$.

1. $2(\cos 30° + i\sin 30°) \cdot 3(\cos 60° + i\sin 60°)$
2. $5(\cos 120° + i\sin 120°) \cdot 4(\cos 60° + i\sin 60°)$
3. $3(\cos 50° + i\sin 50°) \cdot 6(\cos 70° + i\sin 70°)$
4. $7(\cos 130° + i\sin 130°) \cdot 2(\cos 95° + i\sin 95°)$
5. $9(\cos 175° + i\sin 175°) \cdot 3(\cos 275° + i\sin 275°)$

6. $\dfrac{\cos 130° + i \sin 130°}{\cos 40° + i \sin 40°}$

7. $\dfrac{4(\cos 266° + i \sin 266°)}{2(\cos 86° + i \sin 86°)}$

8. $\dfrac{9(\cos 313° + i \sin 313°)}{3(\cos 268° + i \sin 268°)}$

9. $\dfrac{21(\cos 33° + i \sin 33°)}{3(\cos 93° + i \sin 93°)}$

Express each number in trigonometric form and then perform the indicated operations. Leave the results in trigonometric form.

10. $(1 - i)(1 + i\sqrt{3})$

11. $(\sqrt{3} + i)(1 + i)$

12. $(3 - 3i)(4 - 4i)$

13. $4i(5 + 5i)$

14. $3i(1 - i\sqrt{3})(\sqrt{3} + i)$

15. $(2 + 2i)(-\sqrt{3} + i)(-1 + i\sqrt{3})$

16. $\dfrac{3}{2 - 2i}$

17. $\dfrac{2\sqrt{3} - 2i}{1 - i\sqrt{3}}$

18. $\dfrac{4}{3\sqrt{3} - 3i}$

19. $\dfrac{1 + i}{1 + i\sqrt{3}}$

Perform graphically the indicated operations.

20. $2(\cos 45° + i \sin 45°) \cdot 3(\cos 45° + i \sin 45°)$

21. $4(\cos 30° + i \sin 30°) \cdot (\cos 90° + i \sin 90°)$

22. $5(\cos 90° + i \sin 90°) \cdot 2(\cos 60° + i \sin 60°)$

23. $\dfrac{4(\cos 225° + i \sin 225°)}{2(\cos 90° + i \sin 90°)}$

24. $\dfrac{6(\cos 270° + i \sin 270°)}{3(\cos 135° + i \sin 135°)}$

15-5 POWERS AND ROOTS OF NUMBERS

In this section we shall be interested in finding positive integral powers of complex numbers and roots of complex numbers. The following theorem opens the way for the performance of these operations.

Theorem 15-8 If n is any real number, the nth power of the complex number $r(\cos \theta + i \sin \theta)$ is given by the formula

$$[r(\cos \theta + i \sin \theta)]^n = r^n(\cos n\theta + i \sin n\theta). \tag{1}$$

We shall not prove this theorem. We point out, however, that the theorem may be established at once for the case in which n is a positive integer. This is accomplished simply by letting each of the factors in the formula for the product of n complex numbers be equal to $r(\cos \theta + i \sin \theta)$. When n is a positive integer the preceding theorem is called *De Moivre's theorem* in honor of the French mathematician Abraham De Moivre.

Formula (1) yields readily a positive integral power of a complex number. To find roots of numbers, however, we convert the formula to a more useful form.

The number $s(\cos \alpha + i \sin \alpha)$ is, by definition, an nth root of $r(\cos \theta + i \sin \theta)$ if

$$[s(\cos \alpha + i \sin \alpha)]^n = r(\cos \theta + i \sin \theta)$$

We apply De Moivre's formula (1) to the left member of this equation and, because of the periodic property of the sine and cosine functions, replace θ by $\theta + k \cdot 360°$, where k is any integer. This gives

$$s^n(\cos n\alpha + i \sin n\alpha) = r[\cos(\theta + k \cdot 360°) + i \sin(\theta + k \cdot 360°)]$$

We seek values of s and α so that this equation will be satisfied. Consequently, we equate the moduli and amplitudes of the members of the equation. This gives

$$s^n = r \quad \text{and} \quad n\alpha = \theta + k \cdot 360°$$

whence, solving for s and α, we have

$$s = \sqrt[n]{r} \quad \text{and} \quad \alpha = \frac{\theta + k \cdot 360°}{n}$$

Hence

$$s(\cos \alpha + i \sin \alpha) = \sqrt[n]{r}\left(\cos \frac{\theta + k \cdot 360°}{n} + i \sin \frac{\theta + k \cdot 360°}{n}\right)$$

The right member of this equation may be made to take n distinct values by giving k the values $0, 1, 2, \ldots, n - 1$. Any other integral value of k yields a repetition and not a new root. Thus there are n different nth roots of any non-zero number. And it may be observed that the amplitudes are such that graphically the roots are equally spaced about the circumference of a circle of radius $\sqrt[n]{r}$. Hence we have the following theorem.

Theorem 15-9 A nonzero number $r(\cos \theta + i \sin \theta)$ has n nth roots which are given by the formula

$$\sqrt[n]{r}\left(\cos \frac{\theta + k \cdot 360°}{n} + i \sin \frac{\theta + k \cdot 360°}{n}\right) \tag{2}$$

where $k = 0, 1, 2, \ldots, n - 1$.

EXAMPLE 1. Find the fourth power of $(1 - i\sqrt{3})$.
Solution. First changing to polar form, we obtain

$$1 - i\sqrt{3} = 2(\cos 300° + i \sin 300°)$$

Applying formula (1) with $n = 4$, we get

$$(1 - i\sqrt{3})^4 = [2(\cos 300° + i \sin 300°)]^4$$

$$= 2^4(\cos 1200° + i \sin 1200°)$$

$$= 16(\cos 120° + i \sin 120°)$$

$$= -8 + 8i\sqrt{3}$$

EXAMPLE 2. Find the cube roots of -8.

Solution. We first express -8 in polar form. Thus

$$-8 = 8[\cos(180° + k \cdot 360°) + i\sin(180° + k \cdot 360°)]$$

We now apply formula (2) for roots of a number and have

$$\sqrt[3]{8}\left(\cos\frac{180° + k \cdot 360°}{3} + i\sin\frac{180° + k \cdot 360°}{3}\right)$$

$$= 2[\cos(60° + k \cdot 120°) + i\sin(60° + k \cdot 120°)]$$

Assigning k the values 0, 1, and 2, in succession, we find the three cube roots of -8 to be

$$2(\cos 60° + i\sin 60°) = 1 + i\sqrt{3}$$

$$2(\cos 180° + i\sin 180°) = -2$$

$$2(\cos 300° + i\sin 300°) = 1 - i\sqrt{3}$$

The numbers $1 + i\sqrt{3}$, -2, and $1 - i\sqrt{3}$ are the rectangular forms of the cube roots of -8. These are all the cube roots of -8; any other integral value of k gives a repetition of one of these roots. In particular, if the values 3, 4, and 5 are assigned to k, the roots are repeated in the order just mentioned.

EXAMPLE 3. Find the five fifth roots of $-16 + 16i\sqrt{3}$.

Solution. The polar form of $-16 + 16i\sqrt{3}$ is

$$32[\cos(120° + k \cdot 360°) + i\sin(120° + k \cdot 360°)]$$

From formula (2), we write

$$\sqrt[5]{32}\left(\cos\frac{120° + k \cdot 360°}{5} + i\sin\frac{120° + k \cdot 360°}{5}\right)$$

$$= 2[\cos(24° + k \cdot 72°) + i\sin(24° + k \cdot 72°)]$$

The desired roots are obtainable by assigning k the values 0, 1, 2, 3, 4, in succession, in the last expression. Thus the five fifth roots of $-16 + 16i\sqrt{3}$ are

$$2(\cos\ 24° + i\sin\ 24°)$$

$$2(\cos\ 96° + i\sin\ 96°)$$

$$2(\cos 168° + i\sin 168°)$$

$$2(\cos 240° + i\sin 240°)$$

$$2(\cos 312° + i\sin 312°)$$

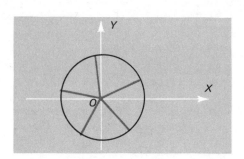

Fig. 15-6

The points representing these roots are equally spaced about the circle of radius 2 in Fig. 15-6.

EXERCISE 15-4

Find the indicated powers in the following problems. Express each result in trigonometric form.

1. $[3(\cos 20° + i \sin 20°)]^2$ **2.** $[2(\cos 35° + i \sin 35°)]^3$

3. $[2(\cos 118° + i \sin 118°)]^3$ **4.** $[5(\cos 220° + i \sin 220°)]^2$

5. $(\cos 10° + i \sin 10°)^8$ **6.** $(\cos 70° + i \sin 70°)^{10}$

7. $(1 + i)^5$ **8.** $(1 - i)^4$ **9.** $(1 + i\sqrt{3})^6$

10. $(\sqrt{3} - i)^4$ **11.** $(-\sqrt{3} + i)^{10}$ **12.** $(2 - 2i)^5$

Find the indicated roots of the given numbers. Leave each result in trigonometric form.

13. The square roots of $16(\cos 80° + i \sin 80°)$

14. The square roots of $25(\cos 134° + i \sin 134°)$

15. The cube roots of $27(\cos 150° + i \sin 150°)$

16. The fourth roots of $81(\cos 160° + i \sin 160°)$

17. The fourth roots of $16(\cos 280° + i \sin 280°)$

18. The fifth roots of $\cos 100° + i \sin 100°$

19. The cube roots of $8 + 8i$ **20.** The cube roots of $27i$

21. The fourth roots of $-i$ **22.** The fourth roots of 81

23. The fifth roots of $-32i$ **24.** The fifth roots of $1 + i$

Find the indicated powers and roots in the following problems, giving each result in algebraic form.

25. $[2(\cos 50° + i \sin 50°)]^6$ **26.** $[3(\cos 36° + i \sin 36°)]^5$

27. $(\cos 210° + i \sin 210°)^4$ **28.** $(\cos 120° + i \sin 120°)^4$

29. $\left(\dfrac{\sqrt{2}}{2} + \dfrac{\sqrt{2}}{2}i\right)^{10}$ **30.** $\left(\dfrac{1}{2} - \dfrac{\sqrt{3}}{2}i\right)^8$

31. The cube roots of $-4\sqrt{3} + 4i$ **32.** The cube roots of $27(4 + 4i\sqrt{3})$
33. The cube roots of $-8 + 8i$ **34.** The fourth roots of $8 + 8i\sqrt{3}$

Solve the following equations.

35. $x^4 + 1 = 0$ **36.** $x^6 - 1 = 0$ **37.** $x^4 - i = 0$
38. $x^2 + (1 - i)x - i = 0$ **39.** $x^2 - 2x + (1 + i) = 0$

40. Prove that complex numbers obey the closure and commutative laws for addition and multiplication. (See Axioms 1 and 2, Sec. 2-2.)

We have stated, Theorem 15-5, that complex numbers obey the distributive law of multiplication with respect to addition. That is:

$$(a + bi)[(c + di) + (e + fi)] = (a + bi)(c + di) + (a + bi)(e + fi)$$

We shall prove the theorem by using Definition 15-2 and Theorem 15-3 to reduce the left member of this equation to the form of the right member. All steps should be followed.

$$(a + bi)[(c + di) + (e + fi)]$$
$$= (a + bi)[(c + e) + (d + f)i]$$
$$= [a(c + e) - b(d + f)] + [b(c + e) + a(d + f)]i$$
$$= [(ac - bd) + (ae - bf)] + [(bc + ad)i + (be + af)i]$$
$$= [(ac - bd) + (bc + ad)i] + [(ae - bf) + (be + af)i]$$
$$= (a + bi)(c + di) + (a + bi)(e + fi)$$

Solve each linear equation for x. Express your result in the form $a + bi$.

41. $2ix = 3 - i$ **42.** $3ix - 8 = i$
43. $(3 + i)x = 4 + i$ **44.** $(5i - 2)x = 5i + 2$
45. $(4 + i)x = 3 - 2i$ **46.** $(6 - i)x = 2i + 3$

Using the quadratic formula to solve the equation

$$2ix^2 - 3x + 4 + i = 0$$

we have here $a = 2i$, $b = -3$, and $c = 4 + i$. Hence, substituting, we find

$$x = \frac{3 \pm \sqrt{9 - 8i(4 + i)}}{4i}$$
$$= \frac{3 \pm \sqrt{17 - 32i}}{4i}$$
$$= \frac{3i \pm \sqrt{32i - 17}}{-4}$$
$$= \frac{-3i \pm \sqrt{32i - 17}}{4}$$

Use the quadratic formula (Theorem 8-1, Sec. 8-2) to solve each equation.

47. $2x^2 - 4ix + i = 0$ **48.** $3x^2 - 2ix - 3i = 0$
49. $4ix^2 + 2ix - 1 = 0$ **50.** $ix^2 + (3 - i)x + 2 = 0$

51. $3ix^2 - 2x - (1 + i) = 0$ **52.** $5ix^2 + 3x + (2 + 3i) = 0$
53. $(1 + i)x^2 + 3x + (2 + 2i) = 0$ **54.** $(2 - i)x^2 - 4x - (3 - 2i) = 0$

Show that the absolute value of a complex number is equal to the absolute value of its conjugate. That is,

$$|a + bi| = |a - bi|$$

Show that $|(a + bi)(c + di)| = |(a + bi)| \cdot |c + di|$

Show that $|(a + bi) + (c + di)| \leq |a + bi| \cdot |c + di|$

In Sec. 7-7 we introduced two ways of expressing a complex number, namely, $a + bi$ and the ordered pair (a, b), where a and b are real numbers and $i = \sqrt{-1}$. So far in this chapter we have used the form $a + bi$, although we could have used the form (a, b). Turning now to this alternative form, we express the sum, difference, product, and quotient of two complex numbers as follows.

Sum: $(a, b) + (c, d) = (a + c, b + d)$

Difference: $(a, b) - (c, d) = (a - c, b - d)$

Product: $(a, b) \cdot (c, d) = (ac - bd, bc + ad)$

Quotient: $\dfrac{(a, b)}{(c, d)} = \left(\dfrac{ac + bd}{c^2 + d^2}, \dfrac{bc - ad}{c^2 + d^2} \right)$ (c, d not both 0).

EXAMPLES.

$$(2, 3) + (4, 1) = (2 + 4, 3 + 1) = (6, 4)$$

$$(5, -2) + (-3, 6) = (5 - 3, -2 + 6) = (2, 4)$$

$$(6, -1) - (2, 2) = (6 - 2, -1 - 2) = (4, -3)$$

$$(3, 5) \cdot (4, -1) = (3 \cdot 4 - 5 \cdot (-1), 5 \cdot 4 + 3(-1)) = (17, 17)$$

$$\frac{(2, 3)}{(5, 6)} = \left(\frac{2 \cdot 5 + 3 \cdot 6}{25 + 36}, \frac{3 \cdot 5 - 2 \cdot 6}{25 + 36} \right) = \left(\frac{28}{61}, \frac{3}{61} \right)$$

Perform the indicated operations, expressing each result as an ordered pair.

55. $(2, 3) + (1, 4)$ **56.** $(-3, 7) + (4, 2)$
57. $(5, 4) + (3, 2)$ **58.** $(3, -1) + (-5, 4)$
59. $(4, 5) - (3, 2)$ **60.** $(2, -3) - (4, -2)$
61. $(-3, 4) - (-4, 3)$ **62.** $(3, -6) - (-4, -2)$
63. $(3, 4) \cdot (1, 2)$ **64.** $(5, 2) \cdot (4, 3)$
65. $(-2, 7) \cdot (3, 2)$ **66.** $(-3, 5) \cdot (-6, -2)$

67. $\dfrac{(2, 4)}{(-1, 2)}$

68. $\dfrac{(-10, 5)}{(2, 1)}$

69. $\dfrac{(26, 13)}{(2, 3)}$

70. $\dfrac{(15, 5)}{(-3, 1)}$

Chapter 16

MATRICES AND DETERMINANTS

16-1 MATRICES AND LINEAR EQUATIONS

In this chapter we shall give a brief introduction to matrices, study determinants, and apply these concepts in solving systems of linear equations. We begin by solving a system of equations by the use of matrices.

Definition 16-1   A rectangular array of numbers forming m rows and n columns is called a *matrix*. Each number of a matrix is called an *element*.

We list here three examples of matrices, denoting each with a capital letter:

$$A = \begin{bmatrix} 2 & 4 \\ -3 & 7 \end{bmatrix}, \qquad B = \begin{bmatrix} 5 & 2 & 1 \\ 3 & 0 & 6 \end{bmatrix}, \qquad C = \begin{bmatrix} 1 & 2 & 4 \\ 0 & -5 & 6 \\ 7 & 8 & 0 \end{bmatrix}$$

It is customary to place brackets, parentheses, or double vertical bars about the elements of a matrix. Matrix A has 2 rows and 2 columns, B has 2 rows and 3 columns, and C has 3 rows and 3 columns. Accordingly they are said to be 2 by 2, 2 by 3, and 3 by 3 matrices, respectively. In general, a matrix of m rows and n columns is an m by n matrix, the number of rows given first.

Let us now see how a system of linear equations can be represented by the use of a matrix. Consider the system

$$2x + 3y + z = 9$$
$$x - 2y - 3z = 1 \qquad (1)$$
$$5x + 4y + 6z = 5$$

We form a 3 by 3 matrix by using the coefficients of the variables, and form a 3 by 4 matrix by joining the constant terms to the coefficients of the variables. Thus we have

$$A = \begin{bmatrix} 2 & 3 & 1 \\ 1 & -2 & -3 \\ 5 & 4 & 6 \end{bmatrix}, \qquad B = \begin{bmatrix} 2 & 3 & 1 & 9 \\ 1 & -2 & -3 & 1 \\ 5 & 4 & 6 & 5 \end{bmatrix}$$

The first matrix is called the *coefficient matrix* and the second matrix the *augmented matrix* for the given system. We note that both matrices of a system such as (1) can be written at once, and that the system of linear equations can be readily written from the corresponding augmented matrix.

Before solving a system of equations by the use of matrices, we recall that the following operations may be performed on a system such as (1) to obtain equivalent systems:

1. The members of any equation may be multiplied by a nonzero constant.
2. The members of one equation may be multiplied by a nonzero constant and added to (or subtracted from) the corresponding members of any other equation.
3. The equations may be written in any chosen order.

Interpreting these operations with respect to the matrix representing the system of equations, we see that

1. The elements of any row may be multiplied by a nonzero constant.
2. The elements of any row may be multiplied by a nonzero constant and added to the corresponding elements of any other row.
3. The rows may be written in any chosen order.

EXAMPLE. Use matrices to solve the system of equations

$$3x + y - 5z = 2$$
$$-x + 4y + 2z = 1$$
$$x + 2y - z = 3$$

Solution. The augmented matrix representing this system is

$$A = \begin{bmatrix} 3 & 1 & -5 & 2 \\ -1 & 4 & 2 & 1 \\ 1 & 2 & -1 & 3 \end{bmatrix}$$

In order to pass to a matrix with two zeros in one of the first three columns, we multiply the elements of the third row of A by 3 and subtract from the corresponding elements of the first row, and next we add the third row to the second row. This gives the matrix

$$B = \begin{bmatrix} 0 & -5 & -2 & -7 \\ 0 & 6 & 1 & 4 \\ 1 & 2 & -1 & 3 \end{bmatrix} \qquad C = \begin{bmatrix} 0 & 7 & 0 & 1 \\ 0 & 6 & 1 & 4 \\ 1 & 8 & 0 & 7 \end{bmatrix}$$

Continuing with B, we multiply the second row by 2 and add to the first row. Then we add the second row to the third row to get C. We now divide the first row of C by 7 and have

$$D = \begin{bmatrix} 0 & 1 & 0 & \frac{1}{7} \\ 0 & 6 & 1 & 4 \\ 1 & 8 & 0 & 7 \end{bmatrix} \qquad E = \begin{bmatrix} 0 & 1 & 0 & \frac{1}{7} \\ 0 & 0 & 1 & \frac{22}{7} \\ 1 & 0 & 0 & \frac{41}{7} \end{bmatrix}$$

The element 1 in the first row of D can now be conveniently used to pass to E with two zeros in the second column. We get E by multiplying the first row of D by 6 and subtracting from the second row, and then multiplying the first row of D by 8 and subtracting from the third row. Finally changing the order of the rows appearing in E, we have

$$F = \begin{bmatrix} 1 & 0 & 0 & \frac{41}{7} \\ 0 & 1 & 0 & \frac{1}{7} \\ 0 & 0 & 1 & \frac{22}{7} \end{bmatrix}$$

The equations corresponding to matrix F are $x = \frac{41}{7}$, $y = \frac{1}{7}$, and $z = \frac{22}{7}$. Since the systems of equations corresponding to the six matrices are all equivalent, we see that the solution set of the given system is $\{(\frac{41}{7}, \frac{1}{7}, \frac{22}{7})\}$.

EXERCISE 16-1

Solve the following systems of equations by the use of matrices.

1. $2x - y + 3z = 9$
$3x + y + 2z = 11$
$x - y + z = 3$

2. $x - 4y + 5z = -4$
$x + 3y + z = 6$
$2x - 3y + 2z = -6$

3. $3x - 4y - z = 1$
$x - y + 3z = 3$
$3x - 2y + 2z = 0$

4. $x - 3y - 7z = 6$
$4x + y = 7$
$2x + 3y + z = 9$

5. $x + y + z = 0$
$3x + y + z = 2$
$5x - 2y + 3z = -8$

6. $3x - 5z = -1$
$2x + 7y = 6$
$x + y + z = 5$

7. $x - y - z = 0$
$2x + 3y + 6z = 3$
$4x + 2y + 2z = 3$

8. $x + 2y + 3z = 4$
$2x + y + z = 1$
$3x + 3y + z = 2$

9. $x + y - z + w = 0$
$3x + 2y - 2z - 3w = 7$
$2x - 4y + 3z + 2w = 8$
$2x + 2y - z - w = 1$

10. $x - y + z - w = 3$
$x + y + z + w = -5$
$x - 3y - z - w = 9$
$x + y - z + w = 1$

16-2 DETERMINANTS OF ORDERS TWO AND THREE

A matrix of n rows and n columns is called a square matrix of order n. With a square matrix we shall associate a number called the *determinant* of the matrix. The determinant of a matrix is said to be of the same order as the matrix. We define the determinant of the second-order matrix

$$\begin{bmatrix} a_1 & b_1 \\ a_2 & b_2 \end{bmatrix}$$

to be the number $a_1 b_2 - a_2 b_1$. To symbolize this definition, we write

$$\det \begin{bmatrix} a_1 & b_1 \\ a_2 & b_2 \end{bmatrix} = \begin{vmatrix} a_1 & b_1 \\ a_2 & b_2 \end{vmatrix} = a_1 b_2 - a_2 b_1$$

It is customary to indicate the determinant of a matrix by enclosing the elements of the matrix by vertical bars instead of brackets.

EXAMPLE 1.

$$\begin{vmatrix} 3 & 4 \\ -2 & 5 \end{vmatrix} = 3(5) - (-2)(4) = 15 + 8 = 23$$

We next show how to express the solution of a system of linear equations by the use of determinants. The system

$$\begin{aligned} a_1 x + b_1 y &= c_1 \\ a_2 x + b_2 y &= c_2 \end{aligned} \tag{1}$$

may be solved by either method of Sec. 6-6 to yield

$$x = \frac{b_2 c_1 - b_1 c_2}{a_1 b_2 - a_2 b_1}, \qquad y = \frac{a_1 c_2 - a_2 c_1}{a_1 b_2 - a_2 b_1} \tag{2}$$

The numerator and denominator of each of these fractions are the numbers associated with matrices of order two. Hence we may express the solution in the symbolic form

$$x = \frac{\begin{vmatrix} c_1 & b_1 \\ c_2 & b_2 \end{vmatrix}}{\begin{vmatrix} a_1 & b_1 \\ a_2 & b_2 \end{vmatrix}}, \qquad y = \frac{\begin{vmatrix} a_1 & c_1 \\ a_2 & c_2 \end{vmatrix}}{\begin{vmatrix} a_1 & b_1 \\ a_2 & b_2 \end{vmatrix}} \tag{3}$$

These formulas may be applied to find the solution of any consistent system of the form (1), that is, systems for which $a_1 b_2 - a_2 b_1 \neq 0$.

EXAMPLE 2. Solve the system of equations

$$3x + 2y = 3$$

$$2x - 5y = -17$$

Solution. Taking $a_1 = 3$, $b_1 = 2$, $c_1 = 3$ and $a_2 = 2$, $b_2 = -5$, $c_2 = -17$ and substituting in formulas (3), we get

$$x = \frac{\begin{vmatrix} 3 & 2 \\ -17 & -5 \end{vmatrix}}{\begin{vmatrix} 3 & 2 \\ 2 & -5 \end{vmatrix}} = \frac{-15 + 34}{-15 - 4} = \frac{19}{-19} = -1$$

$$y = \frac{\begin{vmatrix} 3 & 3 \\ 2 & -17 \end{vmatrix}}{\begin{vmatrix} 3 & 2 \\ 2 & -5 \end{vmatrix}} = \frac{-51 - 6}{-15 - 4} = \frac{-57}{-19} = 3$$

We define the determinant of a third-order matrix by the equation

$$\begin{vmatrix} a_1 & b_1 & c_1 \\ a_2 & b_2 & c_2 \\ a_3 & b_3 & c_3 \end{vmatrix} = a_1 \begin{vmatrix} b_2 & c_2 \\ b_3 & c_3 \end{vmatrix} - b_1 \begin{vmatrix} a_2 & c_2 \\ a_3 & c_3 \end{vmatrix} + c_1 \begin{vmatrix} a_2 & b_2 \\ a_3 & b_3 \end{vmatrix}$$

In this equation we have a symbol for the determinant of a third-order matrix and symbols for the determinants of three second-order matrices. Symbols of this kind, though denoting values of determinants of matrices, are often referred to simply as determinants.

EXAMPLE 3. Find the value of the determinant

$$\begin{vmatrix} 2 & 5 & 2 \\ 1 & -3 & 1 \\ -1 & 1 & -5 \end{vmatrix}$$

Solution. Employing the definition of a third-order determinant and evaluating the resulting second-order determinants, we find

$$\begin{vmatrix} 2 & 5 & 2 \\ 1 & -3 & 1 \\ -4 & 1 & -5 \end{vmatrix} = 2 \begin{vmatrix} -3 & 1 \\ 1 & -5 \end{vmatrix} - 5 \begin{vmatrix} 1 & 1 \\ -4 & -5 \end{vmatrix} + 2 \begin{vmatrix} 1 & -3 \\ -4 & 1 \end{vmatrix}$$

$$= 2(15 - 1) - 5(-5 + 4) + 2(1 - 12)$$

$$= 28 + 5 - 22$$

$$= 11$$

Evaluate each of the following determinants.

1. $\begin{vmatrix} 3 & 4 \\ 9 & -7 \end{vmatrix}$

2. $\begin{vmatrix} 5 & -6 \\ 4 & 3 \end{vmatrix}$

3. $\begin{vmatrix} 4 & -8 \\ 7 & 11 \end{vmatrix}$

4. $\begin{vmatrix} a & 5 \\ 3a & 2 \end{vmatrix}$

5. $\begin{vmatrix} 5 & 20 \\ 0 & -1 \end{vmatrix}$

6. $\begin{vmatrix} x & y \\ 5 & 6 \end{vmatrix}$

Solve the following systems of equations for x and y, using determinants.

7. $4x + 3y = 1$
   $2x + 5y = 11$

8. $2x - 4y = -3$
   $4x + 2y = 9$

9. $3x - 4y = 7$
   $5x - 5y = 8$

10. $2x = 7y + 17$
    $4x = 5y + 25$

11. $7x = 4 - 3y$
    $4x = 3 - 2y$

12. $5x + 7y = -2$
    $4x + 6y = -3$

13. $ax - 3by = c$
    $2ax + 2by = 9c$

14. $ax - by = 2b$
    $3ax + 4by = 7a - b$

15. $3x + ay = 4$
    $2x + ay = 2$

Evaluate each of the third-order determinants.

16. $\begin{vmatrix} 3 & 4 & 5 \\ 2 & 1 & 4 \\ 3 & 2 & 1 \end{vmatrix}$

17. $\begin{vmatrix} 1 & -3 & 2 \\ 4 & -1 & 2 \\ 3 & 5 & 2 \end{vmatrix}$

18. $\begin{vmatrix} 0 & 2 & 6 \\ 1 & 5 & 0 \\ 3 & 7 & 1 \end{vmatrix}$

19. $\begin{vmatrix} -3 & 5 & 1 \\ -4 & 1 & 3 \\ 3 & 1 & 0 \end{vmatrix}$

20. $\begin{vmatrix} 6 & 7 & 0 \\ 0 & -9 & 2 \\ 4 & 0 & -5 \end{vmatrix}$

21. $\begin{vmatrix} 0 & 3 & 4 \\ 2 & 0 & 1 \\ 5 & 6 & 0 \end{vmatrix}$

Solve the following equations for x.

22. $\begin{vmatrix} x & 14 \\ 3 & 11 \end{vmatrix} = 2$

23. $\begin{vmatrix} 4 & -10 \\ x & -3 \end{vmatrix} = 8$

24. $\begin{vmatrix} 2 - x & 3 \\ 2 + x & 4 \end{vmatrix} = 0$

25. $\begin{vmatrix} 6 & -x \\ x - 4 & -2 \end{vmatrix} = 0$

26. $\begin{vmatrix} 2 & x & 1 \\ 1 & 0 & 1 \\ 3 & 4 & 2 \end{vmatrix} = 0$

27. $\begin{vmatrix} 1 & 4 & -3 \\ 4 & x & -5 \\ 2 & 0 & -3 \end{vmatrix} = 0$

16-3 DETERMINANTS OF ORDER n

We have defined determinants of orders two and three. We wish now to give a general definition of determinants which applies to those of orders one, two, three, and higher orders. First, however, let us explain the notion of inversions which is helpful in defining and studying the properties of determinants. In a sequence of positive integers, an *inversion* occurs whenever a larger number precedes a smaller number. Thus the sequence 2143 has two inversions because 2 precedes 1, and 4 precedes 3. But 5143 has four inversions; 5 preceding 1, 4, and 3 furnishes three inversions, and 4 preceding 3 gives another inversion.

Definition 16-2 If A is an n by n matrix, the determinant of the matrix is denoted symbolically by

$$\begin{vmatrix} a_{11} & a_{12} & a_{13} \cdots a_{1n} \\ a_{21} & a_{22} & a_{23} \cdots a_{2n} \\ a_{31} & a_{32} & a_{33} \cdots a_{3n} \\ \vdots & & \vdots \\ a_{n1} & a_{n2} & a_{n3} \cdots a_{nn} \end{vmatrix}$$

The subscripts here indicate the row and column of each number, the first subscript giving the row and the second the column. Thus, a_{34} stands for the number in the third row and fourth column, and a_{ij} stands for the number in the *i*th row and *j*th column.

The determinant stands for the number which is the algebraic sum of all possible products obtained:

1. By taking as factors one and only one number from each row and each column, and
2. By giving to each product a plus or minus sign according as there is an even or odd number of inversions of the second (the column) subscripts when the factors of the product are written so that the first (the row) subscripts have no inversions.

Each of the numbers in the square array is called an *element* of the determinant. The elements along the line from the upper left to the lower right corner form the *principal diagonal*. Each of the signed products described in the definition is called a *term* of the determinant and the sum of the terms is the *expansion*, or *value*, of the determinant. There are, of course, n factors in each term since one and only one factor is taken from each row and each column.

According to the preceding definition, the expansion of a third-order determinant is given by the equation

$$\begin{vmatrix} a_{11} & a_{12} & a_{13} \\ a_{21} & a_{22} & a_{23} \\ a_{31} & a_{32} & a_{33} \end{vmatrix} = \begin{aligned} &a_{11}a_{22}a_{33} - a_{11}a_{23}a_{32} - a_{12}a_{21}a_{33} + a_{12}a_{23}a_{31} \\ &+ a_{13}a_{21}a_{32} - a_{13}a_{22}a_{31} \end{aligned}$$

In this expansion each product has exactly one factor from each row and each column, and the first subscripts are in the order 1, 2, 3. The inversions of the second subscripts of the factors of the terms in the order written are respectively zero, one, one, two, two, three. The signs of the terms with an even number of inversions are positive and those with an odd number of inversions are negative.

16-4 PROPERTIES OF DETERMINANTS*

From the definition of determinants a number of interesting and useful properties can be established. In this section we shall consider some of the elementary properties, including those which lead toward a practical method of evaluating determinants.

Theorem 16-1 The expansion of a determinant of order n has $n!$ terms.

Proof. If the n factors of a term are written with the first subscripts in order of magnitude, then any arrangement of the integers 1 to n, inclusive, as the second subscripts yields the factors of one term. But the n subscripts can be assigned in $n!$ different ways (Sec. 12-2). Since there is a term for each arrangement of the second subscripts, the expansion has $n!$ terms. ∎

Theorem 16-2 The value of a determinant is not changed if corresponding rows and columns are interchanged.

Illustrating the theorem with a third-order determinant, we have

$$\begin{vmatrix} a_{11} & a_{12} & a_{13} \\ a_{21} & a_{22} & a_{23} \\ a_{31} & a_{32} & a_{33} \end{vmatrix} = \begin{vmatrix} a_{11} & a_{21} & a_{31} \\ a_{12} & a_{22} & a_{32} \\ a_{13} & a_{23} & a_{33} \end{vmatrix}$$

Proof. We let D stand for a determinant of order n and D' for the determinant after the columns of D are changed to corresponding rows. If the first subscripts in D denote rows and the second subscripts denote columns, then the first subscripts in D' denote columns and the second subscripts denote rows. The definition of a determinant requires that a term of D and the corresponding term of D' have the same factors. But the sign of the term of D is determined by the inversions of the second subscripts when the first subscripts are in order of magnitude; whereas the sign of the corresponding term of D' is determined by the inversions of the first subscripts when the second subscripts are in order of magnitude. Hence the corresponding terms of the determinants are the same except for a possible difference in signs. Now suppose we interchange adjacent factors of the term of D' repeatedly until the first subscripts are in order of magnitude. Such a series of interchanges introduces the same number of inversions of the second subscripts as the first subscripts had originally. Hence the corresponding terms have the same sign as well as the same factors, and consequently the given determinant and the new determinant have the same value. ∎

Note. Since the interchange of corresponding rows and columns of a determinant does not change the value of the determinant, a property of a determinant involving its rows is also a property involving its columns, and vice

* Depending on the time at hand, and other circumstances, it may be advisable to illustrate some of the following theorems and omit the proofs, especially the lengthier ones.

versa. Hence the proof of a property pertaining to rows or columns of a determinant need be made for only one of the two.

Theorem 16-3 If each of the elements of a row (or column) of a determinant is equal to zero, then the value of the determinant is equal to zero.

Proof. This property follows at once from the fact that each term of the expansion has a factor from every row and every column of the determinant. A product in which zero is a factor is, of course, zero. ■

Theorem 16-4 If two columns (or rows) of a determinant are interchanged, the sign of the determinant is changed.

Proof. First let us suppose that the two columns to be interchanged are adjacent to each other. In this case the interchange will leave the first subscripts of the factors of every term of the expansion unchanged but will either increase or decrease the number of inversions of the second subscripts by one. Hence the sign of the determinant will be reversed.

Now if k columns lie between the two columns to be interchanged, the two particular columns can be brought to their new positions by a succession of interchanges of adjacent columns. First, k interchanges will bring the two particular columns next to each other, then an interchange of the two puts one column in its desired position, and k more interchanges brings the other column to its position. This means that the $2k + 1$ interchanges effect the trade in the positions of the two particular columns and leave all other columns in their original positions. Since the interchange of two adjacent columns of a determinant produces a determinant whose sign is opposite that of the first, the succession of interchanges produces $2k + 1$ reversals of signs, and consequently the sign of the final determinant is opposite the sign of the original determinant. ■

Theorem 16-5 If two columns (or rows) of a determinant are identical, the value of the determinant is zero.

Proof. If the two identical columns are interchanged, the value of the determinant is unchanged. But, by Theorem 16-4, the sign of the determinant is reversed. Hence if D is the value of the determinant, the value of the determinant after the interchange of the identical columns is $-D$. Then we have $D = -D$, $2D = 0$, and $D = 0$. ■

Theorem 16-6 If each of the elements of a column (or row) of a determinant is multiplied by a number m, the value of the determinant is multiplied by m.

Proof. Since each term of the expansion has one and only one factor from each row and each column, m is a factor of each term of the expansion of the new determinant. Then the expansion can be written as m times the expansion of the first determinant. ■

EXAMPLE.

$$\begin{vmatrix} 2 & 5m & 1 \\ 4 & m & 2 \\ 3 & 8m & 7 \end{vmatrix} = m \begin{vmatrix} 2 & 5 & 1 \\ 4 & 1 & 2 \\ 3 & 8 & 7 \end{vmatrix}$$

Corollary   If each element of a column (or row) of a determinant is a constant times the corresponding elements of another column (or row), the value of the determinant is zero.

We leave the proof of this Corollary to the student.

Theorem 16-7   If each element of a column, the kth say, of a determinant is expressed as the sum of two terms, the determinant is equal to the sum of two determinants, where

   (a) the elements of each of the two determinants are identical to the corresponding elements of the given determinant except for the elements of the kth column;

   (b) the first terms of the elements of the kth column of the given determinant form the kth column of one of the two determinants and the second terms form the kth column of the other determinant.

Proof. Consider the determinants in the equation

$$\begin{vmatrix} a_{11} + b_{11} & a_{12} & a_{13} \\ a_{21} + b_{21} & a_{22} & a_{23} \\ a_{31} + b_{31} & a_{32} & a_{33} \end{vmatrix} = \begin{vmatrix} a_{11} & a_{12} & a_{13} \\ a_{21} & a_{22} & a_{23} \\ a_{31} & a_{32} & a_{33} \end{vmatrix} + \begin{vmatrix} b_{11} & a_{12} & a_{13} \\ b_{21} & a_{22} & a_{23} \\ b_{31} & a_{32} & a_{33} \end{vmatrix}$$

Each term of the expansion of the determinant on the left side contains the sum of two terms as a factor, which is equal to the sum of the corresponding terms of the determinants on the right side. For example, $(a_{11} + b_{11})a_{22}a_{33}$ is a term of the left side of the equation, and the corresponding terms, $a_{11}a_{22}a_{33}$ and $b_{11}a_{22}a_{33}$, appear on the right side. Hence the expansion of the left side of the equation is equal to the sum of the expansions on the right side.

The reasoning which we have used here applies in the case of a determinant of any order. We conclude, therefore, that the theorem is true. ∎

Theorem 16-8   The value of a determinant is not changed if each element of a column (or row) is multiplied by a number m and added to the corresponding elements of any column (or row).

Proof. Let us show that

$$D = \begin{vmatrix} a_{11} & a_{12} & a_{13} + ma_{11} \\ a_{21} & a_{22} & a_{23} + ma_{21} \\ a_{31} & a_{32} & a_{33} + ma_{31} \end{vmatrix} = \begin{vmatrix} a_{11} & a_{12} & a_{13} \\ a_{21} & a_{22} & a_{23} \\ a_{31} & a_{32} & a_{33} \end{vmatrix}$$

By Theorem 16-7, the determinant on the left may be written as the sum of two determinants. That is,

$$D = \begin{vmatrix} a_{11} & a_{12} & a_{13} \\ a_{21} & a_{22} & a_{23} \\ a_{31} & a_{32} & a_{33} \end{vmatrix} + \begin{vmatrix} a_{11} & a_{12} & ma_{11} \\ a_{21} & a_{22} & ma_{21} \\ a_{31} & a_{32} & ma_{31} \end{vmatrix}$$

But the second of these determinants is equal to zero since it may be expressed as m times a determinant with the first and third columns identical.

Inasmuch as the method of proof for a third-order determinant is applicable for the case of a determinant of any order, we conclude that the theorem is true. ∎

EXAMPLE. Without expanding the determinant find the root of the equation

$$\begin{vmatrix} 3 & 2 & 4 \\ 4 & -1 & -2 \\ 5 & x & 4 \end{vmatrix} = 0$$

Solution. If we replace x by 2, the elements of the third column become twice the corresponding elements of the second column, and the value of the determinant zero. Why zero? Hence $x = 2$ is the root of the equation.

EXERCISE 16-3

Without expanding show that the value of each of the following determinants is zero.

1. $\begin{vmatrix} 3 & -1 & 3 \\ 2 & 4 & 2 \\ 5 & -3 & 5 \end{vmatrix}$ 2. $\begin{vmatrix} 1 & 6 & -6 \\ 5 & 2 & -2 \\ 4 & -3 & 3 \end{vmatrix}$ 3. $\begin{vmatrix} 1 & 2 & 3 \\ 2 & 4 & 6 \\ 0 & 2 & 1 \end{vmatrix}$

4. $\begin{vmatrix} 1 & 2 & -5 \\ 2 & 4 & -4 \\ 0 & 0 & -3 \end{vmatrix}$ 5. $\begin{vmatrix} 0 & 1 & 0 \\ 5 & -6 & -7 \\ 0 & -4 & 0 \end{vmatrix}$ 6. $\begin{vmatrix} 7 & 6 & 5 \\ 6 & 5 & 4 \\ 5 & 4 & 3 \end{vmatrix}$

Without expanding the determinants, show that the following equations are true.

7. $\begin{vmatrix} 4 & 1 & 6 \\ 2 & 3 & 15 \\ 6 & 0 & 3 \end{vmatrix} = 6 \begin{vmatrix} 2 & 1 & 2 \\ 1 & 3 & 5 \\ 3 & 0 & 1 \end{vmatrix}$ 8. $\begin{vmatrix} 5 & -3 & 1 \\ 4 & 3 & 1 \\ 1 & -2 & 6 \end{vmatrix} = \begin{vmatrix} 9 & 0 & 2 \\ 4 & 3 & 1 \\ 1 & -2 & 6 \end{vmatrix}$

9. $\begin{vmatrix} 6 & 4 & 0 \\ 3 & 2 & 1 \\ 5 & 3 & 1 \end{vmatrix} = \begin{vmatrix} 2 & 4 & 4 \\ 1 & 2 & 3 \\ 2 & 3 & 4 \end{vmatrix}$ 10. $\begin{vmatrix} 2 & 4 & 5 \\ -3 & 7 & 1 \\ 5 & 6 & 0 \end{vmatrix} + \begin{vmatrix} 3 & 4 & 5 \\ 5 & 7 & 1 \\ 4 & 6 & 0 \end{vmatrix} = \begin{vmatrix} 5 & 4 & 5 \\ 2 & 7 & 1 \\ 9 & 6 & 0 \end{vmatrix}$

Without expanding the determinants, solve the following equations for x.

11. $\begin{vmatrix} 2 & 1 & 4 \\ 6 & x & 12 \\ 5 & 0 & 3 \end{vmatrix} = 0$ 12. $\begin{vmatrix} 2 & x & -1 \\ 4 & x^2 & 1 \\ 5 & 5 & 5 \end{vmatrix} = 0$

13. $\begin{vmatrix} 2 & 4 & 8 \\ x & x^2 & x^3 \\ -2 & 4 & -8 \end{vmatrix} = 0$ 14. $\begin{vmatrix} x & a & -3 \\ x^2 & a^2 & 9 \\ 2 & 2 & 2 \end{vmatrix} = 0$

16-5 EXPANSION BY COFACTORS

The definition of the determinant of a matrix (Sec. 16-3) describes its expansion and hence provides a way for evaluating the determinant. But evaluating a determinant of order four or higher directly from the definition would be quite tedious. In this section we shall develop a practical method for finding the values of determinants. As a preliminary step, however, we need to introduce the idea of a minor and cofactor corresponding to an element of a determinant.

Definition 16-3 The *minor* of an element of a determinant of order n is the determinant of order $n - 1$ which remains after the row and column of the element are removed.

EXAMPLE 1. Denoting the minors of a_{23} and a_{41} of the determinant

$$\begin{vmatrix} a_{11} & a_{12} & a_{13} & a_{14} \\ a_{21} & a_{22} & a_{23} & a_{24} \\ a_{31} & a_{32} & a_{33} & a_{34} \\ a_{41} & a_{42} & a_{43} & a_{44} \end{vmatrix} \qquad (1)$$

by A_{23} and A_{41}, we have

$$A_{23} = \begin{vmatrix} a_{11} & a_{12} & a_{14} \\ a_{31} & a_{32} & a_{34} \\ a_{41} & a_{42} & a_{44} \end{vmatrix} \quad \text{and} \quad A_{41} = \begin{vmatrix} a_{12} & a_{13} & a_{14} \\ a_{22} & a_{23} & a_{24} \\ a_{32} & a_{33} & a_{34} \end{vmatrix}$$

Definition 16-4 The *cofactor* of an element a_{ij} of a determinant is the product of the minor of a_{ij} and $(-1)^{i+j}$. That is, if the minor and cofactor of a_{ij} are denoted respectively by A_{ij} and C_{ij}, then $C_{ij} = (-1)^{i+j} A_{ij}$.

According to this definition, the cofactors of a_{23}, a_{41}, and a_{42} of the determinant above are respectively $C_{23} = -A_{23}$, $C_{41} = -A_{41}$, and $C_{42} = A_{42}$.

Theorem 16-9 The value of a determinant may be obtained by the following steps:

(a) Multiply each element of a selected row (or column) by its cofactor.

(b) Then take the algebraic sum of the resulting products.

Before proving this theorem, we illustrate its meaning by referring to the fourth-order determinant of Example 1. The value of the determinant may be expressed by

$$a_{11}C_{11} + a_{12}C_{12} + a_{13}C_{13} + a_{14}C_{14}$$

which is called the expansion according to the elements of the first row. Similarly the expansion according to the elements of the third column is

$$a_{13}C_{13} + a_{23}C_{23} + a_{33}C_{33} + a_{43}C_{43}$$

Proof. We first point out that the terms of a determinant D having a_{11} as a factor are given by $a_{11}A_{11}$, where A_{11} is the minor of a_{11}. This is true because each term of the expansion of D has one and only one factor from each row and each column, and consequently the terms of D involving a_{11} can have no other factor from the first row and first column. Now if each term of A_{11} is multiplied by a_{11}, no change is made in the number of inversions of the subscripts of any product. Hence the product of a_{11} and the expansion of A_{11} are terms of D. Thus we see that the product of the element in the first row and column and its minor are terms of D involving that element.

Next we consider the product of an element and its minor if the element is in the ith row and jth column. This element a_{ij} can be transferred to the first row and first column in the following way:

(a) First bring the ith row to row one by $i - 1$ interchanges of adjacent rows.

(b) Then bring the jth column to column one by $j - 1$ interchanges of adjacent columns.

It should be noticed that these interchanges involve adjacent rows or adjacent columns, and therefore do not disturb the minor of a_{ij}. Further, each interchange reverses the sign of the determinant. Since the total number of interchanges is $i + j - 2$, the final determinant D' and the original determinant are related by the equation

$$D' = (-1)^{i+j-2}D = (-1)^{i+j}D$$

We conclude, then, that the terms of the expansion of D with a_{ij} as a factor are given by

$$a_{ij}(-1)^{i+j}A_{ij} = a_{ij}C_{ij}$$

where A_{ij} is the minor of a_{ij} and C_{ij} is the cofactor.

We recall that each term of the expansion of a determinant has as a factor one and only one element from each row and each column. Hence the sum of the products obtained by multiplying each element of any selected row (or column) by its cofactor yields all the terms of the expansion. This completes the proof of the theorem. ∎

Expanding by cofactors, we can express a determinant of order n in terms of determinants of order $n - 1$. Each of the resulting determinants can then be expressed in terms of determinants of order $n - 2$. A continuation of this process leads to determinants of order 2, whose expansions can be easily written. The evaluation of determinants, however, can be greatly facilitated by employing certain theorems which were established in Sec. 16-4. A particularly helpful device is to obtain a determinant whose elements, except one, of some row or column, are zero. We shall illustrate with examples.

EXAMPLE 2. Evaluate the determinant

$$D = \begin{vmatrix} 2 & -3 & 4 \\ 3 & 5 & -1 \\ 7 & 6 & -2 \end{vmatrix}$$

Solution. Let us obtain an equivalent determinant which has all the elements, except one, of some row or column equal to zero. The element -1 in the second row, third column, makes this end easily achieved. We multiply the elements of the second row by 4 and add the products to the corresponding elements of the first row. Next we multiply the elements of the second row by -2 and add the products to the corresponding elements of the third row. These operations do not alter the value of the determinant (Theorem 16-8), and therefore

$$D = \begin{vmatrix} 2 & -3 & 4 \\ 3 & 5 & -1 \\ 7 & 6 & -2 \end{vmatrix} = \begin{vmatrix} 14 & 17 & 0 \\ 3 & 5 & -1 \\ 1 & -4 & 0 \end{vmatrix}$$

Taking advantage of the zero elements, we expand the new determinant according to the elements of the third column and have

$$D = 0 - (-1)\begin{vmatrix} 14 & 17 \\ 1 & -4 \end{vmatrix} + 0 = 14(-4) - 1(17) = -73$$

Notice that the sum of the row number and the column number of the element -1, given by $2 + 3 = 5$, is odd. Consequently the cofactor of this element is the negative of the minor.

EXAMPLE 3. Evaluate the determinant

$$D = \begin{vmatrix} 2 & -12 & 7 & 5 \\ -4 & 6 & 4 & -3 \\ 10 & 0 & 1 & 2 \\ 6 & 9 & 5 & 0 \end{vmatrix}$$

Solution. We first factor 2 from the elements of column one and 3 from the elements of column two, thus getting

$$D = 6 \begin{vmatrix} 1 & -4 & 7 & 5 \\ -2 & 2 & 4 & -3 \\ 5 & 0 & 1 & 2 \\ 3 & 3 & 5 & 0 \end{vmatrix}$$

To obtain additional zeros in the third row, we multiply the elements of the third column by 5 and subtract the products from the corresponding elements of the first column, and next multiply the elements of the third column by 2 and subtract the products from the corresponding elements of the fourth column. This gives the determinant

$$D = 6 \begin{vmatrix} -34 & -4 & 7 & -9 \\ -22 & 2 & 4 & -11 \\ 0 & 0 & 1 & 0 \\ -22 & 3 & 5 & -10 \end{vmatrix}$$

We now expand according to the elements of the third row and have

$$D = 6 \begin{vmatrix} -34 & -4 & -9 \\ -22 & 2 & -11 \\ -22 & 3 & -10 \end{vmatrix}$$

We next subtract the elements of the second row from the corresponding elements of the third row; and, in the resulting determinant, subtract the elements of the third column from the corresponding elements of the second column. These operations give

$$D = 6 \begin{vmatrix} -34 & -4 & -9 \\ -22 & 2 & -11 \\ 0 & 1 & 1 \end{vmatrix} = 6 \begin{vmatrix} -34 & 5 & -9 \\ -22 & 13 & -11 \\ 0 & 0 & 1 \end{vmatrix}$$

and

$$D = 6 \begin{vmatrix} -34 & 5 \\ -22 & 13 \end{vmatrix} = 6(-442 + 110) = -1992$$

Evaluate the following determinants.

1. $\begin{vmatrix} 1 & 2 & 3 \\ 6 & 4 & -5 \\ 4 & 0 & 1 \end{vmatrix}$
  2. $\begin{vmatrix} 6 & -3 & 7 \\ 7 & 5 & -9 \\ 6 & 3 & 7 \end{vmatrix}$
  3. $\begin{vmatrix} 5 & 0 & 1 \\ 3 & 4 & 2 \\ 6 & 1 & 3 \end{vmatrix}$

4. $\begin{vmatrix} 0 & 5 & 4 & 1 \\ 3 & 2 & 3 & -2 \\ 0 & 1 & 2 & 3 \\ 3 & 0 & 1 & 4 \end{vmatrix}$
  5. $\begin{vmatrix} 6 & 8 & 7 & 2 \\ 3 & 2 & 4 & 1 \\ 2 & 3 & 8 & 2 \\ 5 & 4 & 2 & 1 \end{vmatrix}$
  6. $\begin{vmatrix} 3 & 2 & 2 & 3 \\ 3 & 1 & 1 & 1 \\ 6 & 2 & 2 & 1 \\ 9 & 0 & 1 & 1 \end{vmatrix}$

7. $\begin{vmatrix} 2 & -1 & 2 & 0 \\ 0 & -2 & 1 & 6 \\ 6 & 0 & 0 & 1 \\ 4 & 3 & 4 & 2 \end{vmatrix}$
  8. $\begin{vmatrix} 4 & 8 & 8 & 4 \\ 1 & 2 & 0 & 2 \\ 0 & 1 & 2 & 3 \\ 2 & 4 & 1 & 4 \end{vmatrix}$
  9. $\begin{vmatrix} 3 & 2 & 3 & 0 \\ 2 & 3 & 0 & 2 \\ 1 & 0 & 1 & 1 \\ 0 & 1 & 2 & 3 \end{vmatrix}$

10. $\begin{vmatrix} 3 & -9 & 0 & 12 \\ -2 & 1 & 3 & 0 \\ 3 & -2 & 1 & -3 \\ 0 & 2 & -3 & 1 \end{vmatrix}$
  11. $\begin{vmatrix} 6 & 6 & -1 & 1 \\ 3 & -1 & 1 & 1 \\ 3 & 1 & 6 & 1 \\ 6 & -1 & 1 & -6 \end{vmatrix}$

12. $\begin{vmatrix} 1 & -2 & 1 & -1 & 3 \\ 1 & 2 & 2 & 1 & 1 \\ 2 & 1 & -1 & -3 & 2 \\ 1 & 1 & 3 & 2 & 1 \\ 2 & -1 & 1 & 1 & 1 \end{vmatrix}$
  13. $\begin{vmatrix} 1 & 2 & 2 & 1 & 3 \\ 2 & 1 & 1 & 1 & 1 \\ 3 & 2 & 1 & 2 & 1 \\ 1 & 3 & 2 & 1 & 1 \\ 1 & 1 & 3 & 3 & 1 \end{vmatrix}$

14. Let the vertices of a triangle, reading counterclockwise, be $A(x_1, y_1)$, $B(x_2, y_2)$, and $C(x_3, y_3)$, as in Fig. 16-1. Observe that $DECA$, $EFBC$, and $DFBA$ are trapezoids. The sum of the areas of the first two trapezoids minus the area of the third trapezoid is equal to the area of triangle ABC. Recalling that the area of a trapezoid is equal to half the sum of the parallel sides times the altitude, we have area $DECA = \frac{1}{2}(y_1 + y_3)(x_3 - x_1)$. With this start, show that the area S of the triangle is

$$S = \frac{1}{2}[x_1(y_2 - y_3) - y_1(x_2 - x_3) + (x_2 y_3 - x_3 y_2)]$$

By expanding the determinant, show that the area of the triangle can be expressed as

$$S = \frac{1}{2}\begin{vmatrix} x_1 & y_1 & 1 \\ x_2 & y_2 & 1 \\ x_3 & y_3 & 1 \end{vmatrix}$$

What change is made in this result if the vertices are numbered in the clockwise direction?

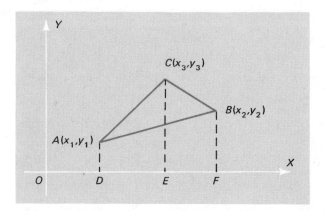

Fig. 16-1

Using the determinant of Problem 14, find the area of each triangle having the given vertices.

15. $A(0,0)$, $B(6,0)$, $C(4,3)$ **16.** $A(-2,4)$, $B(2,-6)$, $C(5,4)$
17. $A(2,7)$, $B(1,1)$, $C(10,8)$ **18.** $A(5,-1)$, $B(-1,4)$, $C(3,6)$
19. $A(1,5)$, $B(6,1)$, $C(8,7)$ **20.** $A(-2,-3)$, $B(3,2)$, $C(-1,-8)$

16-6 SOLUTION OF SYSTEMS OF LINEAR EQUATIONS BY DETERMINANTS

In Sec. 16-2 we solved systems of linear equations in two unknowns by the use of determinants. Determinants may also be used in solving systems of n linear equations in n unknowns. We shall establish the method for the case when $n = 4$. Consider the system

$$a_1x + b_1y + c_1z + d_1w = e_1$$

$$a_2x + b_2y + c_2z + d_2w = e_2$$

$$a_3x + b_3y + c_3z + d_3w = e_3$$

$$a_4x + b_4y + c_4z + d_4w = e_4$$

We shall use D to stand for the determinant of the matrix of the coefficients of the unknowns x, y, z, and w. And N_x, N_y, N_z, and N_w will stand for the four determinants obtained from D by replacing the coefficients of x, y, z, and w, in turn, by the corresponding constant terms. Then the solution of the given system of equations, as we shall show, may be written as

$$x = \frac{N_x}{D}, \qquad y = \frac{N_y}{D}, \qquad z = \frac{N_z}{D}, \qquad w = \frac{N_w}{D}$$

if $D \neq 0$. If we multiply both members of

$$D = \begin{vmatrix} a_1 & b_1 & c_1 & d_1 \\ a_2 & b_2 & c_2 & d_2 \\ a_3 & b_3 & c_3 & d_3 \\ a_4 & b_4 & c_4 & d_4 \end{vmatrix}$$

by x, we get

$$Dx = \begin{vmatrix} a_1 x & b_1 & c_1 & d_1 \\ a_2 x & b_2 & c_2 & d_2 \\ a_3 x & b_3 & c_3 & d_3 \\ a_4 x & b_4 & c_4 & d_4 \end{vmatrix}$$

Now we multiply the second column of the determinant by y, the third column by z, and the fourth column by w and add each of the products to the first column to obtain

$$Dx = \begin{vmatrix} a_1 x + b_1 y + c_1 z + d_1 w & b_1 & c_1 & d_1 \\ a_2 x + b_2 y + c_2 z + d_2 w & b_2 & c_2 & d_2 \\ a_3 x + b_3 y + c_3 z + d_3 w & b_3 & c_3 & d_3 \\ a_4 x + b_4 y + c_4 z + d_4 w & b_4 & c_4 & d_4 \end{vmatrix}$$

The elements of the first column of this determinant are the left members of the given system of equations, and hence may be replaced by their equals as found in the right members. We have then

$$Dx = \begin{vmatrix} e_1 & b_1 & c_1 & d_1 \\ e_2 & b_2 & c_2 & d_2 \\ e_3 & b_3 & c_3 & d_3 \\ e_4 & b_4 & c_4 & d_4 \end{vmatrix} = N_x$$

and $x = N_x/D$.

Similarly, we may find

$$y = \frac{N_y}{D}, \qquad z = \frac{N_z}{D}, \qquad w = \frac{N_w}{D}$$

Then we have a solution of the given system of equations if these values for the unknowns x, y, z, and w satisfy each of the equations. Suppose that we check the first equation. The remaining equations could be checked in the same way. Substituting, we have

$$\frac{a_1 N_x}{D} + \frac{b_1 N_y}{D} + \frac{c_1 N_z}{D} + \frac{d_1 N_w}{D} = e_1$$

Now transposing the constant term to the left side and multiplying by D gives

$$a_1 N_x + b_1 N_y + c_1 N_z + d_1 N_w - e_1 D = 0$$

If this expression on the left is equal to zero, the first equation of the given system is satisfied. The expression is the expansion of a determinant. That is,

$$a_1 N_x + b_1 N_y + c_1 N_z + d_1 N_w - e_1 D = \begin{vmatrix} a_1 & e_1 & b_1 & c_1 & d_1 \\ a_1 & e_1 & b_1 & c_1 & d_1 \\ a_2 & e_2 & b_2 & c_2 & d_2 \\ a_3 & e_3 & b_3 & c_3 & d_3 \\ a_4 & e_4 & b_4 & c_4 & d_4 \end{vmatrix}$$

as may be verified by expanding the determinant according to the elements of the first row. But the determinant is equal to zero since the first and second rows are identical.

The steps which we used in solving the system of four equations may be applied in just the same way to solve a system of n linear equations in n unknowns. Accordingly we shall state a theorem covering the general case. The theorem, known as Cramer's Rule, is due to the Swiss mathematician Gabriel Cramer (1704–1752).

Theorem 16-10 Let a system of n linear equations in n unknowns be denoted by

$$a_{11}x_1 + a_{12}x_2 + a_{13}x_3 + \cdots + a_{1n}x_n = k_1$$
$$a_{21}x_1 + a_{22}x_2 + a_{23}x_3 + \cdots + a_{2n}x_n = k_2$$
$$\vdots$$
$$a_{n1}x_1 + a_{n2}x_2 + a_{n3}x_3 + \cdots + a_{nn}x_n = k_n$$

Let D be the determinant of the coefficients of the unknowns $x_1, x_2, x_3, \ldots, x_n$, and N_i the determinant obtained from D by replacing the coefficients of x_i by the corresponding constant terms and leaving all other elements unchanged. Then, if $D \neq 0$, the solution of the system is given by

$$x_1 = \frac{N_1}{D}, x_2 = \frac{N_2}{D}, x_3 = \frac{N_3}{D}, \ldots, x_n = \frac{N_n}{D}$$

EXAMPLE. Solve the system

$$2x \qquad + z - w = 1$$
$$x + 2y - 2z + 3w = 9$$
$$3x - 2y + 3z - 2w = 2$$
$$4x - 4y - 2z - 3w = 3$$

Solution. The solution may be had from the five determinants D, N_x, N_y, N_z, and N_w. We show how D may be evaluated.

$$D = \begin{vmatrix} 2 & 0 & 1 & -1 \\ 1 & 2 & -2 & 3 \\ 3 & -2 & 3 & -2 \\ 4 & -4 & -2 & -3 \end{vmatrix}$$

We add twice the fourth column to the first column, and add the fourth column to the third column, and then expand by the elements of the first row; whence,

$$D = \begin{vmatrix} 0 & 0 & 0 & -1 \\ 7 & 2 & 1 & 3 \\ -1 & -2 & 1 & -2 \\ -2 & -4 & -5 & -3 \end{vmatrix} = \begin{vmatrix} 7 & 2 & 1 \\ -1 & -2 & 1 \\ -2 & -4 & -5 \end{vmatrix} = 2\begin{vmatrix} 7 & 1 & 1 \\ -1 & -1 & 1 \\ -2 & -2 & -5 \end{vmatrix}$$

Referring to the determinant on the right, we add the third column to the first and also to the second column and obtain

$$D = 2\begin{vmatrix} 8 & 2 & 1 \\ 0 & 0 & 1 \\ -7 & -7 & -5 \end{vmatrix} = -2\begin{vmatrix} 8 & 2 \\ -7 & -7 \end{vmatrix} = 84$$

By evaluating it may be verified that

$$N_x = \begin{vmatrix} 1 & 0 & 1 & -1 \\ 9 & 2 & -2 & 3 \\ 2 & -2 & 3 & -2 \\ 3 & -4 & -2 & -3 \end{vmatrix} = 168 \quad N_y = \begin{vmatrix} 2 & 1 & 1 & -1 \\ 1 & 9 & -2 & 3 \\ 3 & 2 & 3 & -2 \\ 4 & 3 & -2 & -3 \end{vmatrix} = -84$$

$$N_z = \begin{vmatrix} 2 & 0 & 1 & -1 \\ 1 & 2 & 9 & 3 \\ 3 & -2 & 2 & -2 \\ 4 & -4 & 3 & -3 \end{vmatrix} = 0 \quad N_w = \begin{vmatrix} 2 & 0 & 1 & 1 \\ 1 & 2 & -2 & 9 \\ 3 & -2 & 3 & 2 \\ 4 & -4 & -2 & 3 \end{vmatrix} = 252$$

Hence,

$$x = \frac{168}{84} = 2, \quad y = \frac{-84}{84} = -1, \quad z = \frac{0}{84} = 0, \quad w = \frac{252}{84} = 3$$

Solve the following systems of equations by determinants and check your results.

1. $2x + 3y + z = 9$
   $x - 2y - 3z = 1$
   $5x + 4y + 6z = 5$
3. $2x + 3y + z = 9$
   $4x + y = 7$
   $x - 3y - 7z = 6$
5. $3x - 5z = -1$
   $2x + 7y = 6$
   $x + y + z = 5$
7. $-x + 2y + 3z - 2w = 2$
   $3x + y + z - w = 4$
   $2x - y - z + 2w = 0$
   $x + y + z + w = 0$
9. $2x - 3z + w = 1$
   $x - y + 2w = 1$
   $-3y + z + w = 1$
   $x + y + z = 1$
11. $x + y + 2z + 2w = 3$
    $2x - y - z + 2w = 1$
    $2x + 2y - 3z + w = 2$
    $4x - 2y + z + 3w = 6$
13. $x + y + z + w = 2$
    $x - 2y + 2z + 2w = -6$
    $2x + y - 2z + 2w = -5$
    $3x - y + 3z - 3w = -3$

2. $5x + 4y + 3z = 1$
   $2x - y - z = 7$
   $3x + y = 6$
4. $3x + y + z = 2$
   $x + y + z = 0$
   $5x - 2y + 3z = -8$
6. $x + 5y - z = 9$
   $3x - 3y + 2z = 7$
   $2x - 4y + 3z = 1$
8. $2x + 2y - z - w = 1$
   $2x - 4y + 3z + 2w = 8$
   $3x + 2y - 2z - 3w = 7$
   $x + y - z + w = 0$
10. $x - y + z - w = 3$
    $x + y + z + w = -5$
    $x - 3y - z - w = 9$
    $x + y - z + w = 1$
12. $x + y + 2z + 3w = -2$
    $2x + y - 3z - 2w = -1$
    $3x + 2y - z - w = -5$
    $2x + 3y - z - 2w = 1$
14. $2x + 3y + z + w = 2$
    $2x - 2y + z - w = 7$
    $x - y - 3z - w = 2$
    $3x + 2y + 2z + 2w = 5$

16-7 LINEAR SYSTEMS WITH MORE EQUATIONS THAN UNKNOWNS

The systems of linear equations which we have considered thus far have had the same number of equations as unknowns. Systems of equations sometimes arise, however, which do not have this balance between the equations and unknowns. A detailed discussion of this problem is beyond our present stage. Accordingly we shall restrict ourselves to the special case in which the number of equations is one more than the number of unknowns.

To illustrate this case we choose the system

$$a_1x + b_1y = c_1 \qquad (1)$$

$$a_2x + b_2y = c_2 \qquad (2)$$

$$a_3x + b_3y = c_3 \qquad (3)$$

Here we have three equations and two unknowns. A system of this kind may or may not have a solution (i.e. be consistent or inconsistent). However, if any two of the equations are consistent and their solution satisfies the other equation, then the system of three equations is consistent.

Let us assume that equations (1) and (2) are consistent. This means that

$$D = \begin{vmatrix} a_1 & b_1 \\ a_2 & b_2 \end{vmatrix} \neq 0$$

The solution of the two equations is

$$x = \frac{\begin{vmatrix} c_1 & b_1 \\ c_2 & b_2 \end{vmatrix}}{D} \quad \text{and} \quad y = \frac{\begin{vmatrix} a_1 & c_1 \\ a_2 & c_2 \end{vmatrix}}{D}$$

We now substitute these values for x and y in equation (3). This gives

$$\frac{a_3 \begin{vmatrix} c_1 & b_1 \\ c_2 & b_2 \end{vmatrix}}{D} + \frac{b_3 \begin{vmatrix} a_1 & c_1 \\ a_2 & c_2 \end{vmatrix}}{D} = a_3$$

By clearing of fractions, changing signs, and interchanging columns in one of the determinants, we reduce the equation to

$$a_3 \begin{vmatrix} b_1 & c_1 \\ b_2 & c_2 \end{vmatrix} - b_3 \begin{vmatrix} a_1 & c_1 \\ a_2 & c_2 \end{vmatrix} + c_3 \begin{vmatrix} a_1 & b_1 \\ a_2 & b_2 \end{vmatrix} = 0$$

The left side of this equation is the expansion of

$$\begin{vmatrix} a_1 & b_1 & c_1 \\ a_2 & b_2 & c_2 \\ a_3 & b_3 & c_3 \end{vmatrix}$$

Hence this determinant must be equal to zero if the given system of equations is consistent. Conversely, if this determinant is equal to zero and any two of the given equations are consistent, then the three equations form a consistent system.

Systems with one extra equation in more than two unknowns could be treated in an analogous way. In this connection we state a general theorem.

Theorem 16-11 A system of $(n + 1)$ linear equations in n unknowns has a single solution if and only if the following conditions hold:

1. The determinant whose $(n + 1)$ columns are formed from the coefficients of the n unknowns and the constant terms is equal to zero.
2. The determinant of the coefficients of some n of the equations is different from zero.

EXAMPLE. Determine if the following system of equations is consistent.

$$x +\ y +\ z = 0$$

$$x +\ y + 3z = 2$$

$$2x - 3y - 5z = 8$$

$$3x - 2y - 8z = 4$$

Solution. We first evaluate the determinant formed by the coefficients of the unknowns and the constant terms. Thus we find

$$\begin{vmatrix} 1 & 1 & 1 & 0 \\ 1 & 1 & 3 & 2 \\ 2 & -3 & -5 & 8 \\ 3 & -2 & -8 & 4 \end{vmatrix} = 0$$

This shows that one part of the theorem is satisfied. We next determine if the second part of the theorem is satisfied. Trying the first three of the given equations, we find

$$\begin{vmatrix} 1 & 1 & 1 \\ 1 & 1 & 3 \\ 2 & -3 & -5 \end{vmatrix} = 10$$

Since this determinant is different from zero, the three equations have a solution. The solution, as may be verified, is $x = 2$, $y = -3$, $z = 1$. These values also satisfy the fourth equation.

16-8 SYSTEMS OF HOMOGENEOUS LINEAR EQUATIONS

A linear equation is said to be *homogeneous* if the constant term is zero. The equations

$$a_1x + b_1y + c_1z = 0$$

$$a_2x + b_2y + c_2z = 0$$

$$a_3x + b_3y + c_3z = 0$$

constitute a system of homogeneous linear equations. Obviously the equations are satisfied by substituting zero for each unknown. This solution may also be obtained by using determinants if the determinant of the coefficients of the unknown D is not equal to zero. We may write the solution as

$$x = \frac{N_x}{D}, \quad y = \frac{N_y}{D}, \quad z = \frac{N_z}{D}$$

N_x, N_y, and N_z stand for determinants as defined in Sec. 16-6. Each of these numerator determinants, having a column of zeros, is equal to zero. Hence, with $D \neq 0$, this is the only possible solution.

We see at once that any system of homogeneous linear equations is satisfied by the value zero for each unknown. This is called a *trivial solution*. We raise the question as to the existence of solutions other than the trivial solution. Relative to this question, we state a theorem.

Theorem 16-12 A system of n homogeneous linear equations in n unknowns has nontrivial solutions if and only if the determinant of the coefficients is equal to zero.

This theorem is proved in textbooks on theory of equations. Although we do not include a proof, we shall illustrate a method for finding nontrivial solutions.

EXAMPLE. Find nontrivial solutions of the system

$$x - y - 3z = 0$$
$$2x - 2y - 6z = 0$$
$$2x + 3y - z = 0$$

Solution. We first evaluate D. Thus we find

$$D = \begin{vmatrix} 1 & -1 & -3 \\ 2 & -2 & -6 \\ 2 & 3 & -1 \end{vmatrix} = 0$$

Since $D = 0$, there are nontrivial solutions. We next find the solutions. Transposing the z terms in the second and third equations gives

$$2x - 2y = 6z$$
$$2x + 3y = z$$

The solution of these equations for x and y in terms of z is $x = 2z$, $y = -z$. These values for x and y satisfy all three of the given equations. Hence an indefinite number of solutions can be found simply by assigning z any value and using the values $2z$ and $-z$ for x and y. Giving z the literal number k, we write the solution as

$$x = 2k, \quad y = -k, \quad z = k$$

where k can be any number.

EXERCISE 16-6

Test each system of equations for consistency or inconsistency. Find the solution of each system whose equations are consistent.

1. $4x - 3y = 3$
$2x - 3y = 2$
$2x - 6y = 3$

2. $4x - 2y = 3$
$2x - 3y = 2$
$6x - 5y = 4$

3. $3x - 3y = 4$
$2x + 2y = 3$
$7x + y = 10$

4. $x + 7y = 10$
$2x + 5y = 2$
$x + 16y = 28$

5. $2x + 3y = -1$
$3x - 4y = 24$
$x + 10y = -26$

6. $6x - 5y = 14$
$4x + 3y = -13$
$2x + y = 0$

7. $2x - y + 2z = 6$
$3x + y + z = 8$
$x - 2y + z = 0$
$x + y = 1$

8. $2x + y - z = 8$
$3x - 4y - 5z = 8$
$5x + y + z = 10$
$x + y + z = 2$

9. $x - y = 1$
$x + z = -6$
$y - z = 1$
$x + y - 2z = 3$

10. $2x + 3y + z = 9$
$x - 2y - 3z = 1$
$5x + 4y + 6z = 5$
$x + y - z = 6$

11. $x - 2y - z = 3$
$x + 3y + z = 4$
$x - y - 2z = 5$
$x + y + 2z = 6$

12. $2x - y + z = 0$
$2x + y - z = 2$
$2x - y - z = 3$
$2x + y + z = 4$

Test each system of equations for the existence of nontrivial solutions. Solve the systems which have nontrivial solutions.

13. $x + 2y = 0$
$2x - y = 0$

14. $3x - 2y = 0$
$6x - 4y = 0$

15. $3x + 3y = 0$
$2x + 2y = 0$

16. $x + y + z = 0$
$x - y - 5z = 0$
$x + 2y + 4z = 0$

17. $2x - 3y - z = 0$
$x + 3y - 2z = 0$
$x - 3y = 0$

18. $x + 2y - 5z = 0$
$x - y + 4z = 0$
$x + 4y - 9z = 0$

19. $x + y + z + w = 0$
$x - 7y + z - w = 0$
$x + y - 7z - w = 0$
$x - y + 3z + w = 0$

20. $4x - y - 2z + w = 0$
$x - y - z + w = 0$
$5x + y - z - w = 0$
$2x + 2y + 2z - 3w = 0$

16-9 ALGEBRA OF MATRICES

We denote a matrix of m rows and n columns by the array

$$\begin{bmatrix} a_{11} & a_{12} & a_{13} \cdots a_{1n} \\ a_{21} & a_{22} & a_{23} \cdots a_{2n} \\ \vdots \\ a_{m1} & a_{m2} & a_{m3} \cdots a_{mn} \end{bmatrix}$$

which is called an m by n matrix. If $m = n$, the matrix is called *a square matrix of order n.*

Definition 16-5 Two matrices of the same dimensions are said to be equal if all the corresponding elements are equal.

Definition 16-6 The sum of two matrices of the same dimensions is the matrix obtained by adding each element of one of the given matrices to the corresponding element of the other given matrix.

Definition 16-7 The product of a number k and a matrix is the matrix obtained by multiplying each element of the given matrix by k.

EXAMPLE 1.

$$\begin{bmatrix} 2 & -3 \\ 4 & 0 \end{bmatrix} + \begin{bmatrix} 9 & 3 \\ 6 & 2 \end{bmatrix} = \begin{bmatrix} 2+9 & -3+3 \\ 4+6 & 0+2 \end{bmatrix} = \begin{bmatrix} 11 & 0 \\ 10 & 2 \end{bmatrix}$$

EXAMPLE 2.

$$(-3)\begin{bmatrix} 8 & 7 & -2 & 1 \\ 2 & 5 & -8 & 9 \end{bmatrix} = \begin{bmatrix} -24 & -21 & 6 & -3 \\ -6 & -15 & 24 & -27 \end{bmatrix}$$

Definition 16-8 If A is an m by n matrix and B is an n by p matrix, then the product AB is the m by p matrix C ($C = AB$) whose element c_{ij} in the ith row and jth column is given by

$$c_{ij} = a_{i1}b_{1j} + a_{i2}b_{2j} + a_{i3}b_{3j} + \cdots + a_{in}b_{nj}$$

According to this definition, we "multiply" the ith row of A by the jth column of B to find the element in the ith row and jth column of the product AB. Hence the number of columns in A must be equal to the number of rows in B. Matrix multiplication is not defined when this condition fails.

EXAMPLE 3.

$$\begin{bmatrix} 1 & -2 \\ 0 & 6 \\ -4 & 5 \end{bmatrix}\begin{bmatrix} 3 & 6 & -1 & 4 \\ 5 & 2 & 0 & 7 \end{bmatrix} = \begin{bmatrix} -7 & 2 & -1 & -10 \\ 30 & 12 & 0 & 42 \\ 13 & -14 & 4 & 19 \end{bmatrix}$$

The elements -7, 2, -1, -10 of the first row of the product of the two matrices come from the expressions

$$1 \cdot 3 + (-2)5, \quad 1 \cdot 6 + (-2)2, \quad 1(-1) + (-2)0, \quad 1 \cdot 4 + (-2)7$$

That is, we go across the first row of the first matrix and successively down the first, second, and third columns of the second matrix to obtain the elements of the first row of the product of the matrices. Then in a similar way we use the second and third rows of the first given matrix and columns of the second matrix to find the second and third rows of the product of the matrices.

Notice that the first matrix in this example has dimensions 3 by 2 and the second 2 by 3, making the number of columns in the first matrix equal to the number of rows in the second matrix. If we interchange the positions of the matrices, however, the first matrix would than have 4 columns and the second

3 rows, which means the product is not defined. Even though the products *AB* and *BA* of two matrices exist, the results may or may not be the same. The next example illustrates the fact that matrix multiplication is, in general, not commutative.

EXAMPLE 4.

$$\begin{bmatrix} 5 & -1 \\ 6 & 7 \end{bmatrix}\begin{bmatrix} 2 & 1 \\ 3 & 4 \end{bmatrix} = \begin{bmatrix} 5\cdot 2 + (-1)3 & 5\cdot 1 + (-1)4 \\ 6\cdot 2 + 7\cdot 3 & 6\cdot 1 + 7\cdot 4 \end{bmatrix} = \begin{bmatrix} 7 & 1 \\ 33 & 34 \end{bmatrix}$$

$$\begin{bmatrix} 2 & 1 \\ 3 & 4 \end{bmatrix}\begin{bmatrix} 5 & -1 \\ 6 & 7 \end{bmatrix} = \begin{bmatrix} 2\cdot 5 + 1\cdot 6 & 2(-1) + 1\cdot 7 \\ 3\cdot 5 + 4\cdot 6 & 3(-1) + 4\cdot 7 \end{bmatrix} = \begin{bmatrix} 16 & 5 \\ 39 & 25 \end{bmatrix}$$

EXERCISE 16-7

Given the matrices

$$A = \begin{bmatrix} 1 & 3 \\ 4 & 2 \end{bmatrix} \quad \text{and} \quad B = \begin{bmatrix} -2 & 5 \\ 4 & 3 \end{bmatrix}$$

find the matrices resulting from the indicated operations.

1. $A + B$ 2. $B + A$ 3. $A - B$
4. $2A + B$ 5. AB 6. BA

Compute the indicated products.

7. $\begin{bmatrix} 1 & -2 \\ 3 & 2 \end{bmatrix}\begin{bmatrix} 1 & 2 & 3 \\ 3 & 2 & 1 \end{bmatrix}$
   8. $\begin{bmatrix} 2 & 3 & 4 \\ 4 & 1 & 5 \end{bmatrix}\begin{bmatrix} 1 & -1 \\ 0 & -2 \\ 2 & 4 \end{bmatrix}$

9. $\begin{bmatrix} 1 & -1 \\ 0 & -2 \\ 2 & 4 \end{bmatrix}\begin{bmatrix} 2 & 3 & 4 \\ 4 & 1 & 5 \end{bmatrix}$
   10. $\begin{bmatrix} 1 & -2 \\ 4 & -3 \\ 0 & 5 \end{bmatrix}\begin{bmatrix} 1 & 2 & 3 & 4 \\ 4 & 3 & 2 & 1 \end{bmatrix}$

11. $\begin{bmatrix} 3 \\ 2 \\ 1 \end{bmatrix}[1 \quad 3 \quad 5 \quad 6]$
   12. $[-5 \quad -2 \quad 1 \quad 4]\begin{bmatrix} 2 \\ 4 \\ 5 \\ 7 \end{bmatrix}$

13. $\begin{bmatrix} 1 & 2 & 4 \\ 2 & 3 & 5 \\ 1 & 0 & 1 \end{bmatrix}\begin{bmatrix} 2 & -1 & 5 \\ 4 & 0 & 1 \\ 3 & -3 & 0 \end{bmatrix}$
   14. $\begin{bmatrix} 2 & -1 & 5 \\ 4 & 0 & 1 \\ 3 & -3 & 0 \end{bmatrix}\begin{bmatrix} 1 & 2 & 4 \\ 2 & 3 & 5 \\ 1 & 0 & 1 \end{bmatrix}$

Given the matrices

$$A = \begin{bmatrix} a & b \\ c & d \end{bmatrix}, \quad B = \begin{bmatrix} e & f \\ g & h \end{bmatrix}, \quad C = \begin{bmatrix} i & j \\ k & l \end{bmatrix}$$

prove the following equations are true.

15. $A + B = B + A$ **16.** $(A + B) + C = A + (B + C)$
17. $A(B + C) = AB + AC$ **18.** $A(B - C) = AB - AC$
19. $k(A + B) = kA + kB$ **20.** $(m + n)A = mA + nA$

16-10 THE INVERSE OF A SQUARE MATRIX

We recall that any real number a has a multiplicative identity element. That is, if $a \in R$, then $a \cdot 1 = a$ (Axiom 5, Sec. 2-2). This suggests the idea of an identity matrix for matrix multiplication. Some matrices, as we shall show, do have multiplicative identity matrices.

Theorem 16-13   Let A be a square matrix of order n and let I be a square matrix also of order n with all elements on the principal diagonal equal to 1 and all other elements equal to 0. Then it follows that

$$AI = A \quad \text{and} \quad IA = A$$

Proof (for a 3 by 3 matrix). By multiplying, we have

$$\begin{bmatrix} a_1 & b_1 & c_1 \\ a_2 & b_2 & c_2 \\ a_3 & b_3 & c_3 \end{bmatrix} \begin{bmatrix} 1 & 0 & 0 \\ 0 & 1 & 0 \\ 0 & 0 & 1 \end{bmatrix} = \begin{bmatrix} a_1 & b_1 & c_1 \\ a_2 & b_2 & c_2 \\ a_3 & b_3 & c_3 \end{bmatrix}$$

$$\begin{bmatrix} 1 & 0 & 0 \\ 0 & 1 & 0 \\ 0 & 0 & 1 \end{bmatrix} \begin{bmatrix} a_1 & b_1 & c_1 \\ a_2 & b_2 & c_2 \\ a_3 & b_3 & c_3 \end{bmatrix} = \begin{bmatrix} a_1 & b_1 & c_1 \\ a_2 & b_2 & c_2 \\ a_3 & b_3 & c_3 \end{bmatrix}$$

If now A is a square matrix of order n (n any positive integer) and I is the corresponding matrix described in the theorem, we could multiply, just as we have done here, and find that

$$AI = IA = A \quad \blacksquare$$

We recall also that any real number a different from zero has a multiplicative inverse which is expressed by $a(1/a) = 1$. The counterpart of this property of real numbers exists with respect to certain matrices.

Definition 16-9   Let A be a square matrix of order n and I the corresponding identity matrix. If there exists a square matrix A^{-1}, also of order n, such that $AA^{-1} = I$ and $A^{-1}A = I$, then A^{-1} is called the *multiplicative inverse of A*.

EXAMPLE 1. Determine if the matrix

$$A = \begin{bmatrix} 2 & -1 \\ 4 & 5 \end{bmatrix}$$

has an inverse.

Solution. If we let

$$A^{-1} = \begin{bmatrix} x_1 & y_1 \\ x_2 & y_2 \end{bmatrix}$$

then we need to find the unknowns x_1, x_2, y_1, y_2 in the equation

$$\begin{bmatrix} 2 & -1 \\ 4 & 5 \end{bmatrix}\begin{bmatrix} x_1 & y_1 \\ x_2 & y_2 \end{bmatrix} = \begin{bmatrix} 1 & 0 \\ 0 & 1 \end{bmatrix}$$

Performing the multiplication on the left, we have

$$\begin{bmatrix} 2x_1 - x_2 & 2y_1 - y_2 \\ 4x_1 + 5x_2 & 4y_1 + 5y_2 \end{bmatrix} = \begin{bmatrix} 1 & 0 \\ 0 & 1 \end{bmatrix}$$

Since corresponding elements of equal matrices are equal, we write the two systems of equations

$$\begin{cases} 2x_1 - x_2 = 1 \\ 4x_1 + 5x_2 = 0 \end{cases} \quad \begin{cases} 2y_1 - y_2 = 0 \\ 4y_1 + 5y_2 = 1 \end{cases}$$

We observe that the coefficient matrix of each of these systems is the same as the given matrix A. The determinant of A is not equal to zero, and consequently, each system is consistent. The solutions are $x_1 = \frac{5}{14}$, $x_2 = -\frac{2}{7}$ and $y_1 = \frac{1}{14}$, $y_2 = \frac{1}{7}$. Hence

$$A^{-1} = \begin{bmatrix} \frac{5}{14} & \frac{1}{14} \\ -\frac{2}{7} & \frac{1}{7} \end{bmatrix} = \frac{1}{7}\begin{bmatrix} \frac{5}{2} & \frac{1}{2} \\ -2 & 1 \end{bmatrix}$$

Next, reversing the order of multiplication, we find

$$A^{-1}A = \begin{bmatrix} \frac{5}{14} & \frac{1}{14} \\ -\frac{2}{7} & \frac{1}{7} \end{bmatrix}\begin{bmatrix} 2 & -1 \\ 4 & 5 \end{bmatrix} = \begin{bmatrix} 1 & 0 \\ 0 & 1 \end{bmatrix}$$

Since $AA^{-1} = A^{-1}A = I$, we conclude that the given matrix does have a multiplicative inverse.

We shall next establish a theorem by which the inverse of a 3 by 3 matrix may be obtained.

Theorem 16-14 Let

$$A = \begin{bmatrix} a_1 & b_1 & c_1 \\ a_2 & b_2 & c_2 \\ a_3 & b_3 & c_3 \end{bmatrix}$$

be a 3 by 3 matrix whose determinant D is different from zero. Then the inverse matrix exists and is given by

$$A^{-1} = \frac{1}{D} \begin{bmatrix} A_1 & A_2 & A_3 \\ B_1 & B_2 & B_3 \\ C_1 & C_2 & C_3 \end{bmatrix}, \tag{1}$$

where A_1 is the cofactor of a_1 of the determinant of A, B_1 the cofactor of b_1, C_1 the cofactor of c_1, and similarly for the elements of the second and third columns of A^{-1}.

Proof. To find the inverse of A we need to solve the equation

$$\begin{bmatrix} a_1 & b_1 & c_1 \\ a_2 & b_2 & c_2 \\ a_3 & b_3 & c_3 \end{bmatrix} \begin{bmatrix} x_1 & y_1 & z_1 \\ x_2 & y_2 & z_2 \\ x_3 & y_3 & z_3 \end{bmatrix} = \begin{bmatrix} 1 & 0 & 0 \\ 0 & 1 & 0 \\ 0 & 0 & 1 \end{bmatrix} \tag{2}$$

for the elements in the second matrix on the left side. By multiplying we express the left side of the equation in the form

$$\begin{bmatrix} a_1x_1 + b_1x_2 + c_1x_3 & a_1y_1 + b_1y_2 + c_1y_3 & a_1z_1 + b_1z_2 + c_1z_3 \\ a_2x_1 + b_2x_2 + c_2x_3 & a_2y_1 + b_2y_2 + c_2y_3 & a_2z_1 + b_2z_2 + c_2z_3 \\ a_3x_1 + b_3x_2 + c_3x_3 & a_3y_1 + b_3y_2 + c_3y_3 & a_3z_1 + b_3z_2 + c_3z_3 \end{bmatrix}$$

In order for equation (2) to be satisfied the elements in this matrix must be equal to the corresponding elements on the right side of the equation. Equating the elements of the first columns, we have the system

$$a_1x_1 + b_1x_2 + c_1x_3 = 1$$
$$a_2x_1 + b_2x_2 + c_2x_3 = 0$$
$$a_3x_1 + b_3x_2 + c_3x_3 = 0$$

This system has a solution because the determinant of the coefficient matrix is not equal to zero. Applying Cramer's rule, we have

$$x_1 = \frac{\begin{vmatrix} 1 & b_1 & c_1 \\ 0 & b_2 & c_2 \\ 0 & b_3 & c_3 \end{vmatrix}}{\begin{vmatrix} a_1 & b_1 & c_1 \\ a_2 & b_2 & c_2 \\ a_3 & b_3 & c_3 \end{vmatrix}} = \frac{A_1}{D}$$

$$x_2 = \frac{\begin{vmatrix} a_1 & 1 & c_1 \\ a_2 & 0 & c_2 \\ a_3 & 0 & c_3 \end{vmatrix}}{D} = \frac{B_1}{D}, \qquad x_3 = \frac{\begin{vmatrix} a_1 & b_1 & 1 \\ a_2 & b_2 & 0 \\ a_3 & b_3 & 0 \end{vmatrix}}{D} = \frac{C_1}{D}$$

Similarly, we find

$$y_1 = \frac{A_2}{D}, \qquad y_2 = \frac{B_2}{D}, \qquad y_3 = \frac{C_2}{D}$$

and

$$z_1 = \frac{A_3}{D}, \qquad z_2 = \frac{B_3}{D}, \qquad z_3 = \frac{C_3}{D}$$

These values for the unknowns of equation (2) are those contained in equation (1), and consequently we have

$$A^{-1} = \frac{1}{D} \begin{bmatrix} A_1 & A_2 & A_3 \\ B_1 & B_2 & B_3 \\ C_1 & C_2 & C_3 \end{bmatrix} = \begin{bmatrix} \dfrac{A_1}{D} & \dfrac{A_2}{D} & \dfrac{A_3}{D} \\ \dfrac{B_1}{D} & \dfrac{B_2}{D} & \dfrac{B_3}{D} \\ \dfrac{C_1}{D} & \dfrac{C_2}{D} & \dfrac{C_3}{D} \end{bmatrix}$$

To complete the proof of the theorem, we need to show that $A^{-1}A = I$. This may be done by examining the elements of the product $A^{-1}A$. We leave this part for the student and make only the following observations.

$$a_1A_1 + a_2A_2 + a_3A_3 = D$$

which is the determinant matrix A, and

$$b_1A_1 + b_2A_2 + b_3A_3 = 0$$

because it is the determinant of the matrix

$$\begin{bmatrix} b_1 & b_1 & c_1 \\ b_2 & b_2 & c_2 \\ b_3 & b_3 & c_3 \end{bmatrix}$$

which has two identical columns. ∎

We note that the coefficient matrix of each of the systems of equations arising from equation (2) is that of the given matrix A. Each system, then, has a unique solution because the determinant of A is not equal to zero. If the determinant of A were equal to zero, none of the systems would have a solution.

In this case A would have no inverse. Hence we have proved the following theorem for a 3 by 3 matrix. The theorem, however, is true generally.

Theorem 16-15 A square matrix whose determinant is not equal to zero has an inverse. A square matrix whose determinant is equal to zero has no inverse.

EXAMPLE 2. Find the inverse of the matrix

$$A = \begin{bmatrix} 4 & 2 & -7 \\ 2 & 1 & -2 \\ -3 & -5 & 6 \end{bmatrix}$$

Solution. To apply Theorem 16-14, we need to find the value of the determinant D of A and the cofactor of each element of D. Computing we find the value of D to be 21, which ensures the existence of the inverse of A. For the necessary cofactors we have

$$A_1 = \begin{vmatrix} 1 & -2 \\ -5 & 6 \end{vmatrix} = -4, \quad A_2 = -\begin{vmatrix} 2 & -7 \\ -5 & 6 \end{vmatrix} = 23, \quad A_3 = \begin{vmatrix} 2 & -7 \\ 1 & -2 \end{vmatrix} = 3$$

$$B_1 = -\begin{vmatrix} 2 & -2 \\ -3 & 6 \end{vmatrix} = -6, \quad B_2 = \begin{vmatrix} 4 & -7 \\ -3 & 6 \end{vmatrix} = 3, \quad B_3 = -\begin{vmatrix} 4 & -7 \\ 2 & -2 \end{vmatrix} = -6$$

$$C_1 = \begin{vmatrix} 2 & 1 \\ -3 & -5 \end{vmatrix} = -7, \quad C_2 = -\begin{vmatrix} 4 & 2 \\ -3 & -5 \end{vmatrix} = 14, \quad C_3 = \begin{vmatrix} 4 & 2 \\ 2 & 1 \end{vmatrix} = 0$$

From these values we obtain

$$A^{-1} = \tfrac{1}{21} \begin{bmatrix} -4 & 23 & 3 \\ -6 & 3 & -6 \\ -7 & 14 & 0 \end{bmatrix}$$

EXERCISE 16-8

1. Use the method by which Theorem 16-14 was established and prove that if

$$A = \begin{bmatrix} a_1 & b_1 \\ a_2 & b_2 \end{bmatrix}, \quad a_1 b_2 - a_2 b_1 \neq 0$$

then

$$A^{-1} = \frac{1}{\det A} \begin{bmatrix} A_1 & A_2 \\ B_1 & B_2 \end{bmatrix}$$

2. If $A = \begin{bmatrix} 1 & 2 & 1 & 1 \\ 1 & -1 & -1 & 1 \\ 1 & -1 & -1 & -1 \\ 1 & 1 & -1 & -1 \end{bmatrix}$ and $B = \begin{bmatrix} \frac{1}{2} & 0 & \frac{3}{4} & -\frac{1}{4} \\ 0 & 0 & -\frac{1}{2} & \frac{1}{2} \\ \frac{1}{2} & -\frac{1}{2} & \frac{3}{4} & -\frac{3}{4} \\ 0 & \frac{1}{2} & -\frac{1}{2} & 0 \end{bmatrix}$

find AB. Is B the multiplicative inverse of A?

Find the inverse, if it exists, of each of the following matrices. Check your result in the formula $AA^{-1} = I$.

3. $\begin{bmatrix} 3 & 6 \\ 1 & 2 \end{bmatrix}$ **4.** $\begin{bmatrix} 1 & 0 \\ 2 & 0 \end{bmatrix}$ **5.** $\begin{bmatrix} 2 & 1 \\ -4 & 3 \end{bmatrix}$

6. $\begin{bmatrix} 2 & 0 \\ 0 & 2 \end{bmatrix}$ **7.** $\begin{bmatrix} 3 & -1 \\ 4 & 3 \end{bmatrix}$ **8.** $\begin{bmatrix} 1 & 2 \\ -1 & 3 \end{bmatrix}$

9. $\begin{bmatrix} 2 & 1 \\ 1 & -1 \end{bmatrix}$ **10.** $\begin{bmatrix} 4 & -1 \\ 2 & 3 \end{bmatrix}$ **11.** $\begin{bmatrix} 5 & 3 \\ -2 & 2 \end{bmatrix}$

12. $\begin{bmatrix} 1 & 2 & 3 \\ 4 & 5 & 6 \\ 0 & 1 & 2 \end{bmatrix}$ **13.** $\begin{bmatrix} 1 & -1 & 5 \\ 2 & -1 & 3 \\ 3 & 0 & -6 \end{bmatrix}$ **14.** $\begin{bmatrix} 6 & 0 & 0 \\ -6 & 3 & 0 \\ 0 & -3 & 2 \end{bmatrix}$

15. $\begin{bmatrix} 4 & -6 & 1 \\ -1 & -1 & 1 \\ -4 & 11 & -1 \end{bmatrix}$ **16.** $\begin{bmatrix} -2 & 3 & -3 \\ 2 & 2 & 3 \\ 3 & -2 & 2 \end{bmatrix}$ **17.** $\begin{bmatrix} -4 & -6 & -7 \\ 23 & 3 & 14 \\ 3 & -6 & 0 \end{bmatrix}$

LOGARITHMS

I7-I THE LOGARITHMIC FUNCTION

Logarithms are of much value in mathematics. By their use many numerical computations can be made with surprising simplicity. And aside from their computational uses, logarithms are employed to great advantage in calculus and other areas of mathematics.

Definition 17-1 If a is a positive number other than 1 and y is a real number in the equation $a^y = x$, then y is called the *logarithm* of x to the base a. That is, the logarithm of a number x is the exponent which the base a must have to yield a value equal to x.

We write $y = \log_a x$ instead of the lengthier form, y is the logarithm of x to the base a. Hence the equations

$$a^y = x \quad \text{and} \quad y = \log_a x$$

express the same relation among the numbers a, y, and x. The first equation is in *exponential form* and the second in *logarithmic form*.

In the definition of a logarithm we specify that the base $a > 0$ and that the exponent y is a real number (positive, negative, or zero). As a consequence, a^y is a positive number; hence we shall consider logarithms of positive numbers only. We assume (but do not prove) that if a and x are any positive numbers except $a = 1$, then there is one and only one real value of y which will satisfy the equation $a^y = x$. That is, every positive number has a unique logarithm with respect to a given base. We reject unity as a base because 1 with any exponent is still 1.

Since a positive number has a unique logarithm, the equation $y = \log_a x$ defines a function with the set of positive numbers as the domain and the set of real numbers as the range. Hence the function consists of the set of number pairs

$$\{(x, \log_a x)|x > 0\}$$

Suppose we choose $a = 3$ and construct the graph of the equation $y = \log_3 x$. We use the equivalent equation $x = 3^y$, assign integral values to y, and obtain the corresponding values of x as exhibited in the following table.

x	$\frac{1}{9}$	$\frac{1}{3}$	1	3	9
y	-2	-1	0	1	2

The graph (Fig. 17-1) is shown along with the graph of $y = \log_2 x$. Comparatively, we observe that $|\log_3 x| < |\log_2 x|$ except at $x = 1$. Both graphs cross the x axis at $x = 1$, as would be true for any base a because $a^0 = 1$.

As additional illustrations of the meaning of logarithms, we list some logarithmic equations and the corresponding equivalent exponential equations.

LOGARITHMIC FORM	EXPONENTIAL FORM
$\log_2 8 = 3$	$2^3 = 8$
$\log_9 27 = \frac{3}{2}$	$9^{3/2} = 27$
$\log_5 \frac{1}{25} = -2$	$5^{-2} = \frac{1}{25}$
$\log_a a = 1$	$a^1 = a$

EXAMPLE 1. Find x if $\log_5 x = -4$.
Solution. Using the equivalent exponential form, we have

$$x = 5^{-4} = \frac{1}{5^4} = \frac{1}{625}$$

EXAMPLE 2. Find a if $\log_a 36 = \frac{2}{3}$.
Solution. The exponential form is $a^{2/3} = 36$. We take the $\frac{3}{2}$ power of both members of this equation and obtain

$$(a^{2/3})^{3/2} = 36^{3/2} \quad \text{and} \quad a = 216$$

EXAMPLE 3. Find the value of $\log_7 \frac{1}{49}$.

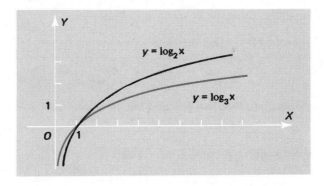

Fig. 17-1

Solution. We let $y = \log_7 \frac{1}{49}$. Then changing to the exponential form, we get

$$7^y = \frac{1}{49} = 7^{-2} \quad \text{and} \quad y = -2$$

EXERCISE 17-1

Give the logarithmic form of each of the following equations.

1. $2^3 = 8$ **2.** $3^4 = 81$ **3.** $5^{-1} = \frac{1}{5}$

4. $7^1 = 7$ **5.** $6^0 = 1$ **6.** $8^{2/3} = 4$

7. $9^{-3/2} = \frac{1}{27}$ **8.** $32^{1/5} = 2$ **9.** $10^3 = 1000$

10. $\left(\frac{3}{2}\right)^{-2} = \frac{4}{9}$ **11.** $\left(\frac{2}{3}\right)^3 = \frac{8}{27}$ **12.** $\left(\frac{1}{25}\right)^{-3/2} = 125$

Give the exponential form of each of the following equations.

13. $\log_4 1 = 0$ **14.** $\log_7 7 = 1$ **15.** $\log_6 36 = 2$

16. $\log_{10} 1 = 0$ **17.** $\log_2 \frac{1}{8} = -3$ **18.** $\log_{16} 4 = \frac{1}{2}$

19. $\log_2 \frac{1}{4} = -2$ **20.** $\log_5 \frac{1}{25} = -2$ **21.** $\log_{1/3} 27 = -3$

22. $\log_{1/5} 125 = -3$ **23.** $\log_{16} 64 = \frac{3}{2}$ **24.** $\log_{1/6} 36 = -2$

Find x, a, or y in each of the following equations.

25. $\log_{10} x = 2$ **26.** $\log_2 x = 3$ **27.** $\log_9 x = -1$

28. $\log_4 x = -2$ **29.** $\log_8 x = \frac{2}{3}$ **30.** $\log_{16} x = \frac{3}{4}$

31. $\log_{25} x = -\frac{3}{2}$ **32.** $\log_{1/4} x = 0$ **33.** $\log_{1/2} x = -4$

34. $\log_a 16 = 2$ **35.** $\log_a 5 = 1$ **36.** $\log_a 8 = 3$

37. $\log_a 25 = \frac{1}{2}$ **38.** $\log_a 4 = -\frac{1}{2}$ **39.** $\log_a \frac{1}{4} = -\frac{1}{2}$

40. $\log_a 64 = -\frac{3}{2}$ **41.** $\log_a 36 = -2$ **42.** $\log_a 2 = -\frac{1}{8}$

43. $y = \log_{125} 25$ **44.** $y = \log_3 27$ **45.** $y = \log_{25} 5$

46. $y = \log_5 \frac{1}{125}$ **47.** $y = \log_{64} 16$ **48.** $y = \log_3 \frac{1}{81}$

17-2 PROPERTIES OF LOGARITHMS

We shall state and prove three properties of logarithms. These properties follow readily from the laws of exponents.

Theorem 17-1 The logarithm of a product is equal to the sum of the logarithms of the factors. That is,

$$\log_a MN = \log_a M + \log_a N$$

Proof. Let $\log_a M = x$ and $\log_a N = y$. Then $M = a^x, N = a^y$,

$$MN = a^x \cdot a^y = a^{x+y}, \quad \text{and} \quad \log_a MN = x + y$$

Substituting for x and y, we obtain

$$\log_a MN = \log_a M + \log_a N$$

For a product of three factors, as MNP, we have

$$\log_a MNP = \log_a (MN)(P)$$
$$= \log_a MN + \log_a P$$
$$= \log_a M + \log_a N + \log_a P$$

Similarly, the theorem may be extended to any number of factors. ■

EXAMPLE 1. $\log_{10} (784)(92.7) = \log_{10} 784 + \log_{10} 92.7$.

Theorem 17-2   The logarithm of a quotient is equal to the logarithm of the dividend minus the logarithm of the divisor. That is,

$$\log_a \frac{M}{N} = \log_a M - \log_a N$$

Proof. If $\log_a M = x$ and $\log_a N = y$, then $M = a^x$ and $N = a^y$. Hence

$$\frac{M}{N} = \frac{a^x}{a^y} = a^{x-y} \quad \text{and} \quad \log_a \frac{M}{N} = x - y$$

By substituting for x and y, we obtain

$$\log_a \frac{M}{N} = \log_a M - \log_a N \quad ■$$

EXAMPLE 2. $\log_5 \frac{1078}{5234} = \log_5 1078 - \log_5 5234$.

Theorem 17-3   The logarithm of a power of a number is equal to the exponent times the logarithm of the number. That is,

$$\log_a M^p = p \log_a M$$

Proof. We let $\log_a M = x$, or $M = a^x$. Then

$$M^p = (a^x)^p = a^{px} \quad \text{and} \quad \log_a M^p = px.$$

Substituting for x gives

$$\log_a M^p = p \log_a M \quad ■$$

EXAMPLE 3. $\log_4 7842^3 = 3 \log_4 7842$.

EXAMPLE 4. $\log_{10} 597^{-5} = -5 \log_{10} 597$.

EXAMPLE 5. $\log_3 \sqrt{95} = \log_3 95^{1/2} = \frac{1}{2} \log_3 95$.

EXAMPLE 6. Express $\log_2 \dfrac{28^2 (6.7)^{1/2}}{(8.4)(17)}$ as the algebraic sum of logarithms.

Solution. We use each of the three preceding theorems and write

$$\log_2 \frac{28^2(6.7)^{1/2}}{(8.4)(17)} = \log_2 28^2(6.7)^{1/2} - \log_2 (8.4)(17)$$

$$= 2\log_2 28 + \tfrac{1}{2}\log_2 6.7 - \log_2 8.4 - \log_2 17$$

EXAMPLE 7. Express $\tfrac{2}{3}\log_6 5 - \log_6 18 - 3\log_6 2 + \log_6 35$ as a single logarithm.

Solution. $\tfrac{2}{3}\log_6 5 - \log_6 18 - 3\log_6 2 + \log_6 35$

$$= \tfrac{2}{3}\log_6 5 + \log_6 35 - (\log_6 18 + 3\log_6 2)$$

$$= \log_6 5^{2/3} + \log_6 35 - (\log_6 18 + \log_6 2^3)$$

$$= \log_6 5^{2/3} \cdot 35 - \log_6 18 \cdot 2^3$$

$$= \log_6 \frac{5^{2/3} \cdot 35}{18 \cdot 2^3}$$

EXERCISE 17-2

Write each of the following expressions as the algebraic sum of logarithms. A base is not indicated. Any positive number except 1 may be used.

1. $\log (7.12)(78)^3$

2. $\log (473)(561)(852)$

3. $\log \dfrac{83}{(47)(51)}$

4. $\log \dfrac{(78)(85)}{69^{2/3}}$

5. $\log \dfrac{(1.44)^{1/2}(8.23)^2}{308}$

6. $\log \dfrac{(33)(14)}{(61)(28)}$

7. $\log \sqrt{\dfrac{(28)(44)(21)}{978}}$

8. $\log \dfrac{\sqrt{57} \cdot \sqrt[3]{91}}{(18.4)^4}$

9. $\log \dfrac{85\sqrt[3]{67}}{(43)\sqrt[2]{65}}$

10. $\log \sqrt{\dfrac{(84)(61)}{37}}$

Given $\log_{10} 2 = 0.3010$ and $\log_{10} 3 = 0.4771$, find the value of each of the following quantities.

11. $\log_{10} 9$ **12.** $\log_{10} 8$ **13.** $\log_{10} 6$

14. $\log_{10} 54$ **15.** $\log_{10} 48$ **16.** $\log \sqrt[3]{12}$

17. $\log_{10} \tfrac{27}{8}$ **18.** $\log_{10} \tfrac{16}{3}$ **19.** $\log_{10} \sqrt[4]{48}$

20. $\log_{10} 144$ **21.** $\log_{10} 81$ **22.** $\log_{10} 32$

17-3 COMMON LOGARITHMS

There are two systems of logarithms which are used extensively in mathematics. In one of these the base is an irrational number, designated by *e*, whose value is approximately 2.718. Logarithms to this base are called *natural*, or *Naperian, logarithms*. This system is convenient in many theoretical considerations and is used in calculus and higher mathematics. The other system employs 10 as a base. Logarithms to the base 10 are called *common*, or *Briggs, logarithms*. Common logarithms are especially suited for making computations.

We shall be interested in logarithms to the base 10; and henceforth, for brevity, this base will not be indicated in logarithmic expressions. Thus we shall write $\log_{10} N$ simply as $\log N$.

The common logarithms of the integral powers of 10 are integers. This is illustrated in the following list of powers of 10 and their logarithms.

$$
\begin{array}{ll}
10^3 = 1000 & \log 1000 = 3 \\
10^2 = 100 & \log 100 = 2 \\
10^1 = 10 & \log 10 = 1 \\
10^0 = 1 & \log 1 = 0 \\
10^{-1} = 0.1 & \log 0.1 = -1 \\
10^{-2} = 0.01 & \log 0.01 = -2 \\
10^{-3} = 0.001 & \log 0.001 = -3
\end{array}
$$

Noticing that $\log 1 = 0$, we may conclude that numbers greater than 1 have positive logarithms and that numbers between 0 and 1 have negative logarithms. Further, since the logarithm of an integral power of 10 is an integer, we conclude that the logarithms of other positive numbers are not integers. For example, 63 is between 10 and 10^2, and therefore its logarithm is between 1 and 2, or 1 plus a fraction (between 0 and 1). In general, then, a logarithm consists of two parts, an integer and a fraction.

17-4 THE CHARACTERISTIC AND MANTISSA

When a logarithm is expressed as an integer plus a decimal (between 0 and 1), the integer is called the *characteristic* and the decimal is called the *mantissa*. Finding the logarithm of a number, then, requires the determination of both the characteristic and the mantissa. We shall consider first the problem of finding the characteristic. Since $\log 1 = 0$ and $\log 10 = 1$, we assume (and it seems reasonable) that the logarithm of a number between 1 and 10 is zero plus a positive fraction. In order to determine the characteristics corresponding to numbers between 0 and 1 and numbers greater than 10, we introduce the idea of the *reference position* of the decimal point.

Definition 17-2 The reference position of the decimal point comes just after the first nonzero digit of the number.

The arrow indicates the reference position of the decimal point of each of the following numbers.

$$
\begin{array}{cccc}
3.741 & 578.3 & 0.456 & 0.000125 \\
\uparrow & \uparrow & \uparrow & \uparrow
\end{array}
$$

The decimal points of these numbers are respectively in the reference position, two places to the right of the reference position, one place to the left, and four places to the left.

A number which is equal to the product of two factors such that the decimal point of one factor is in the reference position and the other factor is an integral power of 10 is said to be expressed in *scientific notation*. Thus the numbers 4832, 0.004832, and 100 are expressed in scientific notation by the equations

$$4832 = 4.832 \times 10^3$$

$$0.004832 = 4.832 \times 10^{-3}$$

$$100 = 1.00 \times 10^2$$

As these illustrations indicate, one factor of a number in scientific notation is equal to or greater than 1 and less than 10. The exponent of 10 in the other factor is equal to plus or minus the number of places the decimal point is removed from the reference position. The exponent is positive if the decimal point is to the right of the reference position and negative if the decimal point is to the left.

Employing Theorems 17-1 and 17-3, we have

$$\log 4832 = \log (4.832 \times 10^3) = 3 + \log 4.832$$

$$\log 0.004832 = \log (4.832 \times 10^{-3}) = -3 + \log 4.832$$

$$\log 100 = \log (1.00 \times 10^2) = 2 + \log 1$$

Thus we see that the characteristic of the logarithm of each of the numbers is the exponent of 10 when the number is expressed in scientific notation.

Consider now the general case in which a number A is expressed in scientific notation as $A = B \times 10^n$, where $1 \le B < 10$ and n is an integer. Taking the logarithm of each member, we write

$$\log A = \log (B \times 10^n) = \log B + n$$

The mantissa of $\log A$ is $\log B$ because $0 \le \log B < 1$, and the characteristic is the integer n. If A is greater than 10, the characteristic is a positive integer; if A is between 0 and 1, the characteristic is a negative integer.

We note also that the characteristic of the logarithm of a number depends solely on the location of the decimal point. Further, to determine the characteristic, it is sufficient to notice what the exponent of 10 would be if the number were expressed in scientific notation. The mantissa, however, is independent of the position of the decimal point; it depends only on the sequence of digits in the number. Accordingly, we state the following.

Theorem 17-4 If the decimal point of a number is in the reference position, the characteristic of its logarithm is zero. If the decimal point is n places to the right of the reference position, the characteristic is n. If the decimal point is n places to the left of the reference position, the characteristic is $-n$.

The mantissa is independent of the position of the decimal point.

EXAMPLE 1. Find the characteristic of the logarithm of each of the numbers 0.0003405, 3.405, and 3405.

Solution. We count the number of places which the decimal point is removed from the reference position in each case. Thus we find that the characteristics are respectively -4, 0, and 3.

$$
\begin{array}{ccc}
0.0003405 & 3.405 & 3045 \\
\uparrow\!\!\!\!\!\underline{\quad}\!\!\!\!\!\uparrow & \uparrow & \uparrow\!\!\!\!\!\underline{\quad}\!\!\!\!\!\uparrow \\
-4 & 0 & 3
\end{array}
$$

EXAMPLE 2. Given log 5.826 = 0.7654, write the logarithms of 582.6, 58260, and 0.05826.

Solution. The mantissas of the logarithms of these numbers are all the same as the given mantissa. Relative to the reference position, the decimal points of 582.6, 58260, and 0.05826 are respectively 2 places to the right, 4 places to the right, and 2 places to the left. Hence the required characteristics are 2, 4, and -2, and we have

$$\log 582.6 = 2.7654,$$

$$\log 58260 = 4.7654$$

$$\log 0.05826 = -2 + 0.7654$$

We have noticed that the logarithm of a number between 0 and 1 is negative. That is, the characteristic is a negative integer, and the mantissa is either zero or a positive fraction. The characteristic may be combined with the mantissa to yield a single negative quantity. For example,

$$\log 0.05826 = -2 + 0.7654 = -1.2346$$

For most computations, however, it is preferable to express a logarithm with the fractional part positive.

The logarithm $-2 + 0.7654$ could not be written as -2.7654 since this would indicate that the fractional part is also negative. A negative characteristic is customarily expressed as the difference of two integers. Thus, since $-2 = 2 - 4$ and also equal to $8 - 10$, we have

$$\log 0.05826 = -2 + 0.7654$$

$$= 2.7654 - 4$$

$$= 8.7654 - 10$$

A negative characteristic may obviously be written as the difference of two integers in an unlimited number of ways. The usual practice, however, is to use a positive integer minus 10 or an integral multiple of 10.

Write the characteristic of the logarithm of each of the following numbers.

1. 38	**2.** 50	**3.** 824.3
4. 7935	**5.** 400.1	**6.** 43780
7. 0.923	**8.** 0.0372	**9.** 0.00347
10. 0.0006	**11.** 0.00008	**12.** 1.800
13. 6.43×10^3	**14.** 2.78×10^{-2}	**15.** 7.85×10^{-5}

Given $\log 86.34 = 1.9362$, find the numbers which have the following logarithms.

16. 2.9362	**17.** 0.9362	**18.** 3.9362
19. $8.9362 - 10$	**20.** $9.9362 - 10$	**21.** 5.9362
22. 4.9362	**23.** $18.9362 - 20$	**24.** $27.9362 - 30$
25. $6.9362 - 10$	**26.** $5.9362 - 10$	**27.** $16.9362 - 20$

17-5 TABLES OF LOGARITHMS

We have shown how to determine immediately the characteristic of the logarithm of a number. Finding mantissas, however, is not simple. But methods are developed in more advanced mathematics which permit the computation of a mantissa to any desired number of decimal places. The mantissas corresponding to many numbers have been computed and arranged in tabular form. Such an arrangement is called a *table of logarithms*, or a *table of mantissas*. Since most mantissas are unending decimal fractions, their values are approximated to two, three, or more decimal places. Accordingly a table is called a *two-place* table, a *three-place* table, and so on, depending on the number of decimal places used.

A four-place table is found on the end papers inside the back cover. This table gives the mantissas of the numbers from 1.00 to 9.99 in steps of 0.01 of a unit. The first column, headed by N, contains the first two digits of a number and the third digit is to the right in the first line of the page. The mantissa corresponding to a number is in the horizontal line of the first two digits of the number and in the column headed by the third digit. As is customary, decimal points are omitted in the table. A decimal point belongs in front of each mantissa. Since the characteristic of the logarithms of 1 and numbers between 1 and 10 is zero, the table gives the complete logarithms of all three-digit numbers from 1.00 to 9.99. In order to use the table for a positive number outside this range, it is necessary to supply the proper nonzero characteristic in accordance with the discussion in Sec. 17-4. We shall explain with examples the use of the table in finding logarithms of numbers and in finding numbers corresponding to given logarithms.

The number corresponding to a given logarithm is called the *antilogarithm* (written "antilog"). That is, if $\log N = x$, then $N = \text{antilog } x$.

EXAMPLE 1.   Find log 4.58.

Solution.   Look in the first column of the table of logarithms, which has N at the top. Go down the column to 45, the first two digits in 4.58. Then look along the horizontal line from 45, stopping at the column which has 8 at the top, the third digit in 4.58. The mantissa .6609 is thus located. Hence

$$\log 4.58 = 0.6609$$

EXAMPLE 2.   Find log 0.00246.

Solution.   The decimal point is three places to the left of the reference position, and this makes the characteristic -3. Although the characteristic is determined from the position of the decimal point, the mantissa is independent of the decimal point. Hence the numbers 0.00246 and 246 have the same mantissa. We now find the mantissa corresponding to 246. In the first column, which has N at the top, locate 24, the first two digits in 246. Then look along the horizontal line from 24, stopping at the column which has 6 at the top, the third digit in 246. The mantissa .3909 is thus located. Hence

$$\log 0.00246 = 7.3909 - 10$$

EXAMPLE 3.   Find log 87,000.

Solution.   Since the decimal point is four places to the right of the reference position, the characteristic is 4. The mantissa may be located by looking across from the number 87, which is in the column headed by N, to the column headed by 0. The mantissa is .9395. Hence

$$\log 87,000 = 4.9395$$

EXAMPLE 4.   Find antilog 3.6284.

Solution.   Here we have the logarithm and wish to find the corresponding number, or antilogarithm. Since the table has only the mantissas of logarithms, we disregard the given characteristic for the moment and look in the columns of mantissas for 6284. The mantissas increase to the right along each row, and from row to row. It is easy, therefore, to locate any mantissa which appears in the table. Having found the entry 6284, we look along its row to the left and find 42 in the column headed by N. The third digit is at the top of the column containing the given mantissa and, in our example, is 5. Hence the sequence of digits in 425 corresponds to the mantissa .6284. To place the decimal point, we notice that the given characteristic is 3; consequently the point belongs three places to the right of the reference position. Hence

$$\text{antilog } 3.6284 = 4250$$

EXAMPLE 5.   If log $N = 8.9741 - 10$, find N.

Solution.   Again we have a logarithm and wish to find the antilogarithm. We first locate 9741 in a column of mantissas. Then along the row of this entry we find 94 in the first column of the page. The given mantissa has 2 at the top of its column. Hence N, except for the decimal point, is 942. The given characteristic is $8 - 10$, or -2, and therefore the decimal point goes two places to the left of the reference position. Accordingly, we have $N = 0.0942$.

Find the logarithm of each of the following numbers.

1. 3.14	**2.** 57.2	**3.** 4310
4. 57,600	**5.** 9.09	**6.** 0.821
7. 0.008	**8.** 0.058	**9.** 0.005
10. 760	**11.** 900	**12.** 0.000087
13. 0.39	**14.** 576,000	**15.** 0.0802

Find the antilogarithm of each of the following logarithms.

16. 1.1847	**17.** 2.4900	**18.** 3.6064
19. 0.9708	**20.** 4.7324	**21.** 0.8927
22. $9.6767 - 10$	**23.** $8.6513 - 10$	**24.** $7.0414 - 10$
25. $6.8476 - 10$	**26.** 0.0043	**27.** $9.4843 - 10$
28. 3.000	**29.** 1.9996	**30.** $7.8692 - 10$

17-6 INTERPOLATION

In finding logarithms and antilogarithms in the preceding section only direct readings from the table were necessary. This was because logarithms of numbers of three or fewer digits (initial and final zeros excluded) were required; and, in the case of a given logarithm, the mantissa appeared in the table. The determination of the logarithm of a four-digit number is not so readily obtained. Also, a given mantissa which is not in the table presents a slight difficulty. These cases, however, can be handled by a process called *interpolation*. According to this process, log N and N are directly proportional for small changes in N. Although this assumption is not entirely true, improved accuracy may be obtained by interpolation.

In the interpolation process we shall have occasion to *round off* numbers, that is, to discard extra digits. In each case the new number will be an approximation to the given number.

EXAMPLES. We round off 2.718 to three digits and have 2.72. We add 1 to the third digit of the given number in order to obtain the best three-figure approximation. For two-figure and one-figure approximations we round off respectively to 2.7 and 3.

In rounding off 2.715 to three digits, we observe that 2.71 and 2.72 are equally valid. In all such cases we shall arbitrarily *round off so that the last digit is an even number*. Accordingly we choose 2.72.

EXAMPLE 1. Find log 2047.

Solution. The number 2047 is 0.7 of the way from 2040 to 2050. The mantissas corresponding to 2040 and 2050 are the same as those for 204 and 205 and can be read directly from the table. These mantissas are 3096 and 3118, except for the decimal points. To obtain a number 0.7 of the way from the smaller to the larger of these numbers, we compute 0.7(3118 − 3096). Our answer is 15, to the nearest integer, and hence 3096 + 15 = 3111 is the required mantissa when the decimal point is supplied. The quantities involved in this interpolation process are tabulated as a further aid in understanding the steps.

$$
10 \left[\; 7 \left[\begin{matrix} 2040 \\ 2047 \\ \end{matrix} \right. \begin{matrix} \text{NUMBER} \end{matrix} \right.
$$

	NUMBER	MANTISSA	
	2040	3096	0.7 × 22 = 15.4
10 [7 [	2047	?] 22	= 15, rounded off
	2050	3118]	3096 + 15 = 3111

Since the required characteristic is 3, we have

$$\log 2047 = 3.3111$$

EXAMPLE 2. Find log 0.07542.

Solution. The characteristic is −2, or 8 − 10. To find the mantissa, we recall that this part of the logarithm of a number is independent of the location of the decimal point. Hence we find the mantissa corresponding to the digits in 7542. The quantities involved are placed here in tabular form.

	NUMBER	MANTISSA	
	7540	8774	0.2 × 5 = 1.0
10 [2 [	7542	?] 5	
	7550	8779]	8774 + 1 = 8775

We see that 7542 is 0.2 of the way across from 7540 to 7550; and we want, therefore, to find the number which is 0.2 of the way from 8774 to 8779. From 8774 to 8779 is 5 units, and 0.2 of 5 is 1. Hence 8774 + 1 = 8775, and

$$\log 0.07542 = 8.8775 - 10$$

The student who is just beginning to learn the method of interpolation will probably need to tabulate the quantities in a way similar to that used in the preceding examples. Having done a few problems, however, he should notice that most, if not all, of the operations can be done mentally. As shown in the examples, three operations are involved in finding a mantissa.

1. A mantissa is to be subtracted from the one immediately following the table.
2. This difference is to be multiplied by an integral multiple of tenths (0.1 or more up to 0.9).
3. The product, rounded off, is to be added to the smaller mantissa.

In most cases the mental performance of these operations can be made easily and quickly.

Let us next consider the problem of finding the number corresponding to a given logarithm whose mantissa does not appear in the table. This is the inverse of the problems just done, but the principle of interpolation is the same.

EXAMPLE 3. Given log $N = 7.8126 - 10$, find N.

Solution. The mantissa 8126 is between the consecutive mantissas 8122 and 8129, and is $\frac{4}{7}$ of the way across. The numbers 6490 and 6500 correspond to the consecutive mantissas. Hence we find the number $\frac{4}{7}$ of the way from 6490 to 6500. This gives 6496 which, except for the decimal point, is the required antilogarithm. Placing the decimal point in accordance with the given characteristic $7 - 10$, or -3, we have

$$N = \text{antilog } 7.8126 - 10 = 0.006496$$

The numbers appearing in this example are presented in the following tabular form.

NUMBER	MANTISSA	
⌈6490	8122⌉	$\frac{4}{7} \times 10 = 6$, rounded off
10⌊ ?	8126⌋ 4 ⌋7	$6490 + 6 = 6496$
⌊6500	8129 ⌋	$N = 0.006496$

Again, the student should observe that, in most cases, the computations required in finding antilogarithms can be done mentally.

EXERCISE 17-5

Find the logarithms of the following numbers.

1. 68.25	**2.** 7.831	**3.** 585.3
4. 0.9872	**5.** 0.09864	**6.** 0.004446
7. 64,050	**8.** 99.09	**9.** 4.827
10. 0.001118	**11.** 0.01001	**12.** 0.00007447
13. 90490	**14.** 10.51	**15.** 234.8
16. 0.8105	**17.** 0.0003077	**18.** 0.4805
19. 1741	**20.** 7621	**21.** 7897
22. 98,890	**23.** 745,500	**24.** 0.2604

Find the antilogarithms of the following logarithms.

25. 2.7743	**26.** 0.8104	**27.** 1.8063
28. 3.0242	**29.** 4.9985	**30.** 0.6090
31. $7.6621 - 10$	**32.** $8.4787 - 10$	**33.** $9.4885 - 10$
34. $9.1336 - 10$	**35.** $8.1731 - 10$	**36.** $6.3011 - 10$

37. 0.2330   **38.** 1.2479   **39.** 2.3182
40. 3.3200   **41.** 4.3346   **42.** 5.2901
43. 9.7375 − 10   **44.** 8.6920 − 10   **45.** 7.9511 − 10
46. 8.0100 − 10   **47.** 7.0350 − 10   **48.** 9.6626 − 10

I7-7 LOGARITHMS USED IN COMPUTATIONS

We are now ready to compute products, powers, roots, or any combination of these operations, by means of logarithms. The three theorems of Sec. 17-2 serve for this purpose. We shall illustrate their use in some examples.

EXAMPLE 1. Compute N by means of logarithms, given

$$N = \frac{(32.41)(81.93)}{(1.854)(0.7949)}$$

Solution. Applying Theorem 17-2, we write

$$\log N = \log (32.41)(81.93) - \log (1.854)(0.7949)$$

And by Theorem 17-1

$$\log N = \log 32.41 + \log 81.93 - (\log 1.854 + \log 0.7949)$$

To help in avoiding errors and also to speed the work, the student should first make an outline form in which the characteristics of the given numbers are written and spaces are left for the quantities to be found. We suggest a form for this problem:

NUMERATOR   DENOMINATOR
$\log 32.41 = 1.$ $\log 1.854 = 0.$
$\log 81.93 = 1.$ $\log 0.7949 = 9.$ − 10
$\log \text{num} =$ $\log \text{denom} =$
$\log \text{denom} =$
$\log N =$

Having made the form, the tables should then be used and the spaces filled in. We show the completed form:

$\log 32.41 = 1.5106$ $\log 1.854 = 0.2681$
$\log 81.93 = 1.9135$ $\log 0.7949 = 9.9003 - 10$
$\log \text{num} = 3.4241$ $\log \text{denom} = 10.1684 - 10$
$\log \text{denom} = 0.1684$ $= 0.1684$
$\log N = 3.2557$
$N = 1802$

EXAMPLE 2. Evaluate $\sqrt[3]{321.4}/\sqrt{208.7}$.

Solution. Letting N stand for the given fraction, we have

$$N = \frac{\sqrt[3]{321.4}}{\sqrt{208.7}} = \frac{(321.4)^{1/3}}{(208.7)^{1/2}}$$

and

$$\log N = \tfrac{1}{3}\log 321.4 - \tfrac{1}{2}\log 208.7$$

$$\log 321.4 = 2.5071 \qquad \log 208.7 = 2.3195$$

$$\tfrac{1}{3}\log 321.4 = 0.8357 \qquad \tfrac{1}{2}\log 208.7 = 1.1598$$

The logarithm of the denominator is greater than the logarithm of the numerator. We wish to obtain the difference, or $\log N$, in the form of a negative integer plus a positive fraction (mantissa). The table may then be used to find N. To obtain $\log N$ in this usable form, we first express the characteristic of the logarithm of the numerator as $10 - 10$. Thus we write

$$\tfrac{1}{3}\log 321.4 = 10.8357 - 10$$
$$\tfrac{1}{2}\log 208.7 = 1.1598$$
$$\overline{\log N = 9.6759 - 10}$$
$$N = 0.4741$$

EXAMPLE 3. Compute $\sqrt[3]{0.8316}$.

Solution. We first write $N = \sqrt[3]{0.8316} = (0.8316)^{1/3}$. Then

$$\log N = \tfrac{1}{3}\log 0.8316$$

$$\log 0.8316 = 9.9199 - 10$$

Dividing $9.9199 - 10$ by 3, we get the result, $3.3066 - 3.3333$, in an undesirable form. A usable form is obtained at once if the characteristic $9 - 10$ is replaced by the equal quantity $29 - 30$. Thus we get

$$\log 0.8316 = 29.9199 - 30$$

$$\tfrac{1}{3}\log 0.8316 = 9.9733 - 10$$

$$N = 0.9404$$

EXERCISE 17-6

Use logarithms and perform the indicated operations.

1. $(3.82)(61.2)$ **2.** $(5.13)(0.837)$ **3.** $(0.0081)(723)$
4. $44.8 \div 81.3$ **5.** $224 \div 68.7$ **6.** $9.07 \div 75.6$
7. $(5.01)(3.77)(9.12)$ **8.** $(10.7)(8.16)(70.7)$ **9.** $(771.3)(31.25)$

10. $(90.92)(0.141)(6.683)$ **11.** $(7.103)(23.09)(1.783)$ **12.** $(12.36)(357.9)(0.167)$

13. $\dfrac{(314)(671)}{(544)(837)}$ **14.** $\dfrac{(7531)(3456)}{(1813)(1339)}$ **15.** $\dfrac{(404)(505)}{(707)(114)}$

16. $(2.718)^2$ **17.** $(0.07184)^4$ **18.** $(3.318)^3$

19. $\sqrt[3]{75.66}$ **20.** $\sqrt[3]{8.082}$ **21.** $\sqrt{0.6118}$

22. $\sqrt[4]{0.04623}$ **23.** $\sqrt[4]{17.81}$ **24.** $\sqrt[3]{66.78}$

25. $(21.08)^{2/3}$ **26.** $(0.1776)^{3/4}$ **27.** $(1.414)^{3/5}$

28. $(11.08)^{-4/3}$ **29.** $(1881)^{-4/5}$ **30.** $(1004)^{-5/2}$

31. $\dfrac{\sqrt[3]{8143}}{\sqrt{5625}}$ **32.** $\dfrac{\sqrt{1357}}{(5.634)^3}$ **33.** $\sqrt{\dfrac{(57.81)(6.183)}{19.78}}$

34. $\sqrt[3]{\dfrac{(4881)(2321)}{87.91}}$ **35.** $\sqrt[4]{\dfrac{(19.03)(30.91)}{(1.456)(2.531)}}$ **36.** $\dfrac{25.73\sqrt{5.183}}{\sqrt[3]{893.2}}$

17-8 EXPONENTIAL EQUATIONS

An equation in which the unknown appears in an exponent is called an *exponential equation*. Some quite simple equations of this kind may be solved by inspection. Certain others are readily solved by the use of logarithms. The following examples illustrate methods of solving exponential equations.

EXAMPLE 1. Solve $5^x = 625^2$.

Solution. First express 625 as 5^4, then $5^x = (5^4)^2 = 5^8$. With the bases equal, the exponents must be the same; hence, $x = 8$.

EXAMPLE 2. Solve $2^x = 29$.

Solution. Since equal quantities have equal logarithms, we may write

$$\log 2^x = \log 29$$

Whence,

$$x \log 2 = \log 29$$

and

$$x = \frac{\log 29}{\log 2} = \frac{1.4624}{0.3010} = 4.858$$

The value of x is obtained by dividing $\log 29$ by $\log 2$. This is the quotient of two logarithms; it should not be confused with the logarithm of a quotient.

EXAMPLE 3. Solve the equation $5(3)^{2x+1} = 7^{1-x}$.

Solution. Taking the logarithm of each member gives

$$\log 5(3)^{2x+1} = \log 7^{1-x}$$

and

$$\log 5 + (2x + 1)\log 3 = (1 - x)\log 7$$

This is a linear equation in x and may be solved in the usual way. Thus

$$(2 \log 3 + \log 7)x = \log 7 - \log 5 - \log 3$$

and

$$x = \frac{\log 7 - \log 5 - \log 3}{2 \log 3 + \log 7} = \frac{-0.3310}{1.7993} = -0.1840$$

Solve the following exponential equations without the use of tables.

1. $5^x = 125$ 2. $3^{x-1} = 9$ 3. $e^{2x} = e^{10}$
4. $4^{x+2} = 16^x$ 5. $10^{3x-3} = 1000$ 6. $25^x = 5$
7. $4^{2x+3} = 4$ 8. $6^x = \frac{1}{36}$ 9. $3^{2x} = 3^0$

Solve the following exponential equations.

10. $4^x = 7$ 11. $9^x = 15$ 12. $(1.27)^x = 2.73$
13. $9^x = 7$ 14. $6^{2x-3} = 4$ 15. $(8.13)^x = 68$
16. $4^x = 5^x$ 17. $11^{2x-1} = 7^x$ 18. $20^x = 4^x$
19. $100^x = 3^x$ 20. $(0.3)^x = 0.5$ 21. $(0.182)^x = 0.93$
22. $(0.718)^x = 1.07$ 23. $3^x = 0.13$ 24. $17^x = 0.19$
25. $6 \cdot 10^x = 11^x$ 26. $9^x = 4 \cdot 19^{x+1}$ 27. $5 \cdot 7^x = 6 \cdot 8^x$

17-9 LOGARITHMS OF A NUMBER TO DIFFERENT BASES

It is sometimes desirable to express the logarithm of a number to one base in terms of its logarithm to another base. In particular, a change from a common logarithm to a natural logarithm, or vice versa, is advantageous in certain calculus problems. We shall derive the relation between the logarithms of a number to any two bases.

Let a and b stand for two positive numbers different from 1, and let

$$\log_a N = x \quad \text{or} \quad N = a^x$$

Taking the logarithm to the base b of both members of the exponential equation, we obtain

$$\log_b N = x \log_b a$$

and

$$x = \frac{\log_b N}{\log_b a}$$

Since $x = \log_a N$, we obtain

$$\log_a N = \frac{\log_b N}{\log_b a}$$

This formula can be used to change from one base to another. By substituting the natural base e (2.718 approximately) for a and 10 for b, we have

$$\log_e N = \frac{\log_{10} N}{\log_{10} e}$$

The value of $\log_{10} e$ is 0.4343 to four decimal places. Making this substitution, we get the relations

$$\log_{10} N = 0.4343 \log_e N$$

and

$$\log_e N = 2.3026 \log_{10} N$$

EXAMPLE 1. $\log_e 46 = 2.3026 \log_{10} 46 = (2.3026)(1.6628) = 3.829$

EXAMPLE 2.

$$\log_4 15 = \frac{\log_{10} 15}{\log_{10} 4} = \frac{1.1761}{0.6021} = 1.953$$

EXERCISE 17-8

Find the following logarithms.

1. $\log_e 100$	**2.** $\log_e 158$	**3.** $\log_e 0.478$
4. $\log_2 54$	**5.** $\log_3 175$	**6.** $\log_4 204$
7. $\log_2 0.416$	**8.** $\log_5 0.067$	**9.** $\log_7 62.4$
10. $\log_{11} 150$	**11.** $\log_e 0.041$	**12.** $\log_{12} 247$
13. $\log_8 9.85$	**14.** $\log_9 0.009$	**15.** $\log_e \sqrt{2}$

17-10 COMPOUND INTEREST

Suppose the interest earned by a given amount of money, called the principal, during a specified period is added to the principal to form new principal for a second period. If this process is repeated for n periods, we say the given principal earned *compound interest* over n interest periods. As we shall show, logarithms may be used in solving certain compound interest problems.

Let r (expressed decimally) be the annual interest rate which P dollars earns at compound interest for n years. At the end of the first year the amount will be

$$A_1 = P + Pr = P(1 + r)$$

At the end of the second year the amount will be

$$A_2 = P(1 + r) + Pr(1 + r) = P(1 + r)^2$$

At the end of the third year the amount will be

$$A_3 = P(1 + r)^2 + Pr(1 + r) = P(1 + r)^3$$

If this process is continued n times, the amount A will be

$$A = P(1 + r)^n \tag{1}$$

EXAMPLE 1. If \$100 is invested at 6% compounded annually, what will be the accumulated amount at the end of 4 years?

Solution. Substituting in formula (1), we get

$$A = 100(1 + .06)^4 = 100(1.06)^4$$
$$\log A = \log 100 + 4 \log (1.06)$$
$$\log 1.06 = 0.0253$$
$$4 \log 1.06 = 0.1012$$
$$\log 100 = 2.0000$$
$$\log A = 2.1012$$
$$A = \$126.25$$

If the interest is compounded t times per year, the interest rate for one period is r/t instead of r, and the number of periods is tn. For this general situation formula (1) becomes

$$A = P\left(1 + \frac{r}{t}\right)^{tn} \tag{2}$$

EXAMPLE 2. If \$500 is invested at 8% per year compounded quarterly, find the accumulated amount at the end of 3 years.

Solution. We use formula (2) with $r = .08$ and $t = 4$. Then we have

$$A = 500 \ (1.02)^{12}$$
$$\log A = \log 500 + 12 \log (1.02)$$
$$\log 1.02 = 0.0086$$
$$12 \log 1.02 = 0.1032$$
$$\log 500 = 2.6990$$
$$\log A = 2.8022$$
$$A = \$634.10$$

EXERCISE 17-9

1. Find the amount if \$200 is invested for 5 years at 6% per annum compounded (a) annually; (b) semiannually.
2. Find the amount if \$300 is invested for 7 years at 7% compounded (a) annually; (b) semiannually.
3. Find the amount if \$400 is invested for 6 years at 6% per annum compounded (a) annually; (b) semiannually; (c) quarterly.
4. How long will it take for \$100 to amount to \$215.80 if interest is paid at the rate of 8% per year compounded annually? [Hint: $215.80 = 100(1.08)^n$, $n = (\log 215.80 - \log 100) \div \log 1.08$.]

5. Find how long it will take for $100 to amount to $179.60 at 5% interest compounded yearly.

6. In 6 years $200 will amount to $300.20 at a certain interest rate compounded annually.   Find   the   interest   rate.   [Suggestion: $300.20 = 200(1 + r)^6$. $\log (1 + r) = \log 300.20 - \log 200) \div 6.$]

7. At what interest rate compounded annually will $100 amount to $229.20 in 17 years?

Chapter 18

PARTIAL FRACTIONS

18-1 RESOLUTION OF FRACTIONS

In Chapter 4 we combined fractions into a single fraction by the process of addition. In certain mathematical situations, particularly in calculus, it is necessary to perform the inverse operation and express a fraction as the sum of simpler fractions. The resulting fractions, obtained by the method described below, are called *partial fractions.*

We shall consider fractions of the form $p(x)/q(x)$, where $p(x)$ and $q(x)$ are polynomials. If the degree of the numerator is less than the degree of the denominator, the fraction is called a *proper fraction.* Otherwise, the fraction is called an *improper fraction.* Clearly, an improper fraction can, by division, be expressed as a polynomial or a polynomial plus a proper fraction. Thus, for example, the left member of the equation

$$\frac{2x^3 - 5x^2 - 3x + 10}{x^2 - 2x + 4} = 2x - 1 + \frac{-13x + 14}{x^2 - 2x + 4}$$

is an improper fraction and the right member is a polynomial plus a proper fraction.

In this chapter we shall restrict ourselves to the consideration of proper fractions. And our problem, then, will be that of expressing such fractions in terms of partial fractions. The process of finding partial fractions is based on a theorem which is proved advanced algebra. The theorem permits us to make the following statements concerning a proper fraction in its lowest terms.

I Corresponding to each linear factor $ax + b$ of the denominator there will be a partial fraction

$$\frac{A}{ax + b}$$

where A is a constant.

II Corresponding to each repeated linear factor $(ax + b)^k$ of the denominator there will be k partial fractions

$$\frac{A_1}{ax + b} + \frac{A_2}{(ax + b)^2} + \cdots + \frac{A_k}{(ax + b)^k}$$

where $A_1, A_2, \ldots, A_k$ are constants and $A_k \neq 0$.

III If $ax^2 + bx + c$ is a factor of the denominator, and not the product of two real linear factors, then corresponding to this quadratic factor there will be a partial fraction

$$\frac{Ax + B}{ax^2 + bx + c}$$

where A and B are constants.

IV If $ax^2 + bx + c$ is not the product of two real linear factors, then corresponding to the repeated factor $(ax^2 + bx + c)^k$ of the denominator there will be k partial fractions

$$\frac{A_1 x + B_1}{ax^2 + bx + c} + \frac{A_2 x + B_2}{(ax^2 + bx + c)^2} + \cdots + \frac{A_k x + B_k}{(ax^2 + bx + c)^k}$$

where $A_1, B_1, A_2, B_2, \ldots, A_k, B_k$ are constants with A_k and B_k not both zero.

We shall show illustrative examples for each of the four types of factors mentioned in the assumptions.

18-2 DISTINCT LINEAR FACTORS

We first illustrate the case in which the factors of the denominator are linear with no factor repeated.

EXAMPLE. Separate $\dfrac{x^2 + 23x - 18}{(x - 1)(x + 2)(2x - 1)}$ into partial fractions.

Solution by Substitution. In accordance with item I the partial fractions have the denominators $x - 1$, $x + 2$, $2x - 1$, and their numerators are constants. Hence we write

$$\frac{x^2 + 23x - 18}{(x - 1)(x + 2)(2x - 1)} = \frac{A}{x - 1} + \frac{B}{x + 2} + \frac{C}{2x - 1} \tag{1}$$

The constants A, B, and C are to be determined so that this equation is an identity. To facilitate their determination we clear equation (1) of fractions and have

$$x^2 + 23x - 18$$
$$= A(x + 2)(2x - 1) + B(x - 1)(2x - 1) + C(x - 1)(x + 2) \tag{2}$$

This equation, also an identity, must be true for any value which is assigned to x. The most convenient values are 1, -2, and $\frac{1}{2}$. The constants are readily found by substituting each of these values in turn in equation (2).

For $x = 1$: $1 + 23 - 18 = A(3)(1)$ and $A = 2$

For $x = -2$: $4 - 46 - 18 = B(-3)(-5)$ and $B = -4$

For $x = \frac{1}{2}$: $\frac{1}{4} + \frac{23}{2} - 18 = C(-\frac{1}{2})(\frac{5}{2})$ and $C = 5$

Therefore,

$$\frac{x^2 + 23x - 18}{(x - 1)(x + 2)(2x - 1)} = \frac{2}{x - 1} - \frac{4}{x + 2} + \frac{5}{2x - 1}$$

Solution by Equating Coefficients. We may evaluate the constants A, B, and C in equation (2) by using the fact that the coefficients of like powers of x in the two members of the equation are equal. To follow this plan we perform the indicated multiplications and collect like terms. Thus we obtain

$x^2 + 23x - 18$

$$= (2A + 2B + C)x^2 + (3A - 3B + C)x + (-2A + B - 2C)$$

Equating the coefficients, we obtain

$2A + 2B + C = 1$, from coefficients of x^2

$3A - 3B + C = 23$, from coefficients of x

$-2A + B - 2C = -18$, from the constant terms

Solving this system for A, B, and C, we find, as before, $A = 2, B = -4, C = 5$.

Either the method of substitution or the method of equating coefficients of the like powers of the variable may be used to determine the unknown constants. The substitution method works particularly well when the factors of the denominator are linear. Frequently, however, it is advantageous to combine the two methods.

EXERCISE 18-1

Resolve each fraction into partial fractions. Check your results by addition.

1. $\dfrac{x - 5}{(x + 1)(x - 1)}$

2. $\dfrac{3}{x(x + 3)}$

3. $\dfrac{7x - 16}{(x - 2)(x - 3)}$

4. $\dfrac{7x - 8}{(2x - 1)(x - 2)}$

5. $\dfrac{3x - 7}{(3x + 1)(3x - 1)}$

6. $\dfrac{5}{(x + 1)(2x - 3)}$

7. $\dfrac{11x + 1}{3x^2 - 2x - 1}$

8. $\dfrac{3x + 1}{2x^2 - x - 6}$

9. $\dfrac{22 - 4x}{3x^2 + 2x - 8}$

10. $\dfrac{6x + 27}{4x^3 - 9x}$

11. $\dfrac{8x - 8}{x^3 - 2x^2 - 8x}$

12. $\dfrac{5x - 2}{6x^3 - 5x^2 + x}$

13. $\dfrac{4x^2 - 27x + 20}{(x - 3)(x - 4)(x + 2)}$

14. $\dfrac{2x + 8}{(x + 2)(x - 2)(x + 1)}$

15. $\dfrac{15}{(2x + 1)(3x - 1)(4x - 1)}$

16. $\dfrac{3x^2 - 15x + 10}{(1 - x)(2 - x)(3 - x)}$

18-3   REPEATED LINEAR FACTORS

When the denominator is the product of linear factors with one or more repeated, we assign denominators to the partial fractions in accordance with Item II, Sec. 18-1.

EXAMPLE.   Resolve $\dfrac{4x^2 - 13x}{(x + 3)(x - 2)^2}$ into partial fractions.

Solution by Substitution. This fraction is the sum of partial fractions with $x + 3$, $x - 2$, and $(x - 2)^2$ as denominators and with constant numerators. Hence,

$$\frac{4x^2 - 13x}{(x + 3)(x - 2)^2} = \frac{A}{x + 3} + \frac{B}{x - 2} + \frac{C}{(x - 2)^2} \tag{1}$$

Clearing of fractions, we obtain

$$4x^2 - 13x = A(x - 2)^2 + B(x + 3)(x - 2) + C(x + 3) \tag{2}$$

We substitute successively the values -3, 2, and 0 for x in this equation.

For $x = -3$: $36 + 39 = 25A$ and $A = 3$

For $x = 2$: $16 - 26 = 5C$ and $C = -2$

For $x = 0$: $0 = 4A - 6B + 3C$

$$6B = 4A + 3C = 12 - 6 = 6$$

$$B = 1$$

Hence,

$$\frac{4x^2 - 13x}{(x + 3)(x - 2)^2} = \frac{3}{x + 3} + \frac{1}{x - 2} - \frac{2}{(x - 2)^2}$$

Solution by Equating Coefficients. To determine the constants A, B, and C by equating the coefficients of like powers of x, we perform the multiplications indicated in equation (2) and collect to get

$$4x^2 - 13x = (A + B)x^2 + (-4A + B + C)x + (4A - 6B + 3C)$$

Hence, from the coefficients of x^2, x, and the constant terms, we obtain

$$A + B = 4$$

$$-4A + B + C = -13$$

$$4A - 6B + 3C = 0$$

The solution of this system of equations is $A = 3$, $B = 1$, $C = -2$, which agrees with the previous determination.

EXERCISE 18-2

Resolve each fraction into partial fractions.

1. $\dfrac{x - 5}{(x - 4)^2}$

2. $\dfrac{7 - 8x}{(2x - 1)^2}$

3. $\dfrac{2x + 7}{(x + 3)^2}$

4. $\dfrac{x^2 - x - 3}{(x - 1)^3}$

5. $\dfrac{18x^2 + 3x + 4}{(3x + 1)^3}$

6. $\dfrac{x^2 - 2x + 3}{(2 - x)^3}$

7. $\dfrac{2x^2 - 2x - 1}{(x - 2)(x - 1)^2}$

8. $\dfrac{6x^2 - 11x - 32}{(x + 6)(x + 1)^2}$

9. $\dfrac{6x + 3}{(x - 1)^2(x - 4)}$

10. $\dfrac{5x^2 + 36x - 27}{x^4 - 6x^3 + 9x^2}$

11. $\dfrac{x^3 - 3x^2 + x - 3}{(x + 1)(x - 1)^3}$

12. $\dfrac{x^2 + 5x - 3}{2x^4 - 3x^3}$

13. $\dfrac{7x^3 + 2x^2 - 27x + 14}{(3x + 2)(2x - 3)(x - 2)^2}$

14. $\dfrac{7x^3 + 32x^2 - 15x - 32}{(2x - 1)(3x + 1)(x + 3)^2}$

15. $\dfrac{2x^3 + 2x^2 + 2x + 4}{x(x + 1)(x + 2)^2}$

16. $\dfrac{6x^3 - 15x^2 + 9x + 3}{(2x - 1)^2(x - 2)^2}$

18-4 DISTINCT QUADRATIC FACTORS

When the denominator contains a quadratic factor which is not the product of two real linear factors, the quadratic factor is the denominator of one of the partial fractions. A linear expression is provided for the numerator of a quadratic denominator (Item III, Sec. 18-1).

EXAMPLE 1. Resolve $\dfrac{x^3 - 25x^2 + 21x - 45}{(x - 3)(x + 2)(2x^2 - x + 3)}$ into partial fractions.

Solution. The quadratic factor $2x^2 - x + 3$ is not the product of two real linear factors. Hence, the partial fractions will have $x - 3$, $x + 2$, and $2x^2 - x + 3$ as denominators. The numerators of the linear denominators are constants, but the numerator of the quadratic denominator may be a linear expression. Thus we have

$$\frac{x^3 - 25x^2 + 21x - 45}{(x - 3)(x + 2)(2x^2 - x + 3)} = \frac{A}{x - 3} + \frac{B}{x + 2} + \frac{Cx + D}{2x^2 - x + 3} \qquad (1)$$

Clearing of fractions, we obtain

$$x^3 - 25x^2 + 21x - 45 = A(x + 2)(2x^2 - x + 3)$$
$$+ B(x - 3)(2x^2 - x + 3) + (Cx + D)(x - 3)(x + 2) \qquad (2)$$

To determine the constants A, B, C, D, we shall combine the method of substitution and the method of equating coefficients. We first substitute the values 3, -2, and 0 for x in equation (2).

$$\text{For } x = 3: \qquad -180 = A(5)(18) \quad \text{and} \quad A = -2$$
$$\text{For } x = -2: \quad -195 = B(-5)(13) \quad \text{and} \quad B = 3$$
$$\text{For } x = 0: \qquad -45 = 6A - 9B - 6D,$$
$$-45 = -12 - 27 - 6D \quad \text{and} \quad D = 1$$

Equating coefficients of x^3 gives

$$1 = 2A + 2B + C,$$
$$1 = -4 + 6 + C \quad \text{and} \quad C = -1$$

Hence,

$$\frac{x^3 - 25x^2 + 21x - 45}{(x - 3)(x + 2)(2x^2 - x + 3)} = \frac{-2}{x - 3} + \frac{3}{x + 2} + \frac{1 - x}{2x^2 - x + 3}$$

EXAMPLE 2. Resolve $\dfrac{4x^3 - 10x^2 - 2x - 9}{(2x^2 - x + 3)(3x^2 + x + 1)}$ into partial fractions.

Solution.

$$\frac{4x^2 - 10x^2 - 2x - 9}{(2x^2 - x + 3)(3x^2 + x + 1)} = \frac{Ax + B}{2x^2 - x + 3} + \frac{Cx + D}{3x^2 + x + 1} \qquad (1)$$

Clearing of fractions gives

$$4x^3 - 10x^2 - 2x - 9 = (Ax + B)(3x^2 + x + 1) + (Cx + D)(2x^2 - x + 3) \qquad (2)$$

To obtain four equations from (2), involving the unknowns A, B, C, and D, let us equate the coefficients of x^3 and then substitute in turn the values 0, 1, and -1 for x.

$$\text{Equate coefficients of } x^3: \qquad 4 = 3A + 2C$$
$$\text{For } x = 0: \qquad -9 = B + 3D$$
$$\text{For } x = 1: \qquad -17 = 5A + 5B + 4C + 4D$$
$$\text{For } x = -1: \quad -21 = -3A + 3B - 6C + 6D$$

From the first two equations of this system, we have

$$C = \frac{4 - 3A}{2} \quad \text{and} \quad B = -9 - 3D$$

Then substituting these results in the other two equations, and solving for A and D, we get

$$A = 2, \quad D = -2,$$

and finally,

$$B = 3, \quad C = -1.$$

Therefore,

$$\frac{4x^3 - 10x^2 - 2x - 9}{(2x^2 - x + 3)(3x^2 + x + 1)} = \frac{2x - 3}{2x^2 - x + 3} - \frac{x + 2}{3x^2 + x + 1}$$

EXERCISE 18-3

Express each fraction in terms of partial fractions.

1. $\dfrac{x^2 + 3x - 1}{(x - 2)(x^2 + 5)}$

2. $\dfrac{x^2 - x + 2}{(x + 1)(x^2 + 3)}$

3. $\dfrac{x^2 - 5x - 3}{(x + 3)(x^2 - 3x + 3)}$

4. $\dfrac{1 - 13x - 10x^2}{(x - 2)(2x^2 + 2x + 1)}$

5. $\dfrac{x^2 + 6x - 12}{(x - 4)(x^2 - x - 5)}$

6. $\dfrac{x^2 + 6x - 12}{(x + 2)(x^2 + x + 2)}$

7. $\dfrac{3x^3 + 8x^2 + 6x - 2}{(x + 2)(x + 1)(x^2 + 2)}$

8. $\dfrac{4x^3 + x^2 + 14x - 4}{x(x - 1)(x^2 + 4)}$

9. $\dfrac{x^2 - 7x + 1}{(x + 1)^2(x^2 - x + 1)}$

10. $\dfrac{3x^3 - 8x^2 + x - 2}{(x - 1)^2(x^2 + x + 1)}$

11. $\dfrac{x^3 + 5x^2 + 4x + 7}{(x^2 + 2)(x^2 + 1)}$

12. $\dfrac{2x^3 + 3x^2 + 3x + 2}{(3x^2 + 2)(2x^2 + 3)}$

13. $\dfrac{x^3 + x^2 - 2x + 1}{(2x^2 + x + 2)(3x^2 - x + 1)}$

14. $\dfrac{3x^3 - 4x^2 + 10x + 4}{(x^2 + 2x + 2)(x^2 - 2x + 2)}$

18-5 REPEATED QUADRATIC FACTORS

When the denominator of the given fraction has a repeated quadratic factor we form the denominators of the corresponding partial fractions as in the case of a repeated linear factor. The numerators, of course, are to be linear expressions.

EXAMPLE. Resolve $\dfrac{x^3 + 4x^2 + 5x + 3}{(x + 1)(x^2 + x + 1)^2}$ into partial fractions.

Solution. We use $x + 1, x^2 + x + 1$, and $(x^2 + x + 1)^2$ as the denominators of the partial fractions. The numerator of $x + 1$ is a constant; the other numerators may be linear expressions. Hence,

$$\frac{x^3 + 4x^2 + 5x + 3}{(x + 1)(x^2 + x + 1)^2} = \frac{A}{x + 1} + \frac{Bx + C}{x^2 + x + 1} + \frac{Dx + E}{(x^2 + x + 1)^2} \qquad (1)$$

Clearing (1) of fractions gives

$$x^3 + 4x^2 + 5x + 3 = A(x^2 + x + 1)^2 + (Bx + C)(x + 1)(x^2 + x + 1)$$
$$+ (Dx + E)(x + 1) \qquad (2)$$

Substituting -1 for x in (2), we obtain at once $A = 1$. To get equations for finding the remaining constants, we next perform the multiplications in the right member of (2). We have, after putting $A = 1$,

$$x^3 + 4x^2 + 5x + 3 = x^4 + 2x^3 + 3x^2 + 2x + 1 + B(x^4 + 2x^3 + 2x^2 + x)$$
$$+ C(x^3 + 2x^2 + 2x + 1) + D(x^2 + x) + E(x + 1)$$

Collecting gives

$$x^3 + 4x^2 + 5x + 3 = (1 + B)x^4 + (2 + 2B + C)x^3$$
$$+ (3 + 2B + 2C + D)x^2 + (2 + B + 2C + D + E)x + (1 + C + E)$$

Equating coefficients, we obtain

$$1 + B = 0, \quad \text{from coefficients of } x^4$$
$$2 + 2B + C = 1, \quad \text{from coefficients of } x^3$$
$$3 + 2B + 2C + D = 4, \quad \text{from coefficients of } x^2$$
$$1 + C + E = 3, \quad \text{from the constant terms}$$

The first two of these equations give readily $B = -1$ and $C = 1$. Then the last two yield $E = 1$ and $D = 1$. Hence,

$$\frac{x^3 + 4x^2 + 5x + 3}{(x + 1)(x^2 + x + 1)^2} = \frac{1}{x + 1} - \frac{x - 1}{x^2 + x + 1} + \frac{x + 1}{(x^2 + x + 1)^2}$$

EXERCISE 18-4

Resolve each fraction into partial fractions.

1. $\dfrac{3x^3 + 4x - 5}{(x^2 + 2)^2}$

2. $\dfrac{x^3 + x^2 - x + 4}{(x^2 + 1)^2}$

3. $\dfrac{3x^2 + 5x + 3}{(x^2 + x + 1)^2}$

4. $\dfrac{x^3 + x^2 + 4x + 1}{(x^2 + 2)^3}$

5. $\dfrac{x^4 - 2x^3 + 9x^2 - 11x + 21}{(x^2 - x + 4)^3}$

6. $\dfrac{2x^3 - x^2 + 6x - 11}{(x^2 + x + 3)^3}$

7. $\dfrac{x^4 + 4x^2 - x + 3}{(x^2 + 1)^3}$

8. $\dfrac{3x^4 - 13x^3 + 30x^2 - 36x + 27}{x(x^2 - 2x + 3)^2}$

9. $\dfrac{5x^2 + 18}{x(x^2 + 3)^2}$

10. $\dfrac{x^3 - 2x^2 + 1}{x^2(2x^2 + 1)^2}$

11. $\dfrac{x^5 + 11x^3 + 10x^2 + 25x + 50}{x^2(x^2 + 5)^2}$

12. $\dfrac{x^4 + 4x^2 - x + 2}{(x + 2)(x^2 + 2)^2}$

13. $\dfrac{x^4 - 6x^3 + 9x^2 - 10x + 5}{(x - 3)(x^2 - 2x + 2)^2}$

14. $\dfrac{4x^4 + 7x^3 + 18x^2 - 3x + 11}{(2x - 1)(x^2 + x + 2)^2}$

ANSWERS TO
ODD-NUMBERED PROBLEMS

1. $\{1, 2, 3, 4, 5, 6\}$, $\{x|x$ is an integer between 0 and 8$\}$
3. $\{M, Tu, W, Th, F, S\}$, $\{x|x$ is a day of the week$\}$
5. $\{$Washington, Adams, Jefferson$\}$, $\{x|x$ is one of the first three presidents of the U.S.$\}$
7. $\{$Genesis, Exodus, Leviticus, Numbers, Deuteronomy$\}$, $\{x|x$ is a book of the Pentateuch$\}$
9. None of these   **11.** $S = T = \varnothing$
13. $a = 0, b = 1, c = 4$
15. $\varnothing, \{1\}, \{2\}, \{3\}, \{1, 2\}, \{1, 3\}, \{2, 3\}, \{1, 2, 3\}$

1. $\{0, 1, 2, 3, 4, 5\}$ **3.** $\{0, 1, 3, 4, 5, 6, 7\}$
5. $\{4, 5, 6, 7\}$ **7.** $\varnothing$
9. $\{0, 1, 2, 3, 4, 5\}$ **11.** $\{0, 1, 3, 4, 5\}$
13. $\{r, s\} = \{r, s\}$ **15.** $\varnothing \neq \{p, q, r, s, t, u\}$
25. $P \times Q = \{(1, 2), (1, 3), (2, 2), (2, 3)\}$
$Q \times P = \{(2, 1), (2, 2), (3, 1), (3, 2)\}$

1. Axiom 2   **3.** Axioms 2 and 5   **5.** Axiom 6
7. Axioms 2 and 6   **9.** Axiom 6   **11.** Axiom 4

13. Axioms 2 and 6   **15.** $-1, \frac{1}{1}$ **17.** $-\frac{2}{3}, \frac{1}{2/3}$

19. $-(2x + y), \dfrac{1}{2x + y}$ **21.** $-\left(-\dfrac{3a}{2b}\right), \dfrac{1}{-3a/2b}$

23. (a) $x = 1$, (b) $x = 0$, (c) $x = d$

EXERCISE 2-3   PAGE 24

1. $0.38 = \frac{38}{100}$ **3.** $-11.62 = -\frac{1162}{100}$ **5.** $\frac{5}{6} = 0.8333\cdots$
7. $\frac{34}{11} = 3.090909\cdots$ **9.** $\frac{21}{37} = 0.567567\cdots$ **11.** $\frac{103}{33} = 3.121212\cdots$
15. $3 < 8$ **17.** $5 > 3$ **19.** $3 > -4$
21. $-5 < -2$ **23.** Positive integers
25. Negative integers   **27.** Rational numbers
29. Irrational numbers

EXERCISE 3-1   PAGE 29

1. -8 **3.** -22 **5.** 9 **7.** 11 **9.** $6mn$
11. 12 **13.** $6x$ **15.** -2 **17.** $-8a$ **19.** -5
21. $-a + 9b - 3c; 11a + 9b - 17c$
23. $6x - xy - 3yz - 9, 13xy + 9yz + 17$
25. $a + 15b + 34c; -15a - b + 8c$
27. $-9r + 4rs + 6x; 7r$ **29.** $x + y - 4$
31. y **33.** $1 - 2a + 2b$
35. $3x + 1$ **37.** $6xy + 5x - 3y$
39. $2 - (5x + 6y - 7z), 2 + (-5x - 6y + 7z)$
41. $c - (xy - 3 + yz), c + (-xy + 3 - yz)$
43. $-(x + 4y + 3z), (-x - 4y - 3z)$

EXERCISE 3-2   PAGE 32

1. 40 **3.** -60 **5.** $-18a^3b^3$ **7.** $-6x^2y^2$
9. $6x^5y^4$ **11.** $x^4y - xy^4$ **13.** $x^2 - 4y^2$ **15.** $4 - 5x - 6x^2$
17. $9x^2 - 25y^2$ **19.** $x^3 + 1$
21. $2c^3 + 13ac^2 + 19a^2c - 4a^3$ **23.** $3x^4 - 5x^3 - 27x^2 + 35x + 42$
25. $a^2 + 2ab + b^2 - 1$ **27.** $a^4 - a^3 - 5a^2 + 17a - 12$
29. $6x^4 + x^3 - x^2 - 10x - 24$ **31.** $x^3 - 6x^2 + 11x - 6$
33. $6 - m - 5m^2 + 2m^3$

EXERCISE 3-3   PAGE 36

1. $2x^2$ **3.** -2 **5.** $5ab^2 - 3$ **7.** $m - n - 1$
9. $1 - 2x^2y$ **11.** $3x - 8$ **13.** $x^2 - 2x - 2 + \dfrac{12}{x + 2}$
15. $x^2 - 3x - 1 + \dfrac{-11}{2x + 6}$ **17.** $a^2 - ab + b^2$ **19.** $x + 9$
21. $2x^2 + x - 3$ **23.** $x^2 + xy + y^2$ **25.** $2m^2 - mn + 3n^2$
27. $x^3 - x + 4 + (8x - 2)/(x^2 + x + 2)$ **29.** $x^3 - x^2 + x + 4$

EXERCISE 3-4 PAGE 38

1. $m^2 - 16$
3. $x^4 - y^4$
5. $9a^4 - 121b^2$
7. $r^2 + 8rs + 16s^2$
9. $4x^2 - 12xy + 9y^2$
11. $16x^4 + 24x^2y^2 + 9y^2$
13. $x^2 - 22x + 121 + 2xy - 22y + y^2$
15. $x^2 + 8x + 15$
17. $6m^2 + 7m + 2$
19. $7a^2 - 41ab - 6b^2$
21. $6r^4 - 5r^2s^2 - 6s^4$
23. $m^2n^2 - mny - 2y^2$
25. $x^3 - 3x^2 + 3x - 1$
27. $8x^3 + 36x^2y + 54xy^2 + 27y^3$
29. $-a^3 - 3a^2b - 3ab^2 - b^3$
31. $a^2 + 2ab + b^2 - 1$
33. $1 - 6m + 9m^2 - n^2$
35. $x^2 + 2x + 1 - 25y^2$

CHAPTER 3 REVIEW EXERCISE PAGE 39

1. 0
3. 6
5. -1
7. 12
9. $6a^3b^4$
11. $-14a^3b^6$
13. $-12a^4b^6$
15. $-6a^6b^4$
17. $2a^2b^2 + 6ab^3 - 4a^2b^5$
19. $-5a^5 + 15a^3b^2 + 15a^4b^2$
21. $2a^2 + ab - b^2$
23. $12a^2 + ab - 6b^2$
25. $b^3 + ab^2 - a^2b + 2a^3$
27. $a^4 + 2a^3 - 5a^2 - 14a - 8$
29. $3x^2$
31. $-2xy^2$
33. $3xy^2 - 2$
35. $x - 7$
37. $2x + 3$

39. $2x + 3 + \dfrac{-2}{4x - 1}$
41. $2x^2 + 6x + 9 + \dfrac{-3}{x - 3}$

43. $x - 9 + \dfrac{54}{x^2 - x + 3}$
45. $3x^2 + x + 2 + \dfrac{-8x - 11}{x^2 - 2x + 4}$

47. $2x^3 + 2x^2 - x + \dfrac{1}{2x - 3}$
49. $9x^4 - 16$

51. $16 - 8x + x^2$
53. $x^4 + 4x^2y^2 + 4y^4$
55. $4x^2 - 12xy + 9y^2 + 8x - 12y + 4$
57. $x^2 - x - 30$
59. $6x^2 + 13xy - 28y^2$
61. $14x^2 + 41xy - 28y^2$
63. $x^3 - 9x^2 + 27x - 27$
65. $x^3y^3 - 9x^2y^2z + 27xyz^2 - 27z^3$
67. $x^2 + 2xy + y^2 - 9$
69. $x^4 + x^2 + 1$
71. $x^2 - 25y^2 + 10y - 1$

EXERCISE 4-1 PAGE 42

1. $x(x^2 - xy + 2y^3)$
3. $3x(x - y + 2y^2)$
5. $(a + 2b)(m - n)$
7. $(3x + y)(3x - y)$
9. $(1 + 7y)(1 - 7y)$
11. $2a(x + 5y)(x - 5y)$
13. $(a - 1 + b)(a - 1 - b)$
15. $(2xy + z - 1)(2xy - z + 1)$
17. $(x - 3)^2$
19. $(4a + b)^2$
21. $3a(x + 1)^2$
23. $(a - 4b)(a^2 + 4ab + 16b^2)$
25. $(3pq + 2)(9p^2q^2 - 6pq + 4)$
27. $2xy(y - x)(y^2 + xy + x^2)$
29. $x(x^2 + 6xy + 12y^2)$
31. $(x^2 + 4)(x + 2)(x - 2)$
33. $(a^2 + 25b^2)(a + 5b)(a - 5b)$

35. $(2x + y)(4x^2 - 2xy + y^2)(2x - y)(4x^2 + 2xy + y^2)$

37. $3a^3b^2(a^2 + 1)(a + 1)(a - 1)$ **39.** $(2m + n - 2)(2m - n)$

41. $a^2b^2(2a - 3b)^2$ **43.** $\left(\dfrac{a}{3} - \dfrac{b}{2}\right)^2$

EXERCISE 4-2   PAGE 44

1. $(m + 2)(m + 4)$ **3.** $(m - 4)(m - 5)$ **5.** $(m - 7n)(m + 3n)$

7. $(m + 2n)(m + 9n)$ **9.** $(3x - 2)(x + 2)$ **11.** $(5x + 1)(2x - 3)$

13. $(7 + x)(3 - 4x)$ **15.** $(2x + y)(3x - 7y)$

17. $(a + 2)(a - 2)(a^2 - 3)$ **19.** $(2a^2 + 1)(a + 1)(a - 1)$

21. $(3a^2 + 2)(a + 2)(a - 2)$ **23.** $(a + 1)(a - 1)(3a + 2)(3a - 2)$

25. $(3x - 3y - 1)(x - y - 2)$ **27.** $3xy(x - 2y)(4x - y)$

29. $(2x + 1)(4x^2 - 2x + 1)(x - 1)(x^2 + x + 1)$

EXERCISE 4-3   PAGE 45

1. $(x + y)(a + 4)$ **3.** $(x - y)(a + 2c)$

5. $(y + 3a)(3x - a)$ **7.** $2(2 - m)(1 + m)(1 - m)$

9. $(x + 1)(x + 1 + a)$ **11.** $(x + y)(x - y + 5)$

13. $(x - y + 1)(x - y - 1)$ **15.** $(m + 2n - 3)(m - 2n + 3)$

17. $(x - 1 + y + z)(x - 1 - y - z)$ **19.** $(x + y - 6)(x - y - 2)$

21. $(a + b)(a^2 - ab + b^2 - a - b)$ **23.** $(8a^2 + 1 + 4a)(8a^2 + 1 - 4a)$

25. $(x^2 + xy + y^2)(x^2 - xy + y^2)$ **27.** $(2a^2 + 3b^2 + 2ab)(2a^2 + 3b^2 - 2ab)$

EXERCISE 4-4   PAGE 47

1. $\dfrac{y}{3x}$ **3.** $\dfrac{a^2}{2b^5}$ **5.** $\dfrac{x}{a}$ **7.** $\dfrac{x - 2}{x + 2}$

9. $\dfrac{-1}{x}$ **11.** $\dfrac{b + a}{b - a}$ **13.** $\dfrac{3}{2}$ **15.** $\dfrac{a - b}{2a - b}$

17. $-\dfrac{x + 2}{4x + 1}$ **19.** $\dfrac{3 - x}{3 - 2x}$ **21.** $\dfrac{a - b}{a^2 + ab + b^2}$

23. $\dfrac{x - y + 1}{x - y}$ **25.** $\dfrac{y - 1}{y + 1}$ **27.** $\dfrac{c + 3}{c - 1}$

EXERCISE 4-5   PAGE 49

1. $\frac{4}{3}$ **3.** $6b$ **5.** $\dfrac{bx}{6}$ **7.** $\dfrac{ab}{18}$

9. $\dfrac{24x}{y^3}$ **11.** $-\frac{1}{2}$ **13.** -2 **15.** $2x - 2y$

17. $\dfrac{2 - x}{4 + 2x + x^2}$ **19.** $\dfrac{x + 3}{x + 6}$ **21.** $\dfrac{b(x + 7)}{2a(3x - 2)}$ **23.** $\dfrac{(x - 3)(x - 2)}{(x + 7)(x + 2)}$

25. $(a - 1)(a - 3)$ **27.** 1 **29.** $\dfrac{x}{2(1 - x)}$

31. $\dfrac{(a - 2)(3a + 5)}{(a + 2)(a + 5)}$ **33.** $\dfrac{(3x + y)(x + 5y)}{2(x + 2y)(3x - 2y)}$

EXERCISE 4-6 PAGE 51

1. $180x^3 y^2$ **3.** $12a(a + 2)(a - 2)$ **5.** $(3x - y)(x + 2y)(2x + y)$

7. $\dfrac{5x + y}{3z}$ **9.** $\dfrac{4}{x - 3}$ **11.** $\dfrac{yz + xz + xy}{xyz}$

13. $\dfrac{a^2 - 2a - 10}{a(a + 5)}$ **15.** $\dfrac{a + 39}{(a - 9)(a + 3)}$ **17.** $\dfrac{2y + 1}{y^2 - 4}$

19. $\dfrac{4y^2 - b^2 + 1}{2y + b}$ **21.** $\dfrac{2a^3 - 4a^2 + 3a - 1}{a^2 + 1}$ **23.** $\dfrac{19m - 39}{(m + 3)(m - 5)}$

25. $\dfrac{16 - 13x + 3x^2}{(3 - x)(1 - x)(2 - x)}$ **27.** $\dfrac{-x^3 - 10x^2 + 20}{(x - 1)(x^2 - 4)}$ **29.** $\dfrac{-2x}{2x + 1}$

31. $\dfrac{x}{2x - 3}$ **33.** $\dfrac{m - 3}{m^2 - m + 1}$ **35.** $\dfrac{2a - b}{a + 2b}$

37. $\dfrac{11a - 1}{(2a + 1)(2a - 3)}$

EXERCISE 4-7 PAGE 54

1. $\frac{1}{8}$ **3.** $\dfrac{4 + x}{4 - x}$ **5.** $\dfrac{3 - x}{3 + x}$

7. $\dfrac{3}{x + 1}$ **9.** $\dfrac{5 + x}{6 - 4x}$ **11.** $\dfrac{1 - x}{2 - 5x}$

13. $\dfrac{(3x - 10)(x + 2)}{(2x + 3)(x - 2)}$ **15.** $\dfrac{(x + 1)(x + 3)}{x(x + 4)}$ **17.** $1 - x$

CHAPTER 4 REVIEW EXERCISE PAGE 54

1. $x(x - 3y + 1)$ **3.** $(3a + 1)(3a - 1)$ **5.** $2(2m + n)(2m - n)$

7. $(2 + a)(4 - 2a + a^2)$ **9.** $(2m + 3n)(4m^2 - 6mn + 9n^2)$

11. $(1 - x - y)(1 + x + y + x^2 + 2xy + y^2)$

13. $(x - 4)(x - 5)$ **15.** $(2 - a)^2$ **17.** $(2x + 3)(x + 2)$

19. $(x^2 + 1)(x + 2)(x - 2)$ **21.** $(3x + y)^2$

23. $(a - b)(4 - a)$ **25.** $(x - y)(a - 2b)(a^2 + 2ab + 4b^2)$

27. $(x - y)(x + y + 3)$ **29.** $(x - 2)(x + y + 1)$

31. $(5 + a - 3)(5 - a + 3)$ **33.** $(2x - y + z)(x - 1)$

35. $(2a + 3)^3$ **37.** $(m + n - 3)^2$

39. $(4x^2 + y^2)(2x + y)(2x - y)$ **41.** $(2m^2 + n^2)(4m^4 - 2m^2n^2 + n^4)$

43. $(x^2 + x + 1)(x^2 - x + 1)$

45. $(2x^2 + 3xy - 3y^2)(2x^3 - 3xy - 3y^2)$

47. $(x + y + 4)(x + y - 1)$

49. $(3m - 3n - 2)(3m - 3n - 1)$

51. $\dfrac{3 - x^2}{-4x - 2}, -\dfrac{3 - x^2}{4x + 2}, -\dfrac{x^2 - 3}{-4x - 2}$

53. $\dfrac{x^2 + x - 2}{x^2 - x - 1}, -\dfrac{2 - x - x^2}{x^2 - x - 1}, -\dfrac{x^2 + x - 2}{1 + x - x^2}$

55. $3x + 4y$ **57.** 2 **59.** $\dfrac{x + 5y}{2x - y}$

61. $\dfrac{x(1 - x)}{2(1 + x)^2}$ **63.** $\dfrac{3}{a + 2}$

65. $\dfrac{-m^2 + m - 9}{(m - 1)(m - 2)(2m + 7)}$ **67.** $\dfrac{x(x + 1)}{4}$

69. $\dfrac{x^2 + 2x - 1}{x^2 + 1}$ **71.** $\dfrac{x + y + 3xy - y^2}{x(y + 1)}$

73. $\dfrac{c - 2d}{2c + 3d}$ **75.** $\dfrac{2 + x}{2 - x}$ **77.** Yes

EXERCISE 5-1   PAGE 61

3. (a) on the x axis, (b) on the y axis

5. (a) on the line bisecting the first and third quadrants; (b) on the line bisecting the second and fourth quadrants

7. (a) $(4, 2)$, (b) $(-4, 2)$, (c) $(6, 4)$

9. $\{2, 3, 4\}$ domain, $\{1, -1, -3\}$ range

11. A relation   **13.** A relation

15. $1, 2, -10$ **17.** $-4, 4, \dfrac{2s + 1}{2s + 2}$

19. $x^2 + 2hx + h^2, 2hx + h^2$

21. $\{x|x \geq 0\}$ domain, $\{y|y \geq 0\}$ range

23. $\{x|-4 \leq x \leq 4\}$ domain, $\{y|0 \leq y \leq 4\}$ range

25. $\{x|x \in R\}$ domain, $\{y|y \geq 1\}$ range

EXERCISE 5-3   PAGE 69

1. 5 **3.** $3\sqrt{2}$ **5.** 5

7. $3, 4, 5$ **9.** $\sqrt{29}, \sqrt{13}, 2\sqrt{17}$ **11.** Isosceles

13. Isosceles   **15.** Straight line

17. $x^2 + y^2 + 12x - 4y + 15 = 0$ **19.** $x^2 + y^2 = 36$

21. $x^2 + y^2 - 6x = 0$ **23.** $x^2 + y^2 - 4x - 8y = 14$

25. $x^2 + y^2 + 8x - 2y + 16 = 0$ **27.** $x^2 + y^2 - 6x - 6y + 10 = 0$

EXERCISE 6-1 PAGE 75

1. $x = -1$ **3.** $x = 0$ **5.** $x = -5$ **7.** $x = \frac{1}{2}$

9. $x = \dfrac{4 + 2b}{10 - 3a}$ **11.** $\dfrac{2a - 7}{2 - 3a}$ **13.** $x = \frac{8}{5}$ **15.** $x = \frac{1}{8}$

17. $x = \frac{2}{3}$ **19.** $x = \frac{3}{4}$ **21.** No solution   **23.** No solution

EXERCISE 6-2 PAGE 77

1. 41, 46, 51   **3.** 5 by 11   **5.** $5\frac{5}{11}$ days

7. 3 hr   **9.** 12 oz   **11.** $10\frac{5}{9}$ oz gold, $4\frac{4}{9}$ oz silver

13. $\dfrac{-3}{1}$ **15.** 6% and 7%   **17.** 88

19. 1.5 hr   **21.** 4 mph, 6 mph   **23.** 12 mph

EXERCISE 6-3 PAGE 83

11. $x = 4, y = 2$ **13.** Dependent equations

15. $x = \frac{3}{2}, y = \frac{3}{2}$ **17.** $x = 5, y = -1$

19. $x = b + a, y = b - a$ **21.** $x = \dfrac{b_2 c_1 - b_1 c_2}{a_1 b_2 - a_2 b_1}, y = \dfrac{a_1 c_2 - a_2 c_1}{a_1 b_2 - a_2 b_1}$

23. $x = -2, y = 0$ **25.** $x = \frac{2}{3}, y = -\frac{3}{2}$ **27.** $x = \frac{1}{3}, y = \frac{1}{2}$

EXERCISE 6-4 PAGE 86

1. $x = -2, y = -2, z = 1$ **3.** $u = 0, v = 0, w = 5$

5. $x = \frac{2}{9}, y = \frac{1}{18}, z = \frac{2}{3}$ **7.** $x = 4, y = -3, z = 2$

9. $x = \frac{1}{8}, y = \frac{1}{8}, z = -\frac{5}{8}$ **11.** $u = a + b, v = a - b, w = 2a$

13. $u = 3, v = 2, w = -4$ **15.** $x = \dfrac{9}{5a}, y = \dfrac{-1}{5b}, z = \dfrac{-3}{5c}$

17. $x = -1, y = -1, z = 3$

EXERCISE 6-5 PAGE 88

1. 9, 23   **3.** 80 dimes, 112 quarters

5. $3,000 at 3%, $2,000 at 4%   **7.** A 15 days, B 20 days

9. $\frac{1}{5}, \frac{1}{6}$ **11.** 215 and 55 mph

13. 5 mph, 3 mph   **15.** 15°, 75°, 90°

17. 80 halves, 80 quarters, 100 dimes, 100 nickels

19. A 12 hr, B 36 hr, C 18 hr   **21.** 817

23. A 60 hr, B 72 hr, C 90 hr

EXERCISE 6-6   PAGE 91

1. $\frac{4}{1}$ **3.** $\frac{1}{8}$ **5.** $\frac{9}{2}, \frac{9}{4}$

7. $33, -\frac{121}{3}$ **9.** $x = \frac{14}{3}$ **11.** $a = \frac{-1}{19}$

13. $x = \frac{4}{5}, y = \frac{6}{5}$ **15.** $5\frac{1}{4}$ mi   **17.** 4.8 in., 5.6 in., 7.2 in.

EXERCISE 6-7   PAGE 94

1. $S = kr^2$ **3.** $V = khr^2$ **5.** $s = kt^2$ **7.** 7.68 qt

9. 1080 cu. in., 540 cu. in., 432 cu. in.   **11.** 15 days

13. 1200 ohms   **15.** Ratio 9 to 16   **17.** 675 watts   **19.** Ratio 5 to 3

CHAPTER 6 REVIEW EXERCISE   PAGE 95

1. $x = -\frac{11}{4}$ **3.** $x = 0$ **5.** $x = -\frac{13}{8}$

7. $x = -3$ **9.** No solution   **11.** $x = 2, y = -4$

13. $x = -\frac{3}{5}, y = -\frac{11}{5}$ **15.** $x = -\dfrac{6}{a}, y = 4$ **17.** $x = \frac{13}{17}, y = -\frac{13}{6}$

19. $x = \frac{14}{13}, y = -\frac{28}{3}$ **21.** $\frac{10}{12}$

23. \$7500 at 7%, \$6250 at 8%   **25.** 93

EXERCISE 7-1   PAGE 100

1. 32   **3.** $\frac{1}{4}$ **5.** 1   **7.** $\frac{1}{5}$ **9.** $\frac{1}{36}$

11. $\frac{1}{9}$ **13.** 7   **15.** $\frac{1}{16}$ **17.** $x^4 y^2$ **19.** b

21. $\dfrac{2a^2 b^5}{5}$ **23.** $\dfrac{8x^3}{y^3}$ **25.** $\dfrac{p^6 q^4}{r^2}$ **27.** $\dfrac{r^2}{p^3}$ **29.** -3

31. $\dfrac{1}{xy^2 + x^2 y}$ **33.** $x^2 + y^2$ **35.** $y^2 - xy + x^2$

EXERCISE 7-2   PAGE 103

1. $\frac{1}{4}$ **3.** $\frac{1}{4}$ **5.** $\frac{27}{8}$ **7.** 16

9. 5   **11.** $x^{1/5}$ **13.** $5^{5/6}$ **15.** $\dfrac{y}{x^2}$

17. $x + x^{1/2} y^{1/2} + y$ **19.** $\dfrac{27}{x^2}$ **21.** $5x^2$ **23.** $x^4 y^3$

25. $\dfrac{4x^{5/6}}{y}$ **27.** $\dfrac{3ab^{1/6}}{4}$ **29.** $\dfrac{x^{12} y^6}{36}$

EXERCISE 7-3   PAGE 105

1. $2\sqrt{3}$ **3.** $2a^2 b\sqrt{5}$ **5.** $4xy\sqrt[3]{xy^2}$ **7.** $\frac{1}{3}\sqrt{6}$

9. $\frac{2}{3}\sqrt[3]{3}$ **11.** $\frac{1}{3}\sqrt[4]{6}$ **13.** $\dfrac{1}{3y}\sqrt[3]{18xy^2}$ **15.** $\dfrac{1}{2c}\sqrt[4]{4bc}$

17. $\sqrt{3}$ **19.** $xy\sqrt{2y}$ **21.** $\frac{1}{x}\sqrt{3x}$ **23.** $\frac{1}{2x}\sqrt{2x^2 + 4x}$

25. $\sqrt{12}$ **27.** $\sqrt{4 - x}$ **29.** $\sqrt{3ab}$ **31.** $\sqrt[6]{3}$

33. $2\sqrt[6]{2}$ **35.** $\sqrt[12]{128}$

EXERCISE 7-4   PAGE 106

1. $4\sqrt{2}$ **3.** $7\sqrt{7}$ **5.** $5\sqrt{2} + 5\sqrt{7}$ **7.** $2\sqrt{3}$

9. $7\sqrt[3]{2} + \sqrt[3]{50}$ **11.** $(2x - 8)\sqrt{2x}$

13. $(2x^2 + x - 15x^3y)\sqrt{3y}$ **15.** $(1 + 2ab - 3b^2)\sqrt[3]{3a^2}$

17. $(1 + a + 2b)\sqrt[3]{2ab^2}$ **19.** $a\sqrt{3}$ **21.** $\frac{x - 3}{2x}\sqrt{2x}$

EXERCISE 7-5   PAGE 108

1. $\sqrt{14}$ **3.** $15\sqrt{2}$ **5.** $3\sqrt[3]{2}$ **7.** $4a\sqrt[3]{b}$

9. $\sqrt[6]{72}$ **11.** $x\sqrt[12]{x}$ **13.** $2\sqrt[3]{3}$ **15.** $\sqrt{9 - a}$

17. $6 - 3\sqrt{2}$ **19.** 3 **21.** $\frac{1}{2}\sqrt{x}$ **23.** $\frac{1}{x}$

25. $\frac{1}{a}\sqrt[3]{28a^2}$ **27.** $\sqrt[6]{3}$ **29.** $\frac{1}{a}\sqrt[6]{a^5b^4}$

31. $\frac{3 + \sqrt{5}}{4}$ **33.** $\frac{5\sqrt{5} - 2}{23}$ **35.** $\frac{7\sqrt{15} - 26}{7}$

EXERCISE 7-6   PAGE 111

1. $2i$ **3.** $i\sqrt{11}$ **5.** $-6i\sqrt{2}$ **7.** $7 + 7i$

9. $-11 - 5i$ **11.** 13 **13.** 7 **15.** $c^2 + d^2$

17. $16i$ **19.** $2c$

CHAPTER 7 REVIEW EXERCISE   PAGE 111

1. $36x^7y^8$ **3.** $\frac{b^3}{a}$ **5.** $2a^2$ **7.** $\frac{1}{8x^6y^3}$

9. $\frac{9q^{12}}{25p^2}$ **11.** $x^2q^3 + x^3q^2$ **13.** $\frac{y^4 - x^4}{y^4 + x^4}$ **15.** $\frac{7}{4}$

17. $\frac{8xy^5z}{5}$ **19.** 9 **21.** $\frac{1}{27}$ **23.** 16

25. 81 **27.** -8 **29.** $\frac{1}{625}$ **31.** $a^{14/15}$

33. $\frac{1}{a^{1/5}}$ **35.** $a^5 - 1$ **37.** $\frac{6^{1/2}}{x^{5/3}y^3}$ **39.** $4x^{5/6}y^{1/6}$

41. $\frac{4}{9x^7y^6}$ **43.** $4\sqrt{2}$ **45.** $3a^2\sqrt{2}$ **47.** $2x\sqrt[3]{4xy^2}$

49. $3xy^3\sqrt{xy^2}$ **51.** $2x^5\sqrt{y}$ **53.** $\frac{1}{3}\sqrt{15a}$ **55.** $\frac{1}{4}\sqrt[3]{9}$

57. $\frac{1}{5}\sqrt[4]{375}$ **59.** $\frac{1}{a}\sqrt[4]{4a^3b^3}$ **61.** $\frac{1}{3x}\sqrt[3]{9x^2(x+1)}$

63. $\frac{1}{bc}\sqrt{bc(a-1)}$ **65.** $\sqrt{16x^2y}$ **67.** $\sqrt{a^2-9}$

69. $\sqrt[3]{12a}$ **71.** $\sqrt[6]{a^5}$ **73.** $\sqrt[6]{75}$
75. $2\sqrt[8]{2}$ **77.** $2\sqrt{5}-18\sqrt{3}$ **79.** $-2\sqrt{15}$
81. $3\sqrt[3]{3}+\sqrt[3]{2}$ **83.** $(2ab+1-3b^2)\sqrt[3]{3a^2}$

85. $\sqrt{3a}$ **87.** $\frac{2a}{b}\sqrt{ab}$ **89.** $-\frac{1}{2}\sqrt[3]{xyz^2}$

91. $78\sqrt{2}$ **93.** $3xy^2\sqrt{6xy}$ **95.** $4a^3\sqrt[3]{b}$
97. $\sqrt[6]{288}$ **99.** $2x\sqrt[6]{27x}$ **101.** $\sqrt[6]{864}$
103. $\sqrt{16-a}$ **105.** $2\sqrt[4]{2}$ **107.** 2

109. $\dfrac{\sqrt[3]{a^2}}{a}$ **111.** $\dfrac{\sqrt[6]{a^2b}}{a}$ **113.** $\dfrac{\sqrt[4]{x}}{x}$

115. $-\dfrac{\sqrt{7}+\sqrt{3}}{4}$ **117.** $2\sqrt{42}-13$ **119.** $\dfrac{10+\sqrt{91}}{3}$

121. $4-\sqrt{14}$ **123.** $\dfrac{33-\sqrt{35}}{62}$

EXERCISE 8-1   PAGE 115

1. $\pm\sqrt{3}$ **3.** $\pm5i$ **5.** $\pm\dfrac{b}{a}$ **7.** $-1,2$

9. $-8,-1$ **11.** $-1,-\frac{1}{2}$ **13.** $\pm\frac{1}{9}$ **15.** $\pm\frac{7}{2}$
17. $-\frac{1}{2},3$ **19.** $\frac{3}{2},\frac{1}{9}$ **21.** $-\frac{2}{3},\frac{1}{4}$ **23.** $-b,a$

25. $-\dfrac{1}{a},b$

EXERCISE 8-2   PAGE 118

1. $1,3$ **3.** $\frac{1}{2},2$ **5.** $2\pm\sqrt{3}$ **7.** $4\pm3i$

9. $\dfrac{3\pm3i}{2}$ **11.** -4 **13.** $\frac{3}{2}$ **15.** $\dfrac{1\pm\sqrt{13}}{2}$

17. $\dfrac{-1\pm i\sqrt{3}}{2}$ **19.** $\dfrac{9\pm\sqrt{33}}{8}$ **21.** $\dfrac{1\pm3i}{2}$

23. $-1,-\dfrac{1}{m}$ **25.** $\dfrac{-1}{a+b},\dfrac{2}{a+b}$ **27.** $11,13;-13,-11$

29. (a) 50 yd by 200 yd, or 100 yd by 100 yd
(b) 75 yd by 150 yd
31. 200 mph and 700 mph
33. (a) 50 yd by 225 yd, or 75 yd by 150 yd
(b) 100 yd by 150 yd
35. 4 sec, 3 sec, 2 sec   **37.** (a) 2.5 sec; (b) 2 sec, 3 sec
39. 1 sec and 2 sec

EXERCISE 8-3   PAGE 121

1. $\pm 1, \pm 3$ **3.** $\pm\sqrt{2}, \pm i$ **5.** 1, 9   **7.** $\pm 1, \pm 8$

9. $\pm\frac{1}{3}, \pm i$ **11.** $\frac{1}{9}, \frac{1}{4}$ **13.** $1, 4, \dfrac{5 \pm \sqrt{37}}{2}$ **15.** $-1 \pm \sqrt{5}$

17. $-1, \frac{1}{2}$ **19.** $\frac{13}{3}$

EXERCISE 8-4   PAGE 123

1. ± 2 **3.** 2, 4   **5.** 4   **7.** No solution
9. 6   **11.** 1   **13.** 4   **15.** 3, 4
17. No solution   **19.** No solution

EXERCISE 8-5   PAGE 125

1. Real, unequal, rational; $-6, 5$ **3.** Real, unequal, irrational; $\frac{1}{2}, -\frac{7}{2}$
5. Real, equal, rational; $3, \frac{9}{4}$ **7.** Imaginary, unequal; $0, \frac{3}{5}$
9. Real, unequal, irrational; $0, -\frac{4}{7}$ **11.** Real, unequal, irrational; $\frac{4}{3}\sqrt{3}, \frac{1}{3}\sqrt{6}$
13. Real, unequal, irrational; $\frac{1}{2}\sqrt{10}, -3$ **15.** $x^2 - 7x + 12 = 0$
17. $9x^2 + 6x - 8 = 0$ **19.** $10x^2 + 7x - 12 = 0$
21. $x^2 + 4 = 0$ **23.** $x^2 - 6x + 6 = 0$
25. $x^2 - 2\sqrt{3}x + 10 = 0$ **27.** 3   **29.** 3
31. 21   **33.** $\frac{27}{64}$ **35.** 10
37. 4   **39.** 4

EXERCISE 8-6   PAGE 131

1. min $(4, -6)$ **3.** max $(-3, 10)$ **5.** min $(-\frac{1}{2}, 0)$
7. max $(-3, 18)$ **9.** min $(\frac{1}{2}, -\frac{19}{4})$ **11.** 110 by 110 rd
13. 3750 sq rd   **15.** 35

CHAPTER 8 REVIEW EXERCISE   PAGE 131

1. $\frac{3}{2}, -1$ **3.** $-\frac{1}{4}, 3$ **5.** $\frac{2}{3}, -\frac{1}{3}$
7. $2a, -2b$ **9.** $-4, 5$ **11.** $-5, 6$

13. $-\frac{1}{3}, -\frac{1}{4}$ **15.** $\dfrac{2 \pm \sqrt{2}}{2}$ **17.** $3 \pm 2i$

19. $-\frac{1}{2}, 2$ **21.** $-\frac{2}{3}, 1$ **23.** $\frac{1}{5}, \frac{1}{4}$

25. $\dfrac{-5 \pm \sqrt{17}}{8}$ **27.** $\dfrac{-5 \pm \sqrt{7}}{9}$ **29.** $\pm 3i, \pm 2i$

31. $\pm\sqrt{3}, \pm 2$ **33.** $\pm 2, \dfrac{\pm i\sqrt{2}}{2}$ **35.** 25

37. $-2, 1$ **39.** $1, 27$ **41.** $-1, -\frac{1}{2}, 1, \frac{3}{2}$

43. $-\frac{1}{3}, -\frac{6}{5}$ **45.** $\dfrac{-3 \pm i\sqrt{71}}{8}, \dfrac{-3 \pm i\sqrt{7}}{8}$

47. ± 3 **49.** 7   **51.** 0
53. No solution   **55.** 7   **57.** 5
59. 2   **61.** 3   **63.** $\frac{7}{2}, 3$
65. 12   **67.** No solution   **69.** $(-4, -4)$
71. $(2, 3)$ **73.** $(1, \frac{1}{2})$ **75.** $(\frac{1}{4}, -\frac{1}{2})$

EXERCISE 9-1   PAGE 139

Solutions obtained graphically should differ, at most, only slightly from the results given here.

1. $(0, 0), (4, 4)$ **3.** $(6, 2), (-4, -3)$ **5.** $(2, 1), (-1.3, -2.3)$
7. $(2.8, 1.4), (-2.8, -1.4), (1.4, 2.8), (-1.4, -2.8)$
9. $(\pm 1, 4), (\pm 1, -4)$ **11.** $(\pm 2, 2)$

EXERCISE 9-2   PAGE 142

1. $(-4, \pm 3), (4, \pm 3)$ **3.** $(-\sqrt{7}, \pm\sqrt{3}), (\sqrt{7}, \pm\sqrt{3})$
5. $(-3, \pm 2i), (3, \pm 2i)$ **7.** $(-1, \pm i\sqrt{2}), (1, \pm i\sqrt{2})$
9. $(-2, \pm 1), (2, \pm 1)$ **11.** $(-1, 3), (-21, -7)$
13. $\left(\dfrac{8 + i\sqrt{11}}{5}, \dfrac{-4 + 2i\sqrt{11}}{5}\right), \left(\dfrac{8 - i\sqrt{11}}{5}, \dfrac{-4 - 2i\sqrt{11}}{5}\right)$
15. $(4, 3), (-2, -5)$ **17.** $(\frac{5}{2}, \frac{1}{2}), (1, -1)$
19. $\left(\dfrac{-3\sqrt{5}}{4}, \dfrac{-3 - 3\sqrt{5}}{2}\right), \left(\dfrac{3\sqrt{5}}{4}, \dfrac{-3 + 3\sqrt{5}}{2}\right)$
21. $(1, 0), (0, -\frac{1}{2})$

EXERCISE 9-3   PAGE 143

1. $(-2, -1), (2, 1), (\pm\sqrt{3}, 0)$ **3.** $(\sqrt{2}, -3\sqrt{2}), (-\sqrt{2}, 3\sqrt{2})$
5. $(-2, 3), (2, -3), (-\frac{1}{3}, \frac{5}{3}), (\frac{1}{3}, -\frac{5}{3})$
7. $(\pm\sqrt{5}, 0), (-3, 4), (3, -4)$ **9.** $(\pm\sqrt{15}, 0), (1, -2), (-1, 2)$
11. $(3, -2), (-3, 2), (8\sqrt{7}/7, -4\sqrt{7}/7), (-8\sqrt{7}/7, 4\sqrt{7}/7)$
13. $(-3i, -2i), (3i, 2i), (2/\sqrt{29}, 3/\sqrt{29}), (-2/\sqrt{29}, -3/\sqrt{29})$
17. $(2, 2), (\sqrt{2}, -\sqrt{2}), (-\sqrt{2}, \sqrt{2})$
19. $(3, 0), (0, 3), (-2, -5), (-5, -2)$

21. 8 by 12   **23.** 10, 13, 13
25. 30 by 40 inches   **27.** 60 cents, 40
29. 40 mph, 60 mph, 100 mi.   **31.** 10 acres, $500 per acre
33. 8 by 10 feet   **35.** 1000 bu. at $3 per bu.
37. 24 ft. and 16 ft.   **39.** 12 by 16 in.

EXERCISE 10-2   PAGE 150

1. $x > 3$ **3.** $x > 4$ **5.** $x < -\frac{1}{2}$
7. $-2 < x < 2$ **9.** $-5 \le x \le 5$ **11.** $1 < x < 3$
13. $-2 < x < 2.5$ **15.** $1 < x < 3.5$ **17.** $x < 0$ or $x > 1$
19. $x < -4$ or $x > 4$ **21.** $x < 0$

EXERCISE 10-4   PAGE 155

1. $-1 < x < 3$ **3.** $-1 < x < 2$
5. No solution, or the empty set $\varnothing$
7. $1 < x < 5$ **9.** $2 < x < 7$ **11.** $-\frac{9}{5} \le x \le 3$
13. The angular region below the line $2x - y - 1 = 0$ and above the line $x - 8y + 1 = 0$, as shown in the figure.
15. The region below the line $2x - y = 2$ and above the line $2x + y = 2$
17. The region above the line $4x + 3y - 6 = 0$
19. The angular region below the line $3x + 4y + 14 = 0$ and also below the line $2x - 5y - 6 = 0$
21. No solution
23. The region inside the circle $x^2 + y^2 = 16$ and below the line $y - 3x + 9 = 0$
25. The region inside the circle $x^2 + y^2 = 16$ and inside the parabola $x^2 - 4y = 0$
27. The region inside $y = x^2 - 4$ and inside $y = -x^2 - 4x$.
29. The region inside $y = x^2$ and inside $y = -x^2 + 4$.
31. The region inside $y = x^2$ and outside $y = 2x^2$.
33. The region outside $x^2 + y^2 = 4$ and inside $x^2 + y^2 = 9$.
35. The whole plane.
37. The region inside the triangle with vertices at $(0, 0)$, $(2, 0)$, $(1, 4)$.

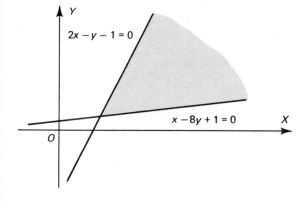

Problem 13
Exercise 10-4

EXERCISE 11-1   PAGE 158

1. $3, 5, 7, 9; a_{20} = 41$

3. $\frac{1}{2}, \frac{2}{3}, \frac{3}{4}, \frac{4}{5}; a_{20} = \frac{20}{21}$

5. $\frac{1}{2}, \frac{1}{6}, \frac{1}{12}, \frac{1}{20}; \frac{1}{420}$

7. $1, \frac{4}{5}, \frac{4}{5}, \frac{16}{17}; a_{20} = \dfrac{2^{20}}{401}$

9. $-2, 4, -8, 16; a_{20} = 2^{20}$

11. $3, 10, 24, 52$

13. $4, \frac{1}{4}, 4, \frac{1}{4}$

15. $-3, -\frac{3}{2}, -\frac{1}{2}, -\frac{1}{8}$

EXERCISE 11-2   PAGE 161

1. $23, 100$

3. $156.4, 4023.9$

5. $-182, -1462$

7. $12x + 9y, 77x + 44y$

9. 1530

11. 1683

15. $2n(n + 1)$

17. $-9, -6, -3, 0, 3, 6, 9$

19. $d = 3, S_n = 195$

21. $a_1 = -40, S_n = 198$

23. $d = 5, a_1 = -16$

25. $n = 6, a_1 = 8$ or $n = 9, a_1 = -4$

27. $a_1 = 8, a_{10} = -19$

29. 208 ft, 784 ft

31. $5445

33. $6405

EXERCISE 11-3   PAGE 164

1. $486, 728$

3. $16, 211$

5. $324, 220$

7. $r = 3$

9. $r = -2$

11. $n = 5, S_n = 726$

13. $n = 6, S_n = 2457$

15. $n = 6, S_n = 94.5$

17. $a_1 = \frac{32}{5}, a_n = -\frac{108}{5}$

19. $651.85

21. $256N$

23. $34.8\%, 29$

25. $252\frac{16}{27}$ ft

27. The first proposition pays \$5800 for the fifth year and \$24,500 for the five-year period. The second proposition pays \$5856.40 for the fifth period and \$24,420.40 for the five-year period.

EXERCISE 11-4   PAGE 167

1. $\frac{3}{2}$

3. $41\frac{1}{4}$

5. $6\frac{2}{3}$

7. -4.9

9. $\frac{8}{9}$

11. $\frac{7}{33}$

13. $\frac{103}{33}$

15. $\frac{91}{111}$

17. $\frac{1}{7}$

19. $2, 1\frac{511}{512}$

21. 1080 ft

23. $\frac{1}{34}$

25. $\frac{1}{3}, \frac{5}{26}, \frac{5}{37}, \frac{5}{48}, \frac{5}{59}, \frac{1}{14}$

27. $\frac{1}{17}, \frac{1}{13}, \frac{1}{9}, \frac{1}{5}, 1$

29. $\dfrac{1}{a}, \dfrac{3}{2a + b}, \dfrac{3}{a + 2b}, \dfrac{1}{b}$

31. $5060

37. ± 12

39. $\pm x\sqrt{2}$

41. $4, \frac{24}{7}, \frac{24}{8}, \frac{24}{9}, \frac{24}{10}, \frac{24}{11}, 2$

EXERCISE 12-1   PAGE 170

1. $16, 12$

3. $100, 48$

5. $30,000$

7. 1024

9. $120, 120$

11. 210

13. 144

15. 420

17. $72, 144$

19. $30,240; 17,280; 8640$

EXERCISE 12-2 PAGE 175

1. 7, 210, 840, 5040
9. 5040
15. 210; 3360; 34,650
21. 28,800; 2880

5. 5040; 40,320; 604,800
11. 4320, 720
17. 90,090
23. 1440, 3600

7. 60
13. 34,560
19. 13,860

EXERCISE 12-3 PAGE 178

1. 465; 2600; 10,626; 1287
7. 35
13. 20,592
19. 70

3. 2380
9. 126, 256
15. 800
21. 792

5. 36, 84
11. 1960
17. 42

EXERCISE 12-4 PAGE 182

1. $\frac{1}{5}, \frac{4}{5}$
7. $\frac{1}{5}, \frac{2}{5}$
13. $\frac{1}{221}, \frac{1}{17}, \frac{4}{17}$

3. $\frac{1}{4}, \frac{1}{2}$
9. $\frac{4}{32}, \frac{2}{33}, \frac{5}{11}$
15. $\frac{1}{14}, \frac{1}{4}$

5. $\frac{1}{32}, \frac{5}{32}, \frac{5}{16}$
11. $\frac{10}{21}, \frac{121}{126}, \frac{125}{126}$
17. $\frac{1}{126}$

EXERCISE 12-5 PAGE 185

1. $\frac{5}{6}$
7. $\frac{1}{3}, \frac{1}{6}, \frac{1}{3}$
13. $\frac{1}{462}, \frac{1}{462}$
19. $\frac{3}{4}$, $375

3. $\frac{1}{6}, \frac{5}{18}, \frac{5}{18}$
9. $\frac{21}{506}, \frac{63}{253}, \frac{175}{2024}$
15. $\frac{3}{4}, \frac{3}{16}$
21. $3.60, $4.29

5. $\frac{1}{2197}, \frac{27}{2197}, \frac{1}{64}$
11. $\frac{19}{30}$
17. $\frac{17}{28}, \frac{11}{18}$

EXERCISE 12-6 PAGE 187

1. $\frac{3}{8}, \frac{1}{4}, \frac{5}{16}$
5. 0.2401, 0.4116, 0.3483

3. $\frac{81}{2401}, \frac{864}{2401}, \frac{1024}{2401}$
7. 2, 1

EXERCISE 13-1 PAGE 191

1. $8a^3 + 12a^2y + 6ay^2 + y^3$

3. $a^5 - 5a^4x + 10a^3x^2 - 10a^2x^3 + 5ax^4 - x^5$

5. $243m^5 + 405m^4n^2 + 270m^3n^4 + 90m^2n^6 + 15mn^8 + n^{10}$

7. $a^8 + 8a^7x + 28a^6x^2 + 56a^5x^3 + 70a^4x^4 + 56a^3x^5 + 28a^2x^6 + 8ax^7 + x^8$

9. $x^8 - 8x^7y + 28x^6y^2 - 56x^4y^3 + 70x^4y^4 - 56x^3y^5 + 28x^2y^6 - 8xy^7 + y^8$

11. $\dfrac{a^6}{b^6} + \dfrac{6a^4}{b^4} + \dfrac{15a^2}{b^2} + 20 + \dfrac{15b^2}{a^2} + \dfrac{6b^4}{a^4} + \dfrac{b^6}{a^6}$

13. $\dfrac{x^5}{243} - \dfrac{5x^4y}{162} + \dfrac{5x^3y^2}{36} - \dfrac{5x^2y^3}{36} + \dfrac{5xy^4}{48} - \dfrac{y^5}{32}$

15. $\dfrac{a^{12}}{4096} - \dfrac{3a^9}{256} + \dfrac{15a^6}{64} - \dfrac{5a^3}{2} + 15 - \dfrac{48}{a^3} + \dfrac{64}{a^6}$

17. $-35a^4x^3$ **19.** $-330x^4$ **21.** $7920a^8$ **23.** $-\dfrac{7a^5b^5}{216}$

25. $\dfrac{231a^6b^6}{16}$ **27.** $90720x^8y^4$ **29.** 84 **31.** $715x^3$

33. 1.126 **35.** 0.886 **37.** 0.851

EXERCISE 13-2   PAGE 194

1. $1 + x + x^2 + x^3 + \cdots$

3. $1 - y + y^2 - y^3 + \cdots$

5. $1 + 3x + 6x^2 + 10x^3 + \cdots$

7. $1 + \dfrac{9x}{2} + \dfrac{135x^2}{8} + \dfrac{945x^3}{16} + \cdots$

9. $1 - \dfrac{16x^2}{3} + \dfrac{224x^4}{9} - \dfrac{8960x^6}{81} + \cdots$

11. $a^{-3} - 6a^{-4}x + 24a^{-5}x^2 - 120a^{-6}x^3 + \cdots$

13. $2^{-11/4}\left(8 - x - \dfrac{3x^2}{16} - \dfrac{7x^3}{128} - \cdots\right)$

15. $3^{-22/3}[3^6 - 4(3)^4y + 14(3)^2y^2 - 140y^3 + \cdots]$

17. 3.04 **19.** 0.99 **21.** 0.50 **23.** 0.97

CHAPTER 13 REVIEW EXERCISE   PAGE 198

1. $81a^4 + 216a^3y + 216a^2y^2 + 96ay^3 + 16y^4$

3. $1 - 12x + 60x^2 - 160x^3 + 240x^4 - 192x^5 + 64x^6$

5. $\dfrac{x^6}{64} + \dfrac{3x^4}{8} + \dfrac{15x^2}{4} + 20 + \dfrac{60}{x^2} + \dfrac{96}{x^4} + \dfrac{64}{x^6}$

7. $\dfrac{y^6}{x^6} + \dfrac{6y^4}{x^5} + \dfrac{15y^2}{x^4} + \dfrac{20}{x^3} + \dfrac{15}{x^2y^2} + \dfrac{6}{xy^4} + \dfrac{1}{y^6}$

9. $945m^4$ **11.** $11{,}430x^8y^{18}$ **13.** $3432x^{14}y^{14}$

15. 61,236 **17.** 1.172 **19.** 1.083

21. $1 + 2x + 3x^2 + 4x^3 + \cdots$

23. $1 - \dfrac{4y}{3} - \dfrac{4y^2}{9} - \dfrac{32y^3}{81} - \cdots$

25. $\dfrac{1}{3^4} + \dfrac{4y}{3^5} + \dfrac{10y^2}{3^6} + \dfrac{20y^3}{3^7} + \cdots$

27. 5.10 **29.** .020 **31.** 1.16

EXERCISE 14-1   PAGE 202

1. 3 **3.** 0 **5.** 67
7. Is a factor **9.** Not a factor **11.** Is a factor
13. Is a factor **15.** Is a factor **17.** Is not a factor
19. Is a factor **21.** Is a factor **23.** Is a factor

25. Is a factor
29. $x^3 + 6x^2 + 11x + 6 = 0$

27. $x^3 - 2x^2 - 5x + 6 = 0$
31. $3x^3 - x^2 - 3x + 1 = 0$

EXERCISE 14-2 PAGE 205

1. $2x^2 - 9x + 12, -11$
5. $4x^2 - 8x + 25, 0$
9. $x^2 + 3, 4$
13. $x^4 + x^3 + 2x^2 + 3x + 6, 15$
17. $x^4 - 2x^3 + 4x^2 - 8x + 16, 0$
19. $x^6 - x^5y + x^4y^2 - x^3y^3 + x^2y^4 - xy^5 + y^6, -2y^7$
21. $-58, 2, 2$

3. $x^2 + x - 7, -3$
7. $5x^2 - 2x + 11, -26$
11. $2x^3 + x^2 - 3x + 10, -19$
15. $x^4 - 2x^3 + 3x^2 - 12x + 48, -191$

23. $9, 19, 389$

EXERCISE 14-3 PAGE 209

1. $-i, -4$
5. $-\sqrt{7}, 2 \pm 2i$
9. $-3 - i\sqrt{2}, \pm i\sqrt{6}$
13. $x^4 - 10x^3 + 27x^2 + 20x - 58 = 0$
17. $x^4 - 4x^3 + 19x^2 - 30x + 50 = 0$
21. $7, 0$
25. $x^3 + 3x^2 - 6x - 8 = 0$
29. $x^4 - 2x^3 - x^2 + 2x = 0$

3. $5 + i, -2$
7. $1 - i\sqrt{3}, \sqrt{2}, -\sqrt{2}$
11. $x^3 - 16x^2 + 93x - 180 = 0$
15. $x^4 - 12x^3 + 77x - 260x + 500 = 0$
19. $3, 5$
23. $0, 3$
27. $x^3 - 7x - 6 = 0$

35. All real; four real, two imaginary; two real, four imaginary; six imaginary.

EXERCISE 14-4 PAGE 213

1. $-2, -2, 3$
7. $-\frac{1}{2}, \frac{1}{2}, \frac{1}{2}$

3. $2, 2 \pm 3i$
9. $-\frac{2}{3}, -1 \pm 2i$

5. $-\frac{1}{4}, \pm 2i$
11. -2

13. $-\frac{1}{2}, \frac{1}{3}, \pm i$

15. $\frac{1}{2}, 2, 2, 2$

17. $\frac{1}{3}, 2, \dfrac{-1 \pm \sqrt{5}}{2}$

19. $1, 1, 2, 2, 3$
23. $(x + 1)(3x - 4)(2x - 1)$
27. $(x - 1)(x + 1)(x + 2)^2$

21. $\frac{3}{2}$
25. $(2x + 1)(x^2 - 4x + 11)$
29. $(x - 1)^3(x + 2)^2$

EXERCISE 14-6 PAGE 220

1. 1.51
7. 0.33
13. $-1.11, 1.25, 2.86$
19. $-0.47, 1.12, 2.35$

3. 1.70
9. 1.08
15. $-0.11, 2.25, 3.86$
21. 1.45

5. 2.04
11. 1.38
17. $-0.25, 1.45, 2.80$
23. 2.07

EXERCISE 14-7 PAGE 221

1. One positive root and two imaginary roots.
3. One negative root and two imaginary roots.
5. Three roots are positive or one root is positive and the others imaginary.
7. One positive root, one negative root, and two imaginary roots.

9. There is no positive root. There are four, two, or no negative roots. Hence there are zero, two, or four imaginary roots.

11. There are either three positive roots or one positive root. There is one negative root. Hence there are two or four imaginary roots.

EXERCISE 15-1   PAGE 228

1. $13 - i$

3. $-2 - (1 + \sqrt{2})i$

5. $3 - i\sqrt{3}$

7. $-14 - 12i$

9. $-15 - 8i$

11. $78 + 9i\sqrt{3}$

13. $-1 - 5i$

15. $\frac{16}{25} + \frac{12}{25}i$

17. $\frac{11}{37} + \frac{8}{37}i$

19. $4 - 3i$

21. i

23. $8i$

25. $-2 + 9i$

27. 2

29. $x = 2, y = -5$

31. $x = 2, y = 2$

33. $x = -\frac{3}{2}, y = -\frac{7}{2}$

EXERCISE 15-2   PAGE 231

1. $\sqrt{2}(\cos 45° + i \sin 45°)$

3. $4(\cos 0° + i \sin 0°)$

5. $4(\cos 300° + i \sin 300°)$

7. $2(\cos 90° + i \sin 90°)$

9. $\cos 270° + i \sin 270°$

11. $16(\cos 330° + i \sin 330°)$

13. $2\sqrt{7}(\cos 300° + i \sin 300°)$

15. $\sqrt{5}(\cos 153° + i \sin 153°)$

17. $25(\cos 254° + i \sin 254°)$

19. $\sqrt{3} + i$

21. $1 + i$

23. $-\dfrac{5\sqrt{3}}{2} + \dfrac{5}{2}i$

25. $-6i$

27. $-0.616 + 0.788i$

EXERCISE 15-3   PAGE 233

1. $6i$

3. $-9 + 9i\sqrt{3}$

5. $27i$

7. -2

9. $\dfrac{7}{2} - \dfrac{7i\sqrt{3}}{2}$

11. $2\sqrt{2}(\cos 75° + i \sin 75°)$

13. $20\sqrt{2}(\cos 135° + i \sin 135°)$

15. $8\sqrt{2}(\cos 315° + i \sin 315°)$

17. $2(\cos 30° + i \sin 30°)$

19. $\dfrac{\sqrt{2}}{2}(\cos 345° + i \sin 345°)$

EXERCISE 15-4   PAGE 237

1. $9(\cos 40° + i \sin 40°)$

3. $8(\cos 354° + i \sin 354°)$

5. $\cos 80° + i \sin 80°$

7. $4\sqrt{2}(\cos 225° + i \sin 225°)$

9. $64(\cos 0° + i \sin 0°)$

11. $1024(\cos 60° + i \sin 60°)$

13. $4(\cos 40° + i \sin 40°), 4(\cos 220° + i \sin 220°)$

15. $3(\cos 50° + i \sin 50°), 3(\cos 170° + i \sin 170°), 3(\cos 290° + i \sin 290°)$

17. $2[\cos (70° + k \cdot 90°) + i \sin (70° + k \cdot 90°)], k = 0, 1, 2, 3$

19. $2[\cos (15° + k \cdot 120°) + i \sin (15° + k \cdot 120°)], k = 0, 1, 2$

21. $\cos (67.5° + k \cdot 90°) + i \sin (67.5° + k \cdot 90°), k = 0, 1, 2, 3$

23. $2[\cos (54° + k \cdot 72°) + i \sin (54° + k \cdot 72°)], k = 0, 1, 2, 3, 4$

25. $32 - 32i\sqrt{3}$ **27.** $-\frac{1}{2} + \frac{\sqrt{3}}{2}i$ **29.** i

31. $1.29 + 1.53i, -1.97 + 0.348i, 0.684 - 1.88i$

33. $1.59 + 1.59i, -2.17 + 0.58i, 0.58 - 2.17i$

35. $\frac{\sqrt{2}}{2}(1 + i), \frac{\sqrt{2}}{2}(-1 + i), \frac{\sqrt{2}}{2}(-1 - i), \frac{\sqrt{2}}{2}(1 - i)$

37. $\cos(22\frac{1}{2}° + k \cdot 90°) + i\sin(22\frac{1}{2}° + k \cdot 90°), k = 0, 1, 2, 3$

39. $\frac{2 - \sqrt{2}}{2} + \frac{\sqrt{2}}{2}i, \frac{2 + \sqrt{2}}{2} - \frac{\sqrt{2}}{2}i$

41. $-\frac{1 + 3i}{2}$ **43.** $\frac{13 - i}{10}$

45. $\frac{10 - 11i}{17}$ **47.** $\frac{2i \pm \sqrt{-4 - 2i}}{2}$

49. $\frac{-i \pm \sqrt{1 - 4i}}{4}$ **51.** $\frac{-i \pm \sqrt{3i - 2}}{3}$

53. $\frac{-3 + 3i \pm \sqrt{-18i - 32}}{4}$

55. $(3, 7)$ **57.** $(8, 6)$

59. $(1, 3)$ **61.** $(1, 1)$

63. $(-5, 10)$ **65.** $(-20, 17)$

67. $(\frac{6}{5}, -\frac{8}{5})$ **69.** $(7, -4)$

EXERCISE 16-1   PAGE 243

1. $(1, 2, 3)$ **3.** $(-2, -2, 1)$ **5.** $(1, 2, -3)$

7. $(\frac{1}{2}, \frac{1}{3}, \frac{1}{6})$ **9.** $(2, -3, -2, -1)$

EXERCISE 16-2   PAGE 246

1. -57 **3.** 100 **5.** -5

7. $x = -2, y = 3$ **9.** $x = -\frac{3}{5}, y = -\frac{11}{5}$ **11.** $x = -\frac{1}{2}, y = \frac{5}{2}$

13. $x = \frac{29c}{8a}, y = \frac{7c}{8b}$ **15.** $x = 2, y = -\frac{2}{a}$ **17.** 40

19. 47 **21.** 63 **23.** $x = 2$

25. $x = -2$ or 6 **27.** $x = -\frac{8}{3}$

EXERCISE 16-3   PAGE 251

11. $x = 3$ **13.** $x = -2, 0, 2$

EXERCISE 16-4   PAGE 256

1. -96 **3.** 29   **5.** -26 **7.** 406   **9.** -36
11. 1083   **13.** 27   **15.** 9   **17.** 23.5   **19.** 19

EXERCISE 16-5   PAGE 261

1. $x = 1, y = 3, z = -2$
3. $x = 1, y = 3, z = -2$
5. $x = 3, y = 0, z = 2$
7. $x = 1, y = -1, z = 1, w = -1$
9. $x = 1, y = -\frac{2}{7}, z = \frac{2}{7}, w = -\frac{1}{7}$
11. $x = \frac{7}{3}, y = \frac{2}{3}, z = 1, w = -1$
13. $x = -2, y = 3, z = \frac{3}{2}, w = -\frac{1}{2}$

EXERCISE 16-6   PAGE 265

1. $x = \frac{1}{2}, y = -\frac{1}{3}$
3. $x = \frac{17}{12}, y = \frac{1}{12}$
5. $x = 4, y = -3$
7. Inconsistent
9. $x = -2, y = -3, z = -4$
11. Inconsistent
13. $x = 0, y = 0$
15. $x = k, y = -k$
17. $x = k, y = \frac{1}{3}k, z = k$
19. $x = 2k, y = k, z = k, w = -4k$

EXERCISE 16-7   PAGE 267

1. $\begin{bmatrix} -1 & 8 \\ 8 & 5 \end{bmatrix}$

3. $\begin{bmatrix} 3 & -2 \\ 0 & -1 \end{bmatrix}$

5. $\begin{bmatrix} 10 & 14 \\ 0 & 26 \end{bmatrix}$

7. $\begin{bmatrix} -5 & -2 & 1 \\ 9 & 10 & 11 \end{bmatrix}$

9. $\begin{bmatrix} -2 & 2 & -1 \\ -8 & -2 & -10 \\ 20 & 10 & 28 \end{bmatrix}$

11. $\begin{bmatrix} 3 & 9 & 15 & 18 \\ 2 & 6 & 10 & 12 \\ 2 & 3 & 5 & 6 \end{bmatrix}$

13. $\begin{bmatrix} 22 & -13 & 7 \\ 31 & -17 & 13 \\ 5 & -4 & 5 \end{bmatrix}$

EXERCISE 16-8   PAGE 272

3. No inverse

5. $\frac{1}{10}\begin{bmatrix} 3 & -1 \\ 4 & 2 \end{bmatrix}$

7. $\frac{1}{13}\begin{bmatrix} 3 & 1 \\ -4 & 3 \end{bmatrix}$

9. $\frac{1}{3}\begin{bmatrix} 1 & 1 \\ 1 & -2 \end{bmatrix}$

11. $\frac{1}{16}\begin{bmatrix} 2 & -3 \\ 2 & 5 \end{bmatrix}$

13. No inverse

15. $\frac{1}{5}\begin{bmatrix} 2 & -1 & 1 \\ 1 & 0 & 1 \\ 3 & 4 & 2 \end{bmatrix}$

17. $\frac{1}{21}\begin{bmatrix} 4 & 2 & -3 \\ 2 & 1 & -5 \\ -7 & -2 & 6 \end{bmatrix}$

EXERCISE 17-1   PAGE 276

1. $\log_2 8 = 3$ **3.** $\log_5 \frac{1}{5} = -1$ **5.** $\log_6 1 = 0$
7. $\log_9 \frac{1}{27} = -\frac{3}{2}$ **9.** $\log_{10} 1000 = 3$ **11.** $\log_{2/3} \frac{8}{27} = 3$
13. $4^0 = 1$ **15.** $6^2 = 36$ **17.** $2^{-3} = \frac{1}{8}$
19. $2^{-2} = \frac{1}{4}$ **21.** $(\frac{1}{3})^{-3} = 27$ **23.** $(16)^{3/2} = 64$
25. $x = 100$ **27.** $x = \frac{1}{9}$ **29.** $x = 4$
31. $x = \frac{1}{125}$ **33.** $x = 16$ **35.** $a = 5$
37. $a = 625$ **39.** $a = 16$ **41.** $a = \frac{1}{6}$
43. $y = \frac{2}{3}$ **45.** $y = \frac{1}{2}$ **47.** $y = \frac{2}{3}$

EXERCISE 17-2   PAGE 278

1. $\log 7.12 + 3 \log 78$ **3.** $\log 83 - \log 47 - \log 51$
5. $\frac{1}{2} \log 1.44 + 2 \log 8.23 - \log 308$
7. $\frac{1}{2}(\log 28 + \log 44 + \log 21 - \log 978)$
9. $\log 85 + \frac{1}{3} \log 67 - 2 \log 43 - \frac{1}{2} \log 65$
11. 0.9542   **13.** 0.7781   **15.** 1.6811
17. 0.5283   **19.** 0.4203   **21.** 1.9084

EXERCISE 17-3   PAGE 282

1. 1   **3.** 2   **5.** 2   **7.** -1 **9.** -3
11. -5 **13.** 3   **15.** -5 **17.** 8.634   **19.** 0.08634
21. 863,400   **23.** 0.08634   **25.** 0.0008634   **27.** 0.0008634

EXERCISE 17-4   PAGE 284

1. 0.4969   **3.** 3.6345   **5.** 0.9586   **7.** $7.9031 - 10$
9. $7.6990 - 10$ **11.** 2.9542   **13.** $9.5911 - 10$ **15.** $8.9042 - 10$
17. 309   **19.** 9.35   **21.** 7.81   **23.** 0.0448
25. 0.000704   **27.** 0.305   **29.** 99.9

EXERCISE 17-5   PAGE 286

1. 1.8341   **3.** 2.7674   **5.** $8.9941 - 10$ **7.** 4.8066
9. 0.6836   **11.** $8.004 - 10$ **13.** 4.9566   **15.** 2.3707
17. $6.4881 - 10$ **19.** 3.2407   **21.** 3.8975   **23.** 5.8724
25. 594.7   **27.** 64.01   **29.** 99,650   **31.** 0.004593
33. 0.3079   **35.** 0.01490   **37.** 1.710   **39.** 208.0
41. 21,600   **43.** 0.5464   **45.** 0.008935   **47.** 0.001084

EXERCISE 17-6   PAGE 288

1. 233.8   **3.** 5.856   **5.** 3.260   **7.** 172.2
9. 24,100   **11.** 292.5   **13.** 0.4627   **15.** 2.532
17. 0.00002662   **19.** 4.230   **21.** 0.7822   **23.** 2.054
25. 7.632   **27.** 1.231   **29.** 0.002402   **31.** 0.2683
33. 4.250   **35.** 3.555

EXERCISE 17-7   PAGE 290

1. 3   **3.** 5   **5.** 2   **7.** -1

9. 0   **11.** 1.233   **13.** 0.8857   **15.** 2.014

17. 0.8414   **19.** 0   **21.** 0.04257   **23.** -1.857

25. 18.80   **27.** -1.366

EXERCISE 17-8   PAGE 291

1. 4.605   **3.** -0.7382 **5.** 4.701   **7.** -1.265

9. 2.124   **11.** -3.194 **13.** 1.100   **15.** 0.3465

EXERCISE 17-9   PAGE 292

1. (a) \$267.64, (b) \$268.80   **3.** (a) \$567.40, (b) \$570.30, (c) \$572.90

5. 12 years   **7.** 5% per year

EXERCISE 18-1   PAGE 296

1. $\dfrac{3}{x+1} - \dfrac{2}{x-1}$ **3.** $\dfrac{2}{x-2} + \dfrac{5}{x-3}$

5. $\dfrac{4}{3x+1} - \dfrac{3}{3x-1}$ **7.** $\dfrac{2}{3x+1} + \dfrac{3}{x-1}$

9. $\dfrac{5}{3x-4} - \dfrac{3}{x+2}$ **11.** $\dfrac{1}{x} + \dfrac{1}{x-4} - \dfrac{2}{x+2}$

13. $\dfrac{5}{x-3} - \dfrac{4}{x-4} + \dfrac{3}{x+2}$ **15.** $\dfrac{2}{2x+1} + \dfrac{27}{3x-1} - \dfrac{40}{4x-1}$

EXERCISE 18-2   PAGE 298

1. $\dfrac{1}{x-4} - \dfrac{1}{(x-4)^2}$ **3.** $\dfrac{2}{x+3} + \dfrac{1}{(x+3)^2}$

5. $\dfrac{2}{3x+1} - \dfrac{3}{(3x+1)^2} + \dfrac{5}{(3x+1)^3}$ **7.** $\dfrac{3}{x-2} - \dfrac{1}{x-1} + \dfrac{1}{(x-1)^2}$

9. $\dfrac{3}{x-4} - \dfrac{3}{x-1} - \dfrac{3}{(x-1)^2}$ **11.** $\dfrac{1}{x+1} - \dfrac{2}{(x-1)^3}$

13. $\dfrac{1}{2x-3} - \dfrac{1}{3x+2} + \dfrac{1}{x-2} + \dfrac{3}{(x-2)^2}$

15. $\dfrac{1}{x} - \dfrac{2}{x+1} + \dfrac{3}{x+2} - \dfrac{4}{(x+2)^2}$

EXERCISE 18-3   PAGE 300

1. $\dfrac{1}{x-2} + \dfrac{3}{x^2+5}$

3. $\dfrac{1}{x+3} - \dfrac{2}{x^2-3x+3}$

5. $\dfrac{4}{x-4} - \dfrac{3x+2}{x^2-x-5}$

7. $\dfrac{1}{x+2} - \dfrac{1}{x+1} + \dfrac{3x}{x^2+2}$

9. $\dfrac{3}{(x+1)^2} - \dfrac{2}{x^2-x+1}$

11. $\dfrac{3x+2}{x^2+1} - \dfrac{2x-3}{x^2+2}$

13. $\dfrac{x+1}{2x^2+x+2} - \dfrac{x}{3x^2-x+1}$

EXERCISE 18-4   PAGE 301

1. $\dfrac{3x}{x^2+2} - \dfrac{2x+5}{(x^2+2)^2}$

3. $\dfrac{3}{x^2+x+1} + \dfrac{2x}{(x^2+x+1)^2}$

5. $\dfrac{1}{x^2-x+4} - \dfrac{3x-5}{(x^2-x+4)^3}$

7. $\dfrac{1}{x^2+1} + \dfrac{2}{(x^2+1)^2} - \dfrac{x}{(x^2+1)^3}$

9. $\dfrac{2}{x} - \dfrac{2x}{x^2+3} - \dfrac{x}{(x^2+3)^2}$

11. $\dfrac{1}{x} + \dfrac{2}{x^2} + \dfrac{x}{(x^2+5)^2} - \dfrac{2}{x^2+5}$

13. $\dfrac{x-3}{(x^2-2x+2)^2} + \dfrac{2x}{x^2-2x+2} - \dfrac{1}{x-3}$

INDEX

Common Logarithms

N	0	1	2	3	4	5	6	7	8	9
10	0000	0043	0086	0128	0170	0212	0253	0294	0334	0374
11	0414	0453	0492	0531	0569	0607	0645	0682	0719	0755
12	0792	0828	0864	0899	0934	0969	1004	1038	1072	1106
13	1139	1173	1206	1239	1271	1303	1335	1367	1399	1430
14	1461	1492	1523	1553	1584	1614	1644	1673	1703	1732
15	1761	1790	1818	1847	1875	1903	1931	1959	1987	2014
16	2041	2068	2095	2122	2148	2175	2201	2227	2253	2279
17	2304	2330	2355	2380	2405	2430	2455	2480	2504	2529
18	2553	2577	2601	2625	2648	2672	2695	2718	2742	2765
19	2788	2810	2833	2856	2878	2900	2923	2945	2967	2989
20	3010	3032	3054	3075	3096	3118	3139	3160	3181	3201
21	3222	3243	3263	3284	3304	3324	3345	3365	3385	3404
22	3424	3444	3463	3483	3502	3522	3541	3560	3579	3598
23	3617	3636	3655	3674	3692	3711	3729	3747	3766	3784
24	3802	3820	3838	3856	3874	3892	3909	3927	3945	3962
25	3979	3997	4014	4031	4048	4065	4082	4099	4116	4133
26	4150	4166	4183	4200	4216	4232	4249	4265	4281	4298
27	4314	4330	4346	4362	4378	4393	4409	4425	4440	4456
28	4472	4487	4502	4518	4533	4548	4564	4579	4594	4609
29	4624	4639	4654	4669	4683	4698	4713	4728	4742	4757
30	4771	4786	4800	4814	4829	4843	4857	4871	4886	4900
31	4914	4928	4942	4955	4969	4983	4997	5011	5024	5038
32	5051	5065	5079	5092	5105	5119	5132	5145	5159	5172
33	5185	5198	5211	5224	5237	5250	5263	5276	5289	5302
34	5315	5328	5340	5353	5366	5378	5391	5403	5416	5428
35	5441	5453	5465	5478	5490	5502	5514	5527	5539	5551
36	5563	5575	5587	5599	5611	5623	5635	5647	5658	5670
37	5682	5694	5705	5717	5729	5740	5752	5763	5775	5786
38	5798	5809	5821	5832	5843	5855	5866	5877	5888	5899
39	5911	5922	5933	5944	5955	5966	5977	5988	5999	6010
40	6021	6031	6042	6053	6064	6075	6085	6096	6107	6117
41	6128	6138	6149	6160	6170	6180	6191	6201	6212	6222
42	6232	6243	6253	6263	6274	6284	6294	6304	6314	6325
43	6335	6345	6355	6365	6375	6385	6395	6405	6415	6425
44	6435	6444	6454	6464	6474	6484	6493	6503	6513	6522
45	6532	6542	6551	6561	6571	6580	6590	6599	6609	6618
46	6628	6637	6646	6656	6665	6675	6684	6693	6702	6712
47	6721	6730	6739	6749	6758	6767	6776	6785	6794	6803
48	6812	6821	6830	6839	6848	6857	6866	6875	6884	6893
49	6902	6911	6920	6928	6937	6946	6955	6964	6972	6981
50	6990	6998	7007	7016	7024	7033	7042	7050	7059	7067
51	7076	7084	7093	7101	7110	7118	7126	7135	7143	7152
52	7160	7168	7177	7185	7193	7202	7210	7218	7226	7235
53	7243	7251	7259	7267	7275	7284	7292	7300	7308	7316
54	7324	7332	7340	7348	7356	7364	7372	7380	7388	7396
N	0	1	2	3	4	5	6	7	8	9